T0142326

Advances in Intelligent Systems and Computing

Volume 850

Series editor

Janusz Kacprzyk, Polish Academy of Sciences, Warsaw, Poland
e-mail: kacprzyk@ibspan.waw.pl

The series "Advances in Intelligent Systems and Computing" contains publications on theory, applications, and design methods of Intelligent Systems and Intelligent Computing. Virtually all disciplines such as engineering, natural sciences, computer and information science, ICT, economics, business, e-commerce, environment, healthcare, life science are covered. The list of topics spans all the areas of modern intelligent systems and computing such as: computational intelligence, soft computing including neural networks, fuzzy systems, evolutionary computing and the fusion of these paradigms, social intelligence, ambient intelligence, computational neuroscience, artificial life, virtual worlds and society, cognitive science and systems, Perception and Vision, DNA and immune based systems, self-organizing and adaptive systems, e-Learning and teaching, human-centered and human-centric computing, recommender systems, intelligent control, robotics and mechatronics including human-machine teaming, knowledge-based paradigms, learning paradigms, machine ethics, intelligent data analysis, knowledge management, intelligent agents, intelligent decision making and support, intelligent network security, trust management, interactive entertainment, Web intelligence and multimedia.

The publications within "Advances in Intelligent Systems and Computing" are primarily proceedings of important conferences, symposia and congresses. They cover significant recent developments in the field, both of a foundational and applicable character. An important characteristic feature of the series is the short publication time and world-wide distribution. This permits a rapid and broad dissemination of research results.

More information about this series at http://www.springer.com/series/11156

Tatiana Antipova · Alvaro Rocha
Editors

Digital Science

 Springer

Editors
Tatiana Antipova
Institute of Certified Specialists
Perm, Russia

Alvaro Rocha
Information Systems and Software
 Engineering
University of Coimbra
Coimbra, Portugal

ISSN 2194-5357 ISSN 2194-5365 (electronic)
Advances in Intelligent Systems and Computing
ISBN 978-3-030-02350-8 ISBN 978-3-030-02351-5 (eBook)
https://doi.org/10.1007/978-3-030-02351-5

Library of Congress Control Number: 2018958770

This Springer imprint is published by the registered company Springer Nature Switzerland AG
The registered company address is: Gewerbestrasse 11, 6330 Cham, Switzerland

Preface

This book contains a selection of papers accepted for the presentation and discussion at the 2018 International Conference on Digital Science (DSIC'18). This Conference had the support of the Institute of Certified Specialists, Russia, AISTI (Iberian Association for Information Systems and Technologies), and Springer. It will take place at Convention Centre, Budva, Montenegro, October 19–21, 2018.

DSIC'18 is an international forum for researchers and practitioners to present and discuss the most recent innovations, trends, results, experiences, and concerns in the several perspectives of Digital Science. The main idea of this Conference is that the world of science is unified and united allowing all scientists/practitioners to be able to think, analyze, and generalize their thoughts.

DSIC aims efficiently to disseminate original research results in natural, social, art, and humanities sciences. An important characteristic feature of the Conference should be the short publication time and worldwide distribution. This Conference enables fast dissemination, so conference participants can publish their papers in print and electronic format, which is then made available worldwide and accessible by numerous researchers.

The Scientific Committee of DSIC'18 was composed of a multidisciplinary group of 26 experts. One hundred and seven invited reviewers who are intimately concerned with Digital Science have had the responsibility for evaluating, in a "double-blind review" process, the papers received for each of the main themes proposed for the Conference: Digital Art and Humanities; Digital Economics; Digital Education; Digital Engineering; Digital Environmental Sciences; Digital Finance, Business and Banking; Digital Media; Digital Medicine, Pharma and Public Health; Digital Public Administration; Digital Technology and Applied Sciences.

DSIC'18 received 88 contributions from 16 countries around the world. The papers accepted for the presentation and discussion at the Conference are published by Springer (this book) and will be submitted for indexing by ISI, SCOPUS, among others.

We acknowledge all of those that contributed to the staging of DSIC'18 (authors, committees, reviewers, organizers, and sponsors). We deeply appreciate their involvement and support that was crucial for the success of DSIC'18.

October 2018   Tatiana Antipova
                                                                         Alvaro Rocha

Organization

General Chair

Tatiana Antipova

Honorary Chair

Álvaro Rocha

Scientific Committee

Alan Sangster	University of Sussex, UK
Andre Carlos Busanelli de Aquino	University of São Paulo (USP), Brasilia
Elena Pimenova	Perm State Agro-Technological University, Russia
Giuseppe Grossi	Nord University, Norway
Joao Vidal de Carvalho	Polytechnic Institute of Porto, Portugal
John Dumay	Macquarie University, Australia
Julia Belyasova	Public Social Welfare Centre, Belgium
Leonid Yasnitsky	Perm State University, Russia
Linda Kidwell	Nova Southeastern University, USA
Luca Bartocci	University of Perugia, Italy
Lucas Oliveira Gomes Ferreira	Federal Court of Accounts, Brasilia
Marina Gurskaya	Kuban State University, Russia
Mikhail Gitman	Perm National Research Polytechnic University, Russia
Mikhail Kuter	Kuban State University, Russia
Miguel Casquilho	IST, Portugal
Nikolai Ustiuzhantsev	Perm State Medical University, Russia

Contents

Digital Education

Digital Art and Humanities

Towards Data-Driven Machine Translation for Lumasaaba

Peter Nabende[✉]

Department of Information Systems, School of Computing and Informatics
Technology, College of Computing and Information Sciences,
Makerere University, P. O. Box 7062, Kampala, Uganda
pnabende@cis.mak.ac.ug

Abstract. This paper reports on results from initial efforts towards the application of data-driven machine translation for low-resourced East African languages. In particular, the paper evaluates the application of phrase-based
statistical machine translation (PBSMT) and neural machine translation
(NMT) for automating translation between Lumasaaba (an East African Bantu
language) and English. As expected, the PBSMT approach outperforms the
NMT approach on a small Bible-based corpus of parallel sentences. The parallel
corpus and the respective machine translation evaluations presented in this paper
can be used as baselines for future machine translation quality improvement
efforts involving Lumasaaba and other related East African Bantu languages.

Keywords: Phrase-based statistical machine translation
Neural machine translation · Lumasaaba

1 Background

There are now numerous and useful applications of Machine Translation (MT) for not
only business purposes but also for instant consumer use. Most of the advances and
benefits of machine translation have been realized in the context of well-resourced
languages. Although there are efforts towards realizing machine translation for low
resourced languages, MT work on many indigenous East African (EA) languages is
scarce. Initial efforts on MT for low resourced languages involved the use of a pivot
language [14] that has more bilingual resources than between a low resourced language
pair. The pivot-based approach is currently not possible for MT involving many EA
indigenous languages because of lack of bilingual resources between the EA languages
and any other language. Recent efforts aim to leverage on the success of state of the art
data-driven MT. However, most of these efforts rely on the similarities between a high
resource language and a low resource language [8]; consequently, this approach is also
not suitable for EA languages since they are barely related to any highly resourced
languages. Several EA languages are also not well written and thus lack even monolingual text. It is therefore necessary to develop from scratch the required resources for
realizing MT for EA languages. This paper reports on the first effort towards building
such MT resources for Lumasaaba (a low resourced EA language). Specifically, the
paper presents a small Bible-based parallel corpus between Lumasaaba and English

T. Antipova and A. Rocha (Eds.): DSIC 2018, AISC 850, pp. 3–11, 2019.
https://doi.org/10.1007/978-3-030-02351-5_1

that can be utilized as seed data for training MT systems and other bilingual applications. This corpus is used to evaluate the application of two state of the art data-driven MT methods for translating between Lumasaaba and English. The findings from the evaluation results can also be associated with other EA languages that have a considerably high lexical similarity to Lumasaaba. The rest of the paper is organized as follows: In Sect. 2, the Lumasaaba language is introduced; in Sect. 3, state of the art data driven MT approaches are described, in Sect. 4, a Bible-based Lumasaaba-English parallel corpus is described; in Sect. 5, data-driven MT experiments are described and evaluation results are discussed; Sect. 6 concludes the paper with pointers to future work.

2   The Lumasaaba Language

Lumasaaba falls under one main family of Bantu languages in Uganda. The other two main language families in Uganda are Nilotic and Central Sudanic. The Bantu languages in Uganda constitute four subfamilies: Lega-Holoholo, Konzo-Ndandi, Masaba-Luhya and Nyoro-Ganda. Lumasaaba is classified under the Masaba-Luhya subfamily and is mostly spoken in the Eastern part of Uganda around the western side of Mt. Elgon. It is mutually intelligible with the Lubukusu language that is mostly spoken in the Western part of Kenya. Lumasaaba has a number of dialects under two main categories [3]: the 'northern' dialects (for example Ludadiri, Luwalasi and Lufumbo) and the 'southern dialects (for example Lubuya, Lusoba and Lukiende). There are clear differences in both pronunciations and vocabulary between the two categories of dialects which warrant a separate treatment in the context of translation. In this paper, we focus on a representation that is mostly associated with the southern Lumasaaba dialects. A reasonable amount of both monolingual and bilingual text has been generated and can be used as seed data for evaluating data-driven MT techniques.

3   Data-Driven Machine Translation

The field of MT is currently dominated by data-driven (or corpus-based approaches) [6] which can be categorized into three: Example-based Machine Translation, Statistical Machine Translation (SMT), and NMT. Only SMT and NMT were considered in this paper as they offer an inexpensive avenue for developing MT models.

3.1   Phrase-Based Statistical Machine Translation

In the past decade or so, the best performing MT systems for language pairs with adequate corpora have been based on the PBSMT approach. Until 2016, Web-based tools such as Google Translate[1] were using this approach. The PBSMT approach has also been tested for low resourced languages [5, 12]. In their study of applying SMT to three under-resourced Asian languages (Lao, Myanmar and Thai), Pa et al. [12]

---

[1] https://translate.google.com

provisionally concluded that the basic PBSMT approach seemed more robust to training on very limited amounts of data. Because of the scarcity of bilingual text involving indigenous EA languages, the PBSMT approach should be the most plausible approach for developing MT models involving EA languages. In the following paragraphs, we describe the PBSMT approach in the context of Lumasaaba-to-English translation. Let us denote a Lumasaaba input sentence as l and an English sentence as e. In PBSMT, l will be broken up into phrases which are then translated using a phrase translation model; the phrase translations are the reordered according to a reordering model to generate the final target sentences. The best English translation e_{best} given l is defined as [10]:

$$e_{\text{best}} = \underset{e}{\operatorname{argmax}}\, p(e|l) = \underset{e}{\operatorname{argmax}}\, p(l|e) \times p_{\text{LM}}(e) \qquad (1)$$

where $p_{\text{LM}}(e)$ is obtained from an English language model. The language model measures the likelihood or fluency of a sequence of words in the English language and is also used to choose the most likely sequence from a set of propositions. Several natural language modeling methods can be used including n-gram methods and neural network-based methods. For translation in the other direction, we would only need an appropriate amount of Lumasaaba text to produce respective language models.

The other component of Eq. 1, $p(l|e)$ is mainly associated with the actual translation, and for PBSMT, it is decomposed further into translation and reordering components according to Eq. 2 below [10]:

$$p(l_1^n|e_1^n) = \prod_{i=1}^{n} \phi(l_i|e_i) \times \text{d}(\text{start}_i - \text{end}_{i-1} - 1) \qquad (2)$$

Equation 2 is representative of the breakup of l into n phrases and the translation of each ith phrase in l (l_i) to an English phrase (e_i) according to a phrase translation probability $\phi(l_i|e_i)$ [10]. Equation 2 also involves a reordering component $\text{d}(\text{start}_i - \text{end}_{i-1} - 1)$ which is obtained from a distance-based re-ordering model [10].

3.2 Neural Machine Translation

Recent machine translation evaluations (for example from the 2016 [1] and 2017 [2] MT shared tasks show that NMT significantly outperforms other MT approaches on a number of language pairs. Currently, the most common NMT model is referred to as an encoder-decoder model. Using the notation in the previous section where l denotes a Lumasaaba sentence and e denotes an English sentence, the source encoder would map each word in l to a word vector and the word vectors would be processed to a sequence of hidden vectors. There are different implementations for this encoder phase including forward, backward and bidirectional. The target language decoder then combines a Recurrent Neural Network (RNN) hidden representation of previously generated English words with the source hidden vectors to predict scores for each possible next English word [9]. A softmax layer is then used to produce a next English word distribution. Current NMT implementations also incorporate an attention mechanism which is used to calculate the the softmax distribution over the next target English words. Figure 1 illustrates this NMT process in translating a Lumasaaba sentence "*mu bifukhu ibyo*" to an English sentence "*in those days*".

4 Bible-Based Parallel Corpus

There is currently a growing interest in using translations of the Christian Bible to develop parallel corpora for several multilingual and cross lingual applications [4]. For many EA under resourced languages, the Bible serves as a first source for extracting highly reliable translations. The Lumasaaba Bible is a recent translation and thus it reflects the present day style and usage of the Lumasaaba language. Only the New Testament part of this Bible is electronically accessible at https://www.bible.is website. For the experiments reported in this paper, the Lumasaaba Bible was paired with the New International Version which also reflects the present day style and usage of the English language. Parallel sentences can be extracted from bilingual alignments of the Bible verses. 7958 parallel verses of the New Testament were considered as a source for extracting parallel sentences. When extracting parallel sentences, there a number of issues to consider [4] including the need to establish sentence alignments and the occurrence of imperfect translations. These issues often require in-depth knowledge of the language pair.

The extraction of parallel sentences was based on the assumption that sentence alignments can be established at corresponding positions of verses where the number of sentences is the same across a verse pair. This assumption was applicable to 6000 parallel verses where a whole sentence corresponds to a verse or where corresponding verses have the same number of sentences. A simple sentence matching algorithm was implemented to automate the extraction of parallel sentences from the 6000 parallel verses; this resulted into 7141 parallel sentences. A random sample of 200 parallel sentences was then extracted for manual validation by three human judges who understand both Lumasaaba and English. All three human judges gave a 100% mapping correctness score implying that the translations in the random sample were all correctly mapped between Lumasaaba and English. Based on this validation we can assume that all the other parallel sentences are correct.

5 Data-Driven MT Experiments

The Lumasaaba-English parallel corpus was utilized in experimental setups to evaluate the PBSMT and NMT methods. Both methods involve similar phases of training, tuning and translation; the main differences are in the respective tools used for each method and the models used for translation. The parallel corpus is subdivided into three sets: a training set of 5300 parallel sentences; a tuning set of 1200 parallel sentences; and a testing set of 641 parallel sentences.

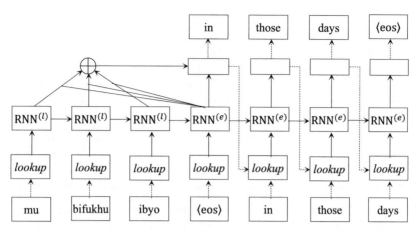

Fig. 1. Illustration of Neural MT from a Lumasaaba sentence to an English sentence. The Lumasaaba words are first mapped to word vectors and then passed to a Recurrent Neural Network. When <eos> symbol is reached, the final time step initializes the target RNN. At each time step, attention is applied over the source RNN and combined with the current hidden state to produce a prediction $p(e_t|e_{1:t-1}, l)$ of the next English word. This prediction is then fed back into the target RNN. Adapted from [9].

5.1 PBSMT Experimental Setup

The freely available moses toolkit [11] which implements the PBSMT method was used for the PBSMT experimental setup. The moses toolkit was the main tool for developing baseline MT systems until 2016. The toolkit has language modeling software [7], tuning software and a decoder software. However, the toolkit utilizes other tools for establishing word alignments. Data preparation, language model construction, training, tuning, and testing were based on the instructions in the moses machine translation user manual and code guide [11]. The data preparation stage involves tokenization (insertion of spaces between words and other symbols), truecasing (conversion of initial words into the most probable casing to reduce on data sparsity), and cleaning (removal of empty and long sentences). Since the size of the corpus is small, all sentences were converted to lower case. The resulting data from the data preparation stage here was also used in the experiments for the NMT method.

Language Modeling. Language modeling is an essential part of PBSMT and it requires a sufficient amount of target language text. For Lumasaaba-to-English MT, English is the target language. The English text used for language modeling is a combination of the English Bible text and English text provided for the MT shared tasks of the 2013 workshop on MT. In total, 164309 English sentences were used for language modeling. The only variation evaluated in the application of the PBSMT method to Lumasaaba-to-English translation is attributed to the different English language models. The KenLM software (which is incorporated in the moses toolkit) was used to implement the language modeling process. Five n-gram models (bi-gram to 6-g) were estimated using Knesser-Ney smoothing without pruning [7].

Training. The training phase involved the following: establishing word alignments between Lumasaaba and English; learning lexical translations; extracting phrases; scoring phrases; learning re-ordering model; learning the generation model and creating a decoding configuration file. The GIZA++ software [15] was used to establish word alignments. The word alignments were then used to estimate a maximum likelihood translation table for building phrase translation pairs which were in turn used to estimate phrase translation probabilities [10].

Tuning. The models from the training phase were tuned using the Minimum Error Rate Training (MERT) implementation in moses. Only mandatory inputs were provided including: the input (Lumasaaba) sentences from the tuning set; the reference (English) translations (also from the tuning set); and where to save the tuned weights.

Translation. The moses decoder was used to automatically translate Lumasaaba sentences to English. The moses decoder implements a beam search algorithm which is designed to reduce computational complexity by pruning away 'bad hypothetical phrase translations' during the translation of 'source input sentences to 'target' output sentences [10]. Default settings of the moses decoder were used along with the configuration file from the tuning phase, the file containing Lumasaaba sentences, and the file that will store the decoder's output of English translations.

5.2 NMT Experimental Setup

The freely available Open NMT [9] Lua/Torch software implementation of an attention-based encoder decoder model was used for the NMT experimental set up. Four different NMT models were configured. The first model was based on the default configuration of OpenNMT where a simple (unidirectional) RNN encoder and Long Short Term Memory units (LSTMs) were used. This first model is denoted as NMT_LSTM_UNI-DIR. To determine which gating mechanism results in better translation quality on the Lumasaaba-English parallel corpus, a second model was configured to use different gating units; a simple RNN encoder was still used but with Gated Recurrent Units (GRUs) instead of LSTMs. Both LSTMs and GRUs are "gating mechanisms" that can be used to learn long term dependencies but they have some key differences. The first difference is in the number of gates used: three gates for LSTMs and two gates for GRUs. The second difference is in whether internal memory is used (LSTMs) or not (GRUs). The third difference is in how the gates are applied: in LSTMs, the input and forget states are combined by an update gate whereas in GRUs, the reset gate is applied directly to the previous hidden state. The second model is denoted as NTM_GRU_UNI-DIR. For the third model, we used a bidirectional RNN encoder with the LSTM gating mechanism. This model is denoted as NMT_LSTM_BI-DIR. So far, the same RNN size was used for the three models. To determine the effect of increasing the RNN size, a fourth model was configured with similar parameters as the third model but where the RNN size is 700. This fourth model is denoted as NMT_LSTM_BI-DIR-700. The standard settings for all the four models are as shown in Table 1 below.

Table 1. Standard hyper parameters for the NMT models.

Parameter	Value
Word embedding size	500
Number of LSTM layers in both encoder and decoder	2
Decoder attention	Global (general)
Maximum sequence size in both source and target	50
Batch size	64
Number of batches (iterations) per epoch	113
Number of epochs	13
Optimizer	SGD
Vocabulary size for Lumasaaba source	11718
Vocabulary size for English target	5145

5.3 Sample of Output from PBSMT and NMT of Lumasaaba-to-English

Table 2 shows a sample of the output translations from the PBSMT and NMT methods. The NMT approach generates reasonably sensible output which is, in most cases, not related to the input sentence. The PBSMT approach outputs some correct words and phrases that are related to the input sentence but the whole output may be incomprehensible. Both approaches should indeed benefit from huge amounts of training data to improve their generalizability on unseen input.

5.4 Automatic Evaluation

The popular BLEU metric [13] was used for evaluating translations from both NMT and PBSMT approaches. The metric considers n-gram matches between the automatically generated translations and reference translations. Table 3 shows the BLEU score results for Lumasaaba-to-English translation for the two approaches.

Table 3 shows that NMT models perform poorly on the small Lumasaaba-to-English parallel corpus. This is to be expected since as observed in Table 2, NMT models mostly generated output that is not related to the input. In this case, the PBSMT models post promising BLEU scores.

6 Conclusion and Future Work

A major requirement in the application of data-driven MT is a sufficient amount of correct parallel sentences. This paper contributes an initial small sentence-aligned Lumasaaba-to-English parallel corpus that can be used not only for MT but other bilingual applications. The small parallel corpus has been used to evaluate two state of the art data-driven MT approaches for translation involving Lumasaaba. As expected, the PBSMT approach results in promising translation quality (according to the BLEU metric). However, the output translations show that while the PBSMT approach generates some words and phrases that are related to input, the NMT approach tends to

Table 2. Translation output from data-driven MT of Lumasaaba-to-English in comparison to reference translations.

1	Lumasaaba	Ne n'arura lundi khu sawa iye khataru, wabona babandu babandi bemile awo mu khatale nga mbaawo sheesi bali khukhola ta.
	Reference	About nine in the morning he went out and saw others standing in the market place doing nothing
	PBSMT	But he went out again on a third, and saw others standing in the market place so as to what they are doing
	NMT	But when they heard this, they did not find him, but they did not find him.

Table 3. BLEU score results for Lumasaaba-to-English translation based on PBSMT and NMT.

Model	BLEU
(a) PBSMT models	
PBSMT-bi-gram	16.67
PBSMT-tri-gram	20.97
PBSMT-4-gram	23.83
PBSMT-5-gram	24.29
PBSMT-6-gram	24.77
(b) NMT models	
NMT_LSTM_UNI-DIR	2.66
NMT_GRU_UNI-DIR	0.00
NMT_LSTM_BI-DIR	3.56
NMT_LSTM_BI-DIR-700	1.92

generalize to output that is in most cases not related to the input. However, it is obvious that most of the limitations for the two data-driven MT approaches can be mostly dealt with by using more adequate amounts of parallel sentences with bigger vocabularies for training and for tuning. Therefore the continuous development of correct parallel sentences should be part of any future efforts for realizing effective data-driven MT applications involving Lumasaaba. Lastly, only sentence alignments were used to learn all data-driven models described in the paper; it should be interesting to evaluate the data-driven approaches when applied to more linguistically inspired features in combination with the sentence alignments.

Acknowledgments. The work in this paper was supported by funds from a Google Faculty award of July 2012.

References

1. Bojar, O., Chatterjee, R., Federmann, C., Graham, Y., Haddow, B., Huck, M., Yepes, A.J., Koehn, P., Logacheva, V., Monz, C., Negri, M., Neveol, A., Neves, M., Popel, M., Post, M., Rubino, R., Scarton, C., Specia, L., Turchi, M., Verspoor, K., Zampieri, M.: Findings of the 2016 conference on machine translation. In: Proceedings of the First Conference on Machine Translation, pp. 131–198. Association for Computational Linguistics, Berlin (2016)
2. Bojar, O., Chatterjee, R., Federmann, C., Graham, Y., Haddow, B., Huang, S., Huck, M., Koehn, P., Liu, Q., Logacheva, V., Monz, C., Negri, M., Post, M., Rubino, R., Specia, L., Turchi, M.: Findings of the 2016 conference on machine translation. In: Proceedings of the First Conference on Machine Translation, pp. 131–198. Association for Computational Linguistics, Berlin (2017)
3. Brown, G.: Phonological Rules and Dialect Variation: A Study of the Phonology of Lumasaaba. Cambridge University Press, Cambridge (1972)
4. Christodouloupoulos, C., Steedman, M.: A massively parallel corpus: the Bible in 100 languages. Lang. Resour. Eval. **49**(2), 375–395 (2015)
5. de Pauw, G., Maajabu, N., Wagacha, P.W.: A knowledge-light approach to Luo machine translation and part-of-speech tagging. In: Proceedings of the Second Workshop on African Lanugage Technology, pp. 15–20. European Language Resources Association, Valletta (2010)
6. Gupta, S.: A survey of data driven machine translation (2012). http://www.cfilt.iitb.ac.in/resources/surveys/MT-Literature%20Survey-2012-Somya.pdf
7. Heafield, K., Pouzyrevsky, I., Clark, J.H., Koehn, P.: Scalable modified Knewser-Ney language model estimation. In: Proceedings of the 51st Annual meeting of the Association for Computational Linguistics (Volume 2: Short papers), pp. 690–696. Association for Computational Linguistics, Sofia (2013)
8. Karakanta, A., Dehdari, J., van Genabith, J.: Neural machine translation for low-resource languages without parallel corpora. Mach. Transl. **32**, 167–189 (2017)
9. Klein, G., Kim, Y., Senellart, J., Rush, A.M.: OpenNMT: open-source toolkit for neural machine translation. ArXiv e-prints 1701.02810 (2017)
10. Koehn, P.: Statistical Machine Translation. Cambridge University Press, Cambridge (2010)
11. Koehn, P.: Moses statistical machine translation user manual and code guide (2017). http://statmt.org/moses/manual/manual.pdf
12. Pa, W.P., Thu, Y.K., Finch, A., Sumita, E.: A study of statistical machine translation methods for under-resourced languages. Procedia Comput. Sci. **81**, 250–257 (2016). SLTU-2016 5th Workshop on Spoken Language Technologies for Under-Resourced Languages, 09–12 May 2016, Yogyakarta, Indonesia (2016)
13. Papineni, K., Roukos, S., Ward, T., Zhu, W-J.: BLEU: a method for automatic evaluation of machine translation. In: Proceedings of the 40th Annual Meeting of the Association for Computational Linguistics, Stroudsburg, PA, USA, pp. 311–318 (2002)
14. Wu, H., Wang, H.: Pivot language approach for phrase-based statistical machine translation. Mach. Transl. **2007**(21), 165–181 (2007)
15. Och, F.J., Ney, H.: A systematic comparison of various statistical alignment models. Comput. Linguist. **29**(1), 19–51 (2003)

Collecting the Database for the Neural Network Deep Learning Implementation

Ekaterina Isaeva[(✉)], Vadim Bakhtin, and Andrey Tararkov

Perm State University, 15, Bukirev Street, 614990 Perm, Russia
ekaterinaisae@gmail.com

Abstract. The paper reviews the work done by the research group in the field of text mining for automating term system construction, i.e., term identification and categorization. The automation is achieved via the usage of binary search trees, a multilayer version of the Rosenblatt's perceptron. The algorithm comprises elements of supervised machine learning with given classes. The paper provides detailed description of the methods and digital resources (online dictionaries, taggers, and corpora) used for manual collection of the database for the subsequent neural network learning. Finally, we present our perspectives on the developed software modification with the transition to deep learning.

Keywords: Terminology · Neural network · Rosenblatt's perceptron
Deep learning · Frame-based terminology · Linguistic analysis
Supervised machine learning

1 Introduction

Modern society is centered around information, its timeliness, precision, completeness, its adaptability into social context, etc. In terms of cognitive sciences information, possessing such properties, is called knowledge. Especially valuable is special knowledge generated in and as a result of professional activity in some industrial and scientific areas, stored in the form of ontologies in the minds of this activity participants, and represented in the form of languages for special communication, terminologies, and term systems. This knowledge is an indispensable prerequisite for the areas development and persistent growth. The rapidness of knowledge generation, modification, and updating motivates research into effective ways of transferring, receiving, processing, and storing special knowledge.

2 Background Knowledge

For this reason, in 2013 our research team initiated a project on development of software for terminology collecting and systemizing. The idea consisted in making an application for the information system (IS) "Semograph" [1], which served as the platform for knowledge base building. The knowledge base included approximately 500 computer security terms in English with their equivalents in Russian, dictionary definitions in English and the contexts of the terms usage in specialized texts. For the

T. Antipova and A. Rocha (Eds.): DSIC 2018, AISC 850, pp. 12–18, 2019.
https://doi.org/10.1007/978-3-030-02351-5_2

term was taken any word (or a two- or three-word combination) encountered in a specialized text and having a special meaning relevant only in a particular subject area (in our case, Computer security). Each term was assigned a field (category), which it belonged to according to the experts' opinion. Starting out from this database the project team developed a program which included some elements of supervised machine learning for text mining implementation and automatic term system enlargement. In the core of our program TSBuilder (State Certificate of Computer Program Registration no. 20166128980) there were two major algorithms, namely the binary search tree and the Rosenblatt's perceptron. The program was aimed at identifying terms (which are already included into the database or new ones), relying on the lemmas (the word stems without inflexions), followed by the categorization of the newly identified terms.

The first module used the method of decision trees for term identification. The procedure consisted of the following steps:

Step 1. Identification of separate words. At this step the text was split into words, similarly to what the programs like CLAWS [2], used in such computer linguistic enterprises as British National Corpus do. The delimiters (space marks) were dropped out. The words were labeled with coordinates, which designated the number of the paragraph and the number of the word in the paragraph.

Step 2. Words lemmatization. As mentioned earlier, word endings were neglected at this step.

Step 3. Retaining lemmas with their coordinates and the flags, which showed whether the words are determined as terms (initially set to false), in the binary search tree.

Step 4. The tree search for terms. When the search was successful, the flag was changed to true.

Step 5. Creating an ordered list of identified terms according to their coordinates in the text.

Step 6. Creating an ordered list of new terms consisting of several words. The word combination was considered a term if the word order coordinates of its elements were sequential and the paragraph coordinates coincided.

The second module dealt with the categorization of the identified terms. This was done with the help of a neural network, which involved the principle of the perceptron, introduced by Frank Rosenblatt in 1957, who applied the mathematical neuron by McCulloch and Pitts and developed a rule for pattern recognition. His task was to separate data into two classes with reference to predefined examples. The algorithm simulated the way the brain neural network operates. The natural process, called synapse enables communication between neurons by means of electric and chemical signals. In brief, the signal is transmitted from one neuron to another if the voltage reaches a certain threshold.

In our database the terms had been classified and the weights had been initialized as 0. One-word terms were classified according to the categories they referred to in the database. For categorization of two- or three-word terms the key parameters were field tags and weights of corresponding fields. Given them the program assigned the category to the whole term.

In TSBuilder the threshold for assigning a certain class tag was determined by the following two conditions:

$$W1 \geq W2 \tag{1}$$

$$\frac{1}{e^{-(w1+w2)}+1} > \frac{1}{2} \tag{2}$$

where W1 and W2 stand for the weights of the first and the second words' categories respectively.

If these conditions were satisfied, the term was assigned the category of the first word, otherwise the second, i.e. the word was categorized in compliance with the class of the word, which has the highest weight.

The terms, which consisted of three words, required a two-layer network for their categorization. These layers could roughly be compared with the convolution layer and the pooling layer respectively. The first layer worked in the same way as in the algorithm above, which refers to the recognition of two-word terms, namely it received as the input the pairs of weights of w1, w2; w1, w3; and w2, w3 and output the highest weight. The tag of the field with the maximum weight was transmitted to the next layer, which worked similarly to the pooling layer. This layer was to decide which category should be assigned to the whole three-word term.

The machine teaching with an expert was achieved by means of the rewards and punishment method, which comprised two options: the initial weight of each word equal to 0.5 was not changed if the classification was correct; otherwise the word with the correct class received the increase in its weight (+1), while the wrong one got the decrease (−1).

3 Current Development

The attempt to alter the algorithm for term identification and categorization is inspired by the research into frame-based terminology, which proves the idea that "trying to find a distinction between terms and words is no longer fruitful or even viable, and that the best way to study specialized knowledge units is by studying their behavior in texts" [3]. By specialized language units this theory disciples understand "compound nominal forms that are used within a scientific or technical field and have meanings specific of this field as well as a syntactic valence or combinatory value" [3]. Taking this into account our multidisciplinary research group have been working on collecting a context-based collection of terms and developing a complex of linguistic rules for underpinning algorithms that will enhance artificial thinking and decision making while implementing computer aided terms identification and categorization.

Thus, our database is built up of terms found by means of the continuous sampling method. We have focused on computer security terminology. The terms have been picked out from specialized texts. This work has been carried out by the students of computer security of Perm State University in the course of English for special purposes while doing tasks on extensive reading of the modern articles, books, and documentations on their specialization.

The database includes the same elements that have been essential in the Semograph collection, namely terms (in English and Russian), definitions, fields, and contexts. To provide a reliable equivalent of an originally English term in Russian we use the following online dictionaries: Multitran [4], Linguee [5], or Reverso context [6]. For definitions we search in online explanatory dictionaries and thesauri, such as The Longman Dictionary of Contemporary English Online [7], Macmillan Dictionary [8], and Your Dictionary [9], as well as specialized dictionaries, glossaries, encyclopedias, and professional forums like SearchSecurityTechTarget [10], Encyclopedia PCMag.-com [11], Technology Dictionary [12], Computer Security Concepts [13], Securelist [14], Computer Hope [15], Tech terms [16], About tech [17], Panda Security [18], etc. Definitions are very important in our research, because they build a bridge between the experts and linguists and facilitate coherence of the linguistic analysis and the expert knowledge.

Beyond that some other linguistic elements have been taken into consideration, in particular, part of speech, morphology, specific semantic categories, general semantic categories, and syntactic roles. We believe that all these factors are subconsciously taken into account in natural cognitive processes of precepting, recognizing, and categorizing words of languages for special purposes.

Let us consider in detail the linguistic aspects, evaluated in our research as the most relevant.

Part of speech (POS) identification can be carried out manually by linguistic experts or automatically with the help of on-line resources like "WordNet", the lexical database for English [19], which differentiates nouns, verbs, adjectives and adverbs, or a more extensive system of POS tagging (61 general categories and subcategories), which is embedded in the British National Corpus [20]. For our purposes we have chosen more concise POS tagging, which includes nouns, verbs, adjectives, and adverbs. Most of the terms in our database are nouns, but we also include other words that make up the frame of the nominal term,

The next linguistic aspect to be tagged is the word morphology or word building, which include the word lemma, root, prefix, suffix, and ending. In our project we tend to manual morphologic tagging by linguistic experts. Each morphological element is analyzed semantically regarding its etymology. For this purpose, we use the online Cambridge dictionary to determine the meanings of prefixes [21] and suffixes [22], alongside with the Online Etymology Dictionary [23].

We also consider if the word only refers to the language of Computer security or can also be used in the language for general purposes with a more common meaning. In the latter case the term is assigned an additional category according to WordNet semantic categorization [19].

Concerning specific characterization of the terms they have been subdivided into 15 classes, including Command/Instruction, Computer Networks, Hardware, Malfunction, Organization, Software/part of software, Virus type, Virus name, Operating systems, Programming languages, File format, Mathematics, Type of malicious activity, Malefactor, and Other. Initially categorization is done manually by the experts in computer security.

The last aspect of our concern is the semantic roles (according to Fillmore [24]) which the terms play in their contexts. Classically, the following semantic roles are highlighted: Agent (initiator of the action), Counter-Agent (the force against which the action is carried out), Objective (something that moves or changes), Result (something that appears as a result of some action), Source (the place where some movement starts from), Goal (the place where something moves to), Experiencer or Patient (someone or something that experiences the effect of an action), Instrument (the stimulus or cause), and others.

Let us take a closer look at an example of the frame-based analysis of the term *botnet*:

At that time the "company" had two key "products": the malicious program, Lurk, and a huge botnet of computers infected with it [25].

The word *botnet* is a noun. In this sample it is used in a singular form. It consists of two grammatical roots, *bot* and *net*. The root *bot* means "a computer program that works automatically, especially one that searches for and finds information on the Internet [26]. The second root *net* denotes "a system of interconnected computer systems, terminals, and other equipment allowing information to be exchanged" [27]. The whole term comprises both semantic meanings and designates "a large number of computers that someone has secretly gained control of and uses to do things such as send spam" [7]. In its frame the term *botnet* plays two semantic roles:

(1) Objective (something the moves or changes): *"the "company" had ... a huge botnet"*. The role is handled by the verb *had*, which means "to own something" [8] and does not imply any changes of the following object.
(2) Instrument (the stimulus or cause): *"and a huge botnet of computers infected with it"*. The verb infect, which means "to affect with a computer virus" [28], determines the role of *botnet*, which is used for affecting *computers*.

The interrelations of all these parameters determine the meaning of the term, its connotation, cognitive model and influence the term's membership in a certain category.

Another method of data preprocessing we are going to add to TSBuilder algorithm is aimed at predicting the probability of multy-word terms categorization. All the data is analyzed with reference to their plausible relation to some category. Presumably, most of the terms are not purely bound to one category, but to different extend can be referred to several categories. At the stage of text mining the degree of this reference will define the weights of the categories in particular cases.

4   Prospects of Further Development

In furtherance of our research we are going to perform a number of updates of the program TSBuilder. First of all, we are introducing "masks" for selection of terms, which will help to exclude noisy data (redundant linguistic units, such as functional words) from the final sample.

Now the program has a two-level hierarchy of term search, namely paragraph number and word number. In the process of program development, it is planned to move to the four-level hierarchy of terms search, namely the numbers of the paragraph, the sentence, the part of the sentence, and the words in the part of the sentence. This will increase the accuracy of detecting terms.

Now, when categorizing, all pairs of words in a term are looked at. In the updated program only pairs of neighboring words in the term are to be viewed, which will increase the importance of neighboring pairs in categorization.

We are determined to identify terms of any length using the term contraction method for categorization. For this purpose, it is planned to use the approach of convolutional networks: first to reduce the term to a mask with a length of 3 words, then to use the current network for categorization. The system will be allowed to include one word in the search term that is not originally a term but is part of a coherent linguistic construction. To do this we are going to implement the analysis of such parameters as POS, gender, number, etc. This will help significantly increase the number of potentially found terms.

5   Conclusion

The article gives an overview of the work done by the research group in the field of text mining for professional communication enhancement. We have reported on the results achieved in the automation of term system development, particularly in terms identification and categorization via supervised machine learning. The program we have developed realizes the algorithm of a multilayer Rosenblatt's perceptron and is able to identify and categorize terms in accordance with the collection of terms' lemmas and predefined classes assigned to the terms. We have also discussed the routine of the database gathering and data preprocessing, which includes morphological, semantic, and syntactic analysis, and makes up the foundation for frame-based terminology management. In the end of the article we have presented our perspectives on embedding the elements of deep learning for upgrading precision of terms identification and categorization.

Finally, we consider the ways the updated program TSBuilder can be used as an operating tool for automated term system building in linguistics and professional language teaching not only in the sphere of computer security, but other sciences as well. This software can also find application in dictionary development and enable automated dictionary appending and updating. In addition, it can be employed by interpreters in the field of professional communication.

Acknowledgments. The reported study was funded by RFBR according to the research project № 18-012-00825 A.

References

1. Semograf: https://semograph.com/. Accessed 30 July 2018
2. CLAWS: http://ucrel.lancs.ac.uk/claws/trial.html. Accessed 30 July 2018
3. LexiCon Research Group: http://lexicon.ugr.es/fbt. Accessed 30 July 2018

4. Multitran: www.multitran.ru. Accessed 30 July 2018
5. Linguee: www.linguee.ru. Accessed 30 July 2018
6. Reverso Context: context.reverso.net. Accessed 30 July 2018
7. The Longman Dictionary of Contemporary English Online: www.ldoceonline.com. Accessed 30 July 2018
8. Macmillan Dictionary: www.macmillandictionary.com. Accessed 30 July 2018
9. Your Dictionary: www.yourdictionary.com. Accessed 30 July 2018
10. SearchSecurityTechTarget: http://searchsecurity.techtarget.com. Accessed 30 July 2018
11. Encyclopedia PCMag.com: www.pcmag.com/encyclopedia. Accessed 30 July 2018
12. Technology Dictionary: www.techopedia.com/dictionary. Accessed 30 July 2018
13. Computer Security Concepts: http://hitachi-id.com/concepts. Accessed 30 July 2018
14. Securelist: https://securelist.com/encyclopedia. Accessed 30 July 2018
15. Computer Hope: www.computerhope.com/jargon.htm. Accessed 30 July 2018
16. Tech Terms: https://techterms.com/definition/api. Accessed 30 July 2018
17. About Tech: http://pcsupport.about.com/od/glossaryterms. Accessed 30 July 2018
18. Panda Security: www.pandasecurity.com/russia/homeusers/security-info/glossary. Accessed 30 July 2018
19. WordNet: http://wordnetweb.princeton.edu/perl/webwn. Accessed 30 July 2018
20. British National Corpus: http://www.natcorp.ox.ac.uk/. Accessed 30 July 2018
21. Cambridge Dictionary: Prefixes from English Grammar Today: https://dictionary.cambridge.org/grammar/british-grammar/word-formation/prefixes. Accessed 30 July 2018
22. Cambridge Dictionary: Prefixes from English Grammar Today: https://dictionary.cambridge.org/grammar/british-grammar/word-formation/suffixes. Accessed 30 July 2018
23. Online Etymology Dictionary: https://www.etymonline.com/. Accessed 30 July 2018
24. Fillmore, Ch.J.: Frame semantics. In: Linguistics in the Morning Calm: Selected Papers from the SICOL-1981, Hanship, Seoul, pp. 111–137 (1982)
25. Stoyanov, R.: The Hunt for Lurk: how we helped to catch one of the most dangerous gangs of financial cybercriminals. https://securelist.com/analysis/publications/75944/the-hunt-for-lurk/. Accessed 30 July 2018
26. Cambridge Dictionary: https://dictionary.cambridge.org/dictionary/english/use. Accessed 30 July 2018
27. ABBYY Lingvo: http://www.lingvo.ru. Accessed 30 July 2018
28. The Free Dictionary by Farlex: https://www.thefreedictionary.com/infect. Accessed 30 July 2018

Digital Economics

Digital Economy: Model for Optimizing the Industry Profit of the Cross-Platform Mobile Applications Market

Ilona Tregub$^{(\boxtimes)}$ ⓘ, Nataliya Drobysheva ⓘ, and Andrey Tregub ⓘ

Financial University under the Government of the RF, Moscow, Russia
ITregub@fa.ru

Abstract. In the conditions of wide distribution of all kinds of mobile devices, the development and use of such applications operating on a variety of platforms from different manufacturers are of particular importance. Such applications are called cross-platform. The paper considers a two-sided market of mobile cross-platform applications. The most important elements of the market are identified. They comprise consumers of the first category - mobile applications consumers - who are simultaneously subscribers of a certain mobile communication operator; consumers of the second category - content providers - that enter into agreements with platform manufacturers to obtain application development tools and mobile operators to ensure the operability of developed applications in the mobile operator networks; and platform-software or a mobile device in which applications can be used. It is noted that the main distinction of the market under study is the interaction between the application producer (provider) and the consumer (subscriber), which is always conducted not only by means of a certain platform, but also through a certain mobile operator. Based on the two-sided mobile applications market model that is developed in the article, a model for optimizing the industry profit of the cross-platform applications market for mobile devices is constructed. This model takes into account the interests of both application manufacturers and intermediaries (platform producers and mobile operators), and consumers of the services via these applications.

Keywords: Two-sided markets · Cross-platform applications
Mobile applications · Pricing models

1 Introduction

In 2017, the Government of the Russian Federation developed and approved a program to create conditions for the country's transition to a digital economy. The Autonomous Non-profit Organization "Digital Economy", created by successful Russian high-tech companies, coordinates the participation of the expert and business communities in the planning of implementation, development and evaluation of the program effectiveness. The key results of this activity should be the elaboration of a general scheme for the development of communication networks and infrastructure for data storage and processing in the Russian Federation until 2024 as well as the concept of creating and developing 5G/IMT-2020 networks. In addition, the roadmap contains an action plan

© Springer Nature Switzerland AG 2019
T. Antipova and A. Rocha (Eds.): DSIC 2018, AISC 850, pp. 21–28, 2019.
https://doi.org/10.1007/978-3-030-02351-5_3

for the development of the state digital platform information systems to provide citizens with state and municipal services in the electronic form. Moreover, by 2024 all data is planned to be made available for the use on digital platforms, including mobile phone terminals with mobile Internet.

The mobile Internet audience has been growing twice as fast as the Internet audience as a whole for the past three years, according to the Yandex data. In November 2017, 87 million people used mobile devices to access the Internet via applications in Russia, which amounted to 71% of the total population of the country, according to Mediascope [1].

Mobile applications are software designed specifically for small portable devices, such as a pocket personal computer, smart phones or cell phones. These applications can be native, i.e. pre-installed in the device in the production process, uploaded by the user via various platforms for distribution, or are web applications processed on the client or server side (JavaScript technology).

The most common mobile operating systems for working and developing applications are iOS from Apple, Android from Google, Windows Phone from Microsoft and Symbian from Nokia. Each of these systems has its own characteristics in the development of applications.

As part of the "Digital Economy" roadmap implementation, the idea of creating platforms for a network of Internet-connected objects capable of collecting data or exchanging data coming from embedded services, the so-called Internet of Things (IoT), has become widespread. Devices included in the Internet of things are any autonomous devices connected to the Internet, which can be monitored and/or managed remotely.

So, the mobile operator "Megafon" announced the change of billing increment. The new billing plan for machine-to-machine (M2M) is based on the basic plan "Remote objects management" having the access to the "Internet of Things for Business" and "IoT without Borders" services. The former will be beneficial for companies that transmit data using communication services (for example, data of hot and cold water meters), and the latter will help companies reduce the dependence on roaming and avoid charging beyond the limit. Basing on the results of the second half of 2017, Megafon connected about 4.5 million SIM cards to M2M, according to the "Kommersant" publication.

MTS plans to launch a platform for deploying IoT at enterprises. It will have an open programming interface and the ability to support work with data networks of various types. For example, you can use the development to monitor the operation of ventilation, water and air conditioning systems. Tele2, VimpelCom and Rostelecom plan to develop similar platforms.

Increasing consumer demand for the mobile application under development is facilitated by the universality of its use for work in all major mobile operating systems. Thus, the elaboration and use of such applications, which are adaptable on a variety of platforms of different manufacturers, is of particular importance in the context of various mobile devices expansion.

This property of applications has been called "cross-platform". Cross-platform applications for mobile devices are applications that can work on more than one hardware platform and/or operating system and are applied in a wide range of mobile devices that use software from various developers.

To increase the effectiveness of the Digital Economy road map activities using cross-platform applications, a prudent pricing policy is necessary. This would ensure maximum profit to platform producers, on the one hand, and as a result, it would lead to maximum allocations to the budget and non-budgetary funds. On the other hand, it would ensure accessibility of the services using digital economy platforms for ordinary consumers. Therefore, the development of a model for optimizing the industry profit of the market for mobile devices cross-platform applications that takes into account the interests of both application producers and intermediaries - platform manufacturers and mobile operators - and consumers of the services via these applications is very timely and relevant.

2   Review of the Literature

The problem of pricing in the two-sided Internet applications market for personal computers, in contrast to the market of mobile device cross-platform applications, is widely discussed in the scientific community.

Ranging from the works on the description of the two-sided market general theory (Armstrong and Wright) [2, 3] to the development of pricing models in these markets as a result of the interaction of both parties, taking into account various network effects, some authors [4] use game-theoretic models to describe the processes of product pricing. Other authors [5] apply models of the logit type for these purposes.

King into consideration the cross-platform in two-sided markets, the research of Choi [6] seems worth mentioning. It resulted in the development of two-sided market models, where both sides have the opportunity to simultaneously use the services of several platforms. The author examined the impact of the measures taken by the platforms to "tie" users, i.e. to compel them to use additional services. In such markets, the users have no opportunity to simultaneously use a number of different platforms.

Evans [7] and Wright [8] considered the issues of "tying" the users to a single platform, as well. The authors conducted the analysis from the standpoint of general consideration of the antimonopoly policy in two-sided markets. However, their discussion was mostly informal and did not deal with the establishment of analytical or quantitative relationships.

Such authors as Rochet and Tirole [9], and Amelio and Julien [10] are notable exceptions in the study of the economic effects of "tying" users to a certain platform in two-sided markets. Roche and Tyrol carried out an economic analysis of the customers' "tying" practice, initiated by the associations of Visa and MasterCard payment cards. Amelio and Jullien [10] made a more general study of tying in two-sided markets. They considered the situation when producers being on one side of the market wanted to set platform prices below zero to solve the problem of demand management in two-sided markets. According to their analysis, "tying" can serve as a mechanism for introducing hidden subsidies on one side of the market to tackle the problem of increasing demand in two-sided markets. As a result, "tying" of users can benefit consumers in case of a monopoly platform. However, in the context of a duopoly, tying also has a strategic impact on competition. Amelio and Jullien have shown that the "tying" influence on consumer surplus and social welfare depends on the asymmetry degree of external factors between the two parties.

The article by Amelio and Julien and the work by Choi are focused on different aspects of "tying" the users. For example, they compare the effects of tying to different market structures (monopolistic and duopolistic), but do not consider the issues of oligopoly when there are more than two participants in the market.

This work is the development of approaches to the study of pricing processes in two-sided markets [6, 10–16] with regard to cross-platform applications developed for mobile devices under the assumption of the oligopolistic market of mobile communication operators.

3 Model Development

Cross-platform applications developed for the interaction of mobile application market participants can be attributed to innovative products, and the mobile applications market itself can be referred to two-sided markets having two groups of users between which network effects arise. The goals of using the network and the role of users in the network are markedly different. Representatives of various groups, being interdependent, outline different requirements on the functionality of a two-sided network. On the other hand, end-users of the services provided via cross-platform applications for mobile devices use mobile Internet through intermediaries (their mobile operators), who in turn establish their billing for the traffic.

Considering the market of mobile applications as a two-sided one, it is necessary to single out such most important elements as: users of the first category - consumers of mobile applications that are simultaneously subscribers of a certain mobile communication operator; users of the second category - mobile application manufacturers - content providers that enter into agreements with platform manufacturers to obtain application development tools and mobile operators to ensure the functionality of developed applications in mobile operator networks; platform - software or a mobile device in which applications can be used.

The functioning of the market has a number of features that are not inherent in any other market. The main difference is the ways of interaction between the service provider and the consumer (subscriber), which are always developed not only via a certain platform, but also through a certain mobile operator [16]. In this case, the operator has the right to choose from a variety of service providing companies it will cooperate with, and what price will be set for this or that service to the end user.

Currently, the market has more than three hundred organizations that develop services to provide subscribers with. The value of the Herfindahl-Hirschman index, calculated for the first fifty largest companies, does not exceed one-tenth (HHI = 0.049 < 0.100) [14], which indicates a high intensity of competition and a low concentration of the market for producers of additional mobile services.

At the same time, the mobile operators' market, represented by three largest companies, the total share of which in the sector of additional mobile services exceeds 90.0%, is oligopolistic (HHI = 0.317 > .200) [14].

Competition in the provider market leads to a decrease in the price of the application, the value of which is established at the equilibrium of the sectoral demand and supply. Consumers and the operator, transporting services to subscribers via a mobile application, are interested in the reduction of prices. To increase its own profit, the operator will choose those providers that offer their services at the lowest price.

Given a high loyalty of subscribers to their operator, which is observed in the Russian mobile communications market, there is practically no redistribution of consumers among operators. Because of this, the operator, in fact, is a monopolist that can dictate its price of both basic and additional services to the subscribers who have already been connected. At the same time, attracting new subscribers requires the operator to set the minimum price for new billing plans in the oligopoly market.

The pricing process is shown schematically in Fig. 1.

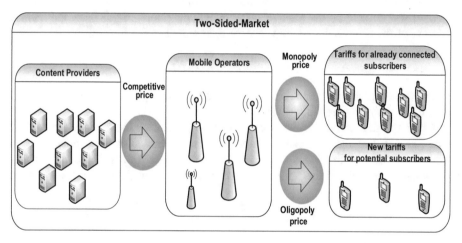

Fig. 1. A pricing model for cross-platform applications for mobile devices on a two-sided market.

Under such a market functioning scheme, it is possible to find conditions for maximizing the profit of mobile operators and industry profits when implementing cross-platform applications for mobile devices. The final purchase price of a cross-platform application for a subscriber, p_i consists of the price formed in the providers' competitive market, p_c and the surcharge established by the mobile operator Δp_i, and $p_i = p_c + \Delta p_i$.

To estimate the operator's premium Δp_i to the price, they consider the conditions for maximizing the operator's profit, defined as the difference between the revenue and costs

$$\pi_i = y_i^d(p_i) \cdot p_i - c\left(y_i^d\right) \tag{1}$$

The revenue of the i-th mobile operator from the provision of applications is equal to the product of the number of applications for the period under consideration and their price. At the same time, the number of the developed applications depends on their price (the law of demand) and is determined by the individual demand function of the mobile operator $y_i^d(p_i)$, represented in quantitative terms [13].

The costs of the i-th mobile operator $c(y_i^d)$ depend on the number of services rendered by using the application and determined by the demand, which, in turn, depends on the price $c(y_i^d(p_i))$.

When determining the price level for cross-platform applications provided to subscribers through their networks (especially when launching new service provision), the i-th mobile operator seeks to increase its profit primarily by attracting new consumers. Therefore, setting the price of a cross-platform application, the i-th mobile operator is guided not only by the price of the service, formed in the competitive provider market and the demand of their subscribers, but also by the prices of similar services implemented in the competitors' networks. So, the price set by the i-th operator will depend on the price of other mobile operators. In this case, the demand function of the i-th mobile operator can be represented in the form of

$$y_i^d = f(p_i, p_j); \ j = 1, \ldots, N; \ j \neq i \tag{2}$$

where j is the number of the mobile operator, and N is the quantity of mobile operators working in the coverage area of the i-th operator.

Given the dependence of the price premium Δp_i, (set by the i-th operator) on the price premiums of their competitors, and the case of an additive dependence of the demand function of the i-th operator on the competitors' prices, the conditions for maximizing the profit of the i-th mobile operator can be put down in a general form

$$\begin{cases} \frac{\partial \pi_i}{\partial \Delta p_i} = y_i^d(p_i) \cdot \frac{\partial p_i}{\partial \Delta p_i} + p_i \cdot \sum_{j=1}^{4} \frac{\partial y_i^d(p_i)}{\partial p_j} \cdot \frac{\partial p_j}{\partial \Delta p_i} - \frac{\partial c(y_i^d)}{\partial y_i^d} \cdot \frac{\partial y_i^d(p_i)}{\partial \Delta p_i} = 0; \\ \Delta p_i = p_i - p_c \\ \frac{\partial^2 \pi_i}{\partial(\Delta p_i)^2} < 0; \quad i = \overline{1, 4}. \end{cases} \tag{3}$$

The first equation in the system (3) reflects the necessary condition for the existence of an extremum, and the inequation in the system (3) is a sufficient condition for the maximum. The solution to the system (3) for each i allows estimating the price extra charges of mobile operators at which their profit is maximum.

The industry profit is determined by the amount of individual companies' profits

$$\Pi = \sum_{i=1}^{4} \pi_i \tag{4}$$

The conditions for optimizing the industry profit are given in the form of

$$\begin{cases} d\Pi(\Delta p_1, \Delta p_2, \Delta p_3, \Delta p_4) = 0; \\ A(d\Delta p_1, d\Delta p_2, d\Delta p_3, d\Delta p_4) = \sum_{i,j=1}^{4} \frac{\partial^2 \Pi \cdot d\Delta p_i \cdot d\Delta p_j}{\partial(\Delta p_i)\partial(\Delta p_j)} \end{cases} \tag{5}$$

The cross-platform usage leads to an increase in the demand for these applications, raising their value and functionality in the mind of a consumer. Undoubtedly, such an application can be considered as the most attractive in the conditions of high competition in the market. In this case, its cost will be higher than the cost of applications that run only on one platform. On the one hand, this will result in an increase in the industry profits and the corresponding deductions to the budget. On the other hand, the drop in the demand for the application due to its price rise, as well as the emergence of two-sided market network effects reflected by the presence of a negative term in the model (3) leads to the possibility of obtaining optimal price values. They take into account both the interests of application producers and the interests of consumers, subscribers of the mobile operators.

In most cases, a mobile application acts as a new product in the market (as an innovation). On the one hand, the revenues earned from its disposal should cover the costs of its development and introduction to the market. On the other hand, the price for this product should contribute to the sufficient intensity of demand, and the goods must be fixed in the minds of consumers.

In addition, there will be a high level of uncertainty or complexity of the product life cycle forecast in the market.

4 Summing Up

Summarizing the research and the results of developing a model for optimizing the industry profit of cross-platform applications for mobile devices, it should be noted that the problems of considering the innovative product pricing in the two-sided mobile application markets have not yet been properly explored either in theory or in practice. This study should be regarded as an attempt to consider the issue of innovative product pricing in the current conditions of increasing competition and the presence of a growing number of developers in the market. Based on the indicated approaches, a further study and modeling of approaches to the pricing of the applications created for various mobile devices can be carried out taking into account the tendency to make use of the cross-platform opportunities.

References

1. Russian Internet Forum. http://2017.russianinternetforum.ru/news/1298. Accessed 02 June 2018
2. Armstrong, M., Wright, J.: Two-sided markets, competitive bottlenecks and exclusive contracts. Econ. Theory **32**, 353–380 (2007)
3. Gale, D., Shapley, L.: College admissions and the stability of marriage. Am. Math. Mon. **1** (69), 9–15 (1962)
4. Moldovanu, B.: Are stable demands vectors in the core of two-sided markets? some graph-theoretical considerations. In: Albers, W., Güth, W., Hammerstein, P., Moldovanu, B., van Damme, E. (eds.) Understanding Strategic Interaction. Springer, Berlin (1997)

5. Grigolon, L., Verboven, F.: Nested logit or random coefficients logit? A comparison of alternative discrete choice models of product differentiation. Rev. Econ. Stat. **96**(5), 916–935 (2014)
6. Choi, J.: Tying in two-sided markets with multi-homing. J. Ind. Econ. **LVIII**(3), 607–626 (2010)
7. Evans, D.: The antitrust economics of multi-sided platform markets. Yale J. Regul. **20**, 325–381 (2003)
8. Wright, J.: One-sided logic in two-sided markets. Rev. Netw. Econ. **3**(1), 44–64 (2004)
9. Rochet, J.-C., Tirole, J.: Tying in Two-Sided Markets and the Impact of the Honor All Cards Rule. Unpublished manuscript (2003)
10. Amelio, A., Julien, B.: Tying and Freebie in Two-Sided Markets. Unpublished manuscript (2006)
11. Laffont, J.-J., Rey, P., Tirole, J.: Competition between telecommunications operators. Eur. Econ. Rev. **41**, 701–711 (1997)
12. Tregub, I.V.: On the applicability of the random walk model with stable steps for forecasting the dynamics of prices of financial tools in the Russian market. J. Math. Sci. **5**(216), 716–721 (2016)
13. Akhmetchina, A.V., Tregub, I.V.: Social investments and goodwill. Int. J. Appl. Fundam. Res. **2**, 1–3 (2014)
14. Belikov, V.V., Gataullina, R.I., Tregub, I.V.: The methodology for the mid-term forecasting of the financial results of firms. http://www.freit.org/WorkingPapers/Papers/FirmLevelGeneral/FREIT711.pdf. Accessed 02 June 2018
15. Tregub, I.V.: International diversification. In: PSTM, Moscow (2015)
16. Tregub, I.V.: Mathematical models of development the world information society. In: Forum for Research in Empirical International Trade, San Rafael, pp. 1–4 (2015)

On Digital Development of Russian Urban Transport Infrastructure

Zhanna Mingaleva[✉]

Perm National Research Polytechnic University, Perm 614000, Russia
mingall@psu.ru

Abstract. The increasing size of the cities requires important of urban environment. It is urgent to optimize urban transport infrastructure. The aim of the work is to show the role of public transport in the harmonious progress of towns and the prospects of transport infrastructure development on the digital technology. The problems which concern different aspects of public transport working and development, namely its work effectiveness improving, particularity of financing and economics of the public transport sector, new approach to the public transport management and its security, introduction of intellectual systems in transport and many others have been examined in the article. The author came the next conclusion: cities are doomed to degradation and worsening of life quality without public transport development. All-round introduction of digital urban transport control technologies including transfer to the new kinds of public and individual urban transport is of great importance today. At the same time substitution of individual motor transport by the public one permits to improve the efficiency of the roads utilization two-three times more.

Keywords: Digital technology · Urban development · Urban environment
Urban transport infrastructure · Public transport · Digital approach

1 Introduction

Since 60s of the twentieth century the process of urbanization has grown in the world. In 2013 the share of urban population in all countries of the world made up 53.0% of the total population size, and in the number of countries achieved 90% (there were such countries as Singapore, Belgium, Kuwait, Iceland, Australia, Luxemburg, Israel, Malta) [1]. In 2018 in Russia there were 74.43% of urban population and 15 cities with a million population [2]. The city of Perm is among them occupying the 13th place in the rating. According to the experts of the World Union Organization urban population will make up 66% of all countries of the world in 2050 [3].

The growth of urban population makes the governments of many states solve the problem of normal transport supply in order to give all citizens an opportunity to reach the places of work and social destination (educational institutions, hospitals), as well as sales outlets, etc. [4–6].

The issues of social justice in transport [7, 8] and the sociological perspectives on labor and educational mobilities [9] are very important.

© Springer Nature Switzerland AG 2019
T. Antipova and A. Rocha (Eds.): DSIC 2018, AISC 850, pp. 29–35, 2019.
https://doi.org/10.1007/978-3-030-02351-5_4

Fundamental system of public passenger transportation is the main factor of social, political and economic stability of towns and cities and rural households warrantee for their harmonious progress [10–13].

2   The Method of Research

The paper presents complex interdisciplinary analysis of social, psychological, legal and economic aspects of urban transport infrastructure improvement.

Comparative analysis methods were also used in this research: the method of comparative analysis of mobility of older persons in Norway, Sweden and Denmark [14] and the benchmarking methods of transport decision-making [15].

Also we examined the results of the correlation analysis between transport, unmet activity needs and wellbeing in later life, which were proposed in the works of foreign researchers [16, 17] and the methods of spatial and demographic analysis of trip making by the example of Hamilton CMA in Canada [18].

Also the paper used the methodology of study the quality of service desired by users of a public transport system [19], the methods of research the problems of road safety, enforcement of the social conditions for drivers in road transport as well as supporting the fair competition between road transport undertakings [20], the methods of research the possibility of using a system which gathers the data from the digital television transport stream, stores all the information in the database on the server and hands out this information to the clients [21], the results of study of transport stream in the satellite digital TV Time Service System [22].

In the course of this study social, economic and institutional determinants of the demand for public transport [23, 24] and the assessment of public transport demand elasticities [25, 26] were used.

3   Public Transport Priority Development

In recent decades one of the main tendencies of Western European towns' development has become the refusal from "adjustment to automobile" concept for the benefit of public transport infrastructure growth. It is remarkable that the emphasis is made on rebirth of rail public transport, namely on the light rail train. Just in this direction French, German, Belgium and other European countries' towns is being developed.

The goal of the public transport systems' development has the priority of modernization of transport infrastructure in Russian towns. According to the approved in 2014 "Transport Strategy of the Russian Federation for the period until 2030 [27], there is considerable attention given to the growth of municipal transport and municipal transport infrastructure. At the same time solution of this problem legislatively transferred to the state power bodies of the RF subjects and urban municipality.

As the experience of foreign countries shows joint activity and initiatives of urban municipalities and central government is indispensable for solving the most urgent problems, connected with transport accessibility in big towns, for legislative solution of conflicts, impeded the development of urban transport infrastructure. World practice shows that in democratic states urban municipality is conservative by its nature as far as any managerial or investment decisions are connected with taxpayers costs and are limited by the budget size. In these conditions town as the subject of the social system begins to make its actual problems only in three cases: (1) when traditional decisions have already been inefficient and further worsening the situation and could be seriously disruptive for the town; (2) when town begins to lose its position in global competition (and for Russian conditions – in regional one) for investments and better personnel; (3) when the state itself initiate the programs of urban modernization and achievement new, higher quality of life in towns.

Thus, the development of transport infrastructure and the modern public transport in many Russian towns, mainly in Moscow and St.-Petersburg, as well as in other megalopolises, which allows preventing the lowering of the life quality in towns, is possible only due to the initiatives and support of the state government.

4   ICT Utilization for Traffic Infrastructure Control

The second important direction of transport infrastructure development both in the countries of the world and in Russia is the effective utilization of the already existed objects of transport infrastructure instead of a large scaled construction of new ones. This could be done at the expense of wide introduction of modern ICT for the urban traffic flows control (creation of the so called intellectual transport systems).

ICT utilization for traffic infrastructure control is one of the most effective and the least expensive ways of solution the problems mentioned above. Introduction of intellectual transport systems permits to increase significantly the capacity of the existed road network (thereby to reduce expenditures on the construction of new roads, the cost of which is overwhelming), to improve transport accessibility of many objects, to reduce and make predictable the time of movement around the town due to the absence of traffic congestions, to improve the urban logistics and the work of public transport, to decrease harmful emissions, to bate cars and roads service expenses, to rise traffic safety for all its participants, to reduce time for accident response as well as to solve some other problems connected with road traffic. Thus, the evaluation of the efficiency of intellectual transport systems, already working abroad, has proved the fact, that these systems increase the capability of the roads by 15–50%, decrease the accident rate by 20–40%, and reduce the traffic violation number by 1, 5–2 times.

Moreover, introduction of intellectual transport systems allows reducing ecological damage. Only at the expense of traffic flows control optimization fuel consumption could be lessen and, correspondingly, the bulk of harmful emission is possible to decrease by 10–20%. According to the predictions of Euro commission, creation of all-European intellectual transport system will result in the reduction of accidental death rate by 1.7 times [28]. ICT are not only an efficient means of transport flows' optimization and improvement of traffic safety, but they enable considerable increase of cost efficiency for road network development and operation.

In some cases introduction of the elements of intellectual transport systems could change the construction of new objects of road infrastructure (for example, new road junctions). Meanwhile, according to the estimations the average cost of one crossing capital repairs is 5–6 times higher the cost of equipment of the same crossing in Moscow [28]. Experts prove that only at the expense of optimization of the traffic lights work on the ordinary Moscow crossing it could be saved 300 h of dead time every day. This indicator is equal to the loss of 950 man-hours.

Optimization of traffic flows control on the base of intellectual transport systems is carried out not only thanks to the "clever" traffic lights, road signs and other elements of road equipment, but also with the help of computer-based data exchange between the elements of road infrastructure and automobile (for example, the systems of automated speed limitation, systems of warning about conflicting traffic lane, etc.).

Utilization of intellectual transport systems in order to reduce traffic jams is very important for their introduction.

5 The Experience of Digital Approach to the Development of Transport Infrastructure in Russia

Russia makes only the first steps of introduction the elements of intellectual transport system into the development of urban transport infrastructure.

The first project of such kind, namely the project of complex intellectual transport system creation has begun to be realized in Moscow. The developed system consists of number of subsystems, providing data gathering and their real time transfer to the center of traffic control, information processing, informing traffic participants about the situation in the streets of the town and efficient traffic flows regulation.

Complex experiment consisting of utilization of the traffic movement adaptive control systems is being carried out in Nizhny Novgorod. In the frame of this experiment the main phases of complete urban intellectual transport system creation have been worked out.

In Perm the direction of public transport development has been determined by municipal plan of general development of the town. Perm general plan is quite unusual document for Russian megalopolises. It is not an official legal document, but just by this plan the main vectors of the town's growth in the long-term outlook have been defined. And namely this master-plan has become the basis for the new municipal General Plan. Creation of effective urban transport infrastructure, accounting the modern and perspective levels of citizens' auto mobilization is being one of the main goals of the General Plan of Perm [29].

Also at present municipal government of Perm is working out the concept of public transport development. The goal of this concept is the definition of directions and principles of urban public transport development, accounting the needs of citizens, for the period till 2015. In the frame of this document the elaboration of single standard of the Perm public transport work quality is supposed to be done. This standard should take into consideration the citizens' demands [30]. The concept is being worked out accounting the proposals and recommendations concerning the development and reforming the urban public transport, mentioned in the Transport Strategy of the Russian Federation for the period till 2020, as well as in the Transport Strategy of Russian Federation for the period till 2030 [31]. In the Strategy the main strategic goals of the state in the sphere of transport policy have been defined. They are as follows:

- development of modern and effective infrastructure;
- improvement of accessibility of public transport complex services;
- increase of competitive ability of the Russian transport system and realization of transit potential of the country;
- growth of complex safety and transport system stability;
- improvement of investment climate and market relations development in transport complex.

And at last, it is necessary to mention successful experience of Novokuznetsk, Omsk and Tomsk. There the developed and approved urban public transport network with the determined routes and stops of public transport were revealed by the analysis of the passenger traffic investigation. The next elements were taken into consideration:

1. The length of all town streets, including the primary roads of municipal and regional importance.
2. The length of the streets with the public transport movement.
3. The average capacity of one vehicle unit.
4. The total number of passengers in the separate kinds of transport with distribution of them to the paid and preferential ones.

Banning of local (urban, regional) intellectual transport systems' creation on different technological platforms, with different architecture and functionality is being one of the goals of creation the single information field of control and monitoring of the Russian Federation transport complex condition. Otherwise in future, on the stage of unification of these local systems into the national one the problem of separate systems compatibility and impediment for their integration into the intellectual transport system of the higher level – the national - could arise.

6   Conclusion

In the last decade municipal government of many Russian towns began seriously think of public transport work, improvement of its accessibility and of transport ser-vice quality. All-round introduction of new urban transport control technologies including transfer to the new kinds of public and individual urban transport is of great importance today. Digital equipment and digital devices are utilization for traffic infrastructure control is one of the most effective and the least expensive ways of solution the problems. Digital equipment and digital devices are is not only an efficient means of transport flows' optimization and improvement of traffic safety, but they enable considerable increase of cost efficiency for road network development and operation.

It has already been proved that the town where every citizen uses his individual car won't be comfortable. It is impossible to provide enough parking places, roads in the center of the town for every car owner, as the number of automobiles increases annually.

To overcome negative tendencies in the development of urban transport infrastructure in Russia and to provide population with more qualitative transport services it is advisable to use the abroad experience, where the development of public transport has become the first priority in the activity of both the national power and municipal officials.

Substitution of individual motor transport by the public one permits to increase the effectiveness of the road network utilization two-three times.

Acknowledgment. The work is carried out based on the task on fulfilment of government contractual work in the field of scientific activities as a part of base portion of the state task of the Ministry of Education and Science of the Russian Federation to Perm National Research Polytechnic University (topic # 26.6884.2017/8.9 "Sustainable development of urban areas and the improvement of the human environment").

References

1. Human Report Development, pp. 216–219. UNDP, New York (2014)
2. The population of the Russian Federation by sex and age on January 1, 2018. Rosstat. http://www.ru. Accessed 14 June 2018
3. World Urbanization Prospects: The 2014 Revision. United Nations Press, New York. http://esa.un.org/unpd/wup/Highlights/WUP2014-Highlights.pdf. Accessed 14 June 2018
4. Bohte, W., Maat, K., van Wee, B.: Measuring attitudes in research on residential self-selection and travel behaviour: a review of theories and empirical research. Transp. Rev. **29** (3), 325–357 (2009)
5. Ryan, J., Wretstrand, A., Schmidt, S.M.: Exploring public transport as an element of older persons' mobility: a capability approach perspective. J. Transp. Geogr. **48**, 105–114 (2015)
6. Boschmann, E.E., Brady, S.A.: Travel behaviors, sustainable mobility, and transit-oriented developments: a travel counts analysis of older adults in the Denver, Colorado metropolitan area. J. Transp. Geogr. **33**, 1–11 (2013)
7. Beyazit, E.: Evaluating social justice in transport: lessons to be learned from the capability approach. Transp. Rev. **31**(1), 117–134 (2011)
8. Metz, D.: Transport policy for an ageing population. Transp. Rev. **23**(4), 375–386 (2003)
9. Cairns, S., Harmer, C., Hopkin, J., Skippon, S.: Sociological perspectives on travel and mobilities: a review. Transp. Res. Part A Policy Pract. **63**, 107–117 (2014)
10. Davoudi, S.: Polycentricity in European spatial planning: from an analytical tool to a normative agenda. Eur. Plan. Stud. **11**(8), 979–1000 (2003)
11. Anas, A., Arnott, R., Small, K.A.: Urban spatial structure. J. Econ. Lit. **36**(3), 1426–1464 (1998)
12. Casello, J.M.: Transit competitiveness in polycentric metropolitan regions. Transp. Res. Part A Policy Pract. **41**(1), 19–40 (2007)
13. Smith, N., Hirsch, D., Davis, A.: Accessibility and capability: the minimum transport needs and costs of rural households. J. Transp. Geogr. **21**, 93–101 (2012)
14. Hjorthol, R.J., Levin, L., Sirén, A.: Mobility in different generations of older persons. The development of daily travel in different cohorts in Denmark, Norway and Sweden. J. Transp. Geogr. **18**(5), 624–633 (2010)
15. Jones, P., Lucas, K.: The social consequences of transport decision-making: clarifying concepts, synthesising knowledge and assessing implications. J. Transp. Geogr. **21**, 4–16 (2012)
16. Nordbakke, S., Schwanen, T.: Transport, unmet activity needs and wellbeing in later life: exploring the links. Transportation **23**, 1 (2014)

17. Spinney, J.E.L., Scott, D.M., Newbold, K.B.: Transport mobility benefits and quality of life: a time-use perspective of elderly Canadians. Transp. Policy **16**(1), 1–11 (2009)
18. Páez, A., Scott, D., Potoglou, D., Kanaroglou, P., Newbold, K.B.: Elderly mobility: demographic and spatial analysis of trip making in the Hamilton CMA, Canada. Urban Stud. **44**(1), 123–146 (2007)
19. Dell'Olio, L., Ibeas, A., Cecin, P.: The quality of service desired by public transport users. Transp. Policy **18**(1), 217–227 (2011)
20. Rychter, M., Rychter, R.: The impact of European Registers of Road Transport Undertakings on security and enforcement of the system of digital tachograph. In: IOP Conference Series: Materials Science and Engineering, vol. 148, no. 1, p. 012050 (2016)
21. Pijetlovic, S., Jovanovic, N., Jovanov, N., Ocovaj, S.: One solution of a system for data acquisition and storage from the digital television transport stream and its exposure to the clients. In: 21st Telecommunications Forum Telfor, TELFOR 2013 - Proceedings of Papers, pp. 721–723 (2013)
22. Pan, D., Hua, Y., Xiang, Y.: Research of transport stream and PCR timestamp in the satellite digital TV time service system. Appl. Mech. Mater. **411–414**, 799–802 (2013)
23. Souche, St: Measuring the structural determinants of urban travel demand. Transp. Policy **17**(3), 127–134 (2010)
24. Cordera, R., Canales, C., dell' Olio, L., Ibeas, A.: Public transport demand elasticities during the recessionary phases of economic cycles. Transp. Policy **42**, 173–179 (2015)
25. Bresson, G., Dargay, J., Madre, J.-L., Pirotte, A.: The main determinants of the demand for public transport: a comparative analysis of England and France using shrinkage estimators. Transp. Res. Part A Policy Pract. **37**(7), 605–627 (2003)
26. Bresson, G., Dargay, J., Madre, J.-L., Pirotte, A.: Economic and structural determinants of the demand for public transport: an analysis on a panel of French urban areas using shrinkage estimators. Transp. Res. Part A Policy Pract. **38**(4), 269–285 (2004)
27. Transport Strategy of the Russian Federation for the period until 2030, approved by the Government of the Russian Federation of November 22, 2008 N 1734-r, as amended. Order of the Government of the Russian Federation No. 1032-r of 11.06.2014. http://rosavtodor.ru/documents/119. Accessed 14 June 2018
28. Urban Development: Best Practices and Current Trends. National Report, Moscow, p. 44 (2015)
29. The Master Plan of Perm. Administration of Perm City. https://www.kcap.eu/en/projects/v/perm_strategic_masterplan. Accessed 11 May 2018
30. Concept of uniform information space management and monitoring of the transport sector of the Russian Federation (MIS RF LC). Ministry of Transport: Regulatory Documents. http://www.chitaitext.ru/upload/medialibrary/kons.DOC. Accessed 11 May 2018
31. The Transport Strategy of the Russian Federation up to 2030. Ministry of Transport: Regulatory Documents. www.mintrans.ru/documents/detail.php?ELEMENT_ID=13008. Accessed 11 May 2018

Investigation of the Influence of a Number of Socio-Economic Factors on Migration Flows Between the Federal Districts of the Russian Federation

Sergey Rusakov[✉], Olga Rusakova, and Maksim Immis

Perm State University, Bukireva, 15, Perm 614990, Russia
rusakov@psu.ru

Abstract. The paper examines the trends of migration flows between the federal districts of the Russian Federation in the period from 2009 to 2016. It was revealed that in the given period the previously noted trend remains, in accordance with which the balance of migration is positive only for the three federal districts, and in the remaining districts there is an outflow of population. In addition, it is noted that in 2012–2013 migration flows reach their maximum, and then their intensity begins to fall. A number of socio-economic factors are considered from the point of view of their influence on migration. On the basis of statistical analysis, two factors were selected from them, based on which the linear regression model was constructed for each year of the period under study. It is noted that, at a qualitative level, the change in the parameters of the regression model corresponds to the trends of migration flows.

Keywords: Migration flows · Factors affecting migration · Regression model

1 Introduction

Migration plays an important role in shaping the structure of the population, labor resources, social and economic development of the region and the country as a whole. So the central federal district since the beginning of the twenty-one century has a negative natural increase. This means that it is born less than it dies. But due to an increase in the flow of migrants from other federal districts of the country, the population of the region only increases. In the most other federal districts there is a reverse trend - outflow of population from the region. This is a negative trend. The state, realizing the importance of this phenomenon, tries to solve this problem by creating federal services, developing the migration policy. In the conditions of the development of the digital economy, migratory processes acquire a new quality, since many IT professionals can work in a distributed mode without leaving the region of their residence.

T. Antipova and A. Rocha (Eds.): DSIC 2018, AISC 850, pp. 36–39, 2019.
https://doi.org/10.1007/978-3-030-02351-5_5

2 Methods and Methodology

Analyzing the research and the articles of the authors dealing with the research of migration processes in the Russian Federation, one can come to the following conclusions:

- all researchers agree that the centers of attraction of migrants are the regions of the European part of Russia [1];
- the regions of the Urals, Siberia and the Far East are "donor regions" of migrants;
- population migration is conditioned by the goal of raising the standard of living that has worsened throughout the country since the collapse of the Soviet Union.

In this paper we analyzed data taken from an open source - the site of the Federal Service of State Statistics. At the same time, the migration flows between federal districts (FD) of the Russian Federation in the period from 2009 to 2016 were investigated. The choice of this period is due to a number of reasons. First, in earlier works (see, for example, [1–3]), 2000–2010 were considered. Secondly, until 2009, there were 7 federal districts in the Russian Federation and only in 2009 the North Caucasus Federal District was separated from the Southern Federal District and the total number of districts increased to 8. In addition, in 2014, the Crimean Federal District was formed, which in 2016 was part of the Southern Federal District. Accordingly, data on migration flows of the Crimean Federal District for the period 2014–2016. were summed with the values of the Southern Federal District for the same period.

3 Discussion

The data obtained are presented in Table 1, in which the difference between arriving and departing residents is presented for each federal district (migration balance in thousands). The last column of this table shows the ratio of the total migration balance for all 8 years to the population of the corresponding district in 2009 (in percent).

Table 1. The values of migration flows.

	2009	2010	2011	2012	2013	2014	2015	2016	%
CFD	77.9	103.6	113.5	146.0	142.6	119.4	126.7	86.3	2.5
N-WFD	7.6	7.7	33.1	29.5	41.1	35.8	38.4	38.9	1.7
SFD	7.4	10.7	28.7	16.6	35.8	18.2	29.0	39.7	1.4
N-CFD	−14.4	−19.4	−40.7	−48.2	−46.9	−28.0	−31.5	−31.1	−2.8
VFD	−29.1	−37.7	−61.0	−63.5	−64.9	−50.5	−60.8	−46.1	−1.4
UFD	−6.7	−8.4	1.9	−1.7	−16.4	−13.4	−20.5	−15.6	−0.7
SFD	−19.7	−26.1	−43.2	−42.7	−52.1	−45.8	−48.5	−47.1	−1.7
FEFD	−23.0	−30.4	−32.4	−36.1	−39.3	−35.7	−32.8	−25.1	−3.9

Here and below, the abbreviation is used: CFD – Central Federal District, N-WFD – North-West Federal District, SFD – South Federal District, N-CFD – North-Caucasian Federal District, VFD – Volga Federal District, UFD – Ural FD, SFD – Siberian Federal District, FEFD – Far Eastern Federal District.

Analysis of the table shows that the balance of migration has a positive value only for the three federal districts (CFD, N-WFD, SFD). And the maximum modulo value for all districts, with the exception of the Southern Federal District and the Ural Federal District, is reached in 2012 or 2013.

4   Investigation of the Impact on Migration of Some Factors

Eleven socio-economic indicators of federal districts were analyzed to reveal the factors influencing the migration flows: incomes of the population, expenditures of the population, the number of offenses, the incidence of tuberculosis, the number of unemployed, the average size of accrued pensions, the average price of housing in the secondary market, the consumer price index, investments in fixed capital, the average annual amplitude of air temperature, the number of pensioners per 1000 people of the population. For each of the years under consideration, the coefficients of pairwise linear correlation (the Pearson coefficient) for the values of all these factors were calculated. A strong correlation dependence of a number of them was found, which made it possible to reduce the number of factors considered to 5. Further, for the remaining factors, models of multiple linear regression were constructed, the coefficients of which were checked for significance in accordance with the t-test. Finally it was obtained that the level of significance of 0.1 is satisfied by the parameters of the regression model of the form:

$$Y(N) = b_0(N) + b_1(N)X_1(N) + b_2(N)X_2(N) \qquad (1)$$

$N = 2010, \ldots, 2016$ years,
$Y(N)$ - migration rate,
$X_1(N)$ - values of the factor "incidence of tuberculosis",
$X_2(N)$ - values of the factor "investment in fixed assets".

Under the migration factor in this paper we will understand the migration balance attributed to 100 thousand people of the population.

Table 2. The results of the analysis for the coefficients of the regression equation for significance level 0.1.

N	$b_0(N)$	$b_1(N)$	$b_2(N)$	R-square
2010	27.0	−4.52	0.22	0.74
2011	−101.9	−4.74	0.27	0.56
2012	−267.7	−4.49	0.31	0.66
2013	11.4	−6.49	0.18	0.56
2014	144.8	−7.44	0.12	0.67
2015	145.1	−7.38	0.11	0.59
2016	133.2	−6.01	0.07	0.44

Analyzing the obtained Table 2, it can be concluded that the incidence of tuberculosis negatively affects migration flows to the region, and investment in fixed assets is positive. The resulting coefficient of determination is not very high, but it can still indicate the validity of the model obtained, even at a qualitative level.

It can be noted that the incidence of tuberculosis is strongly associated with the level of crime, the standard of living and the climatic conditions of the region, and investments have a serious impact on the income level of the population. Therefore, the presence of these two factors in the resulting regression model is fully understandable. In addition, in the values of the parameters of the model two pronounced trends are observed in the periods of 2010–2012 and 2014–2016: in the second period the value of the modulus of the coefficient $b_1(N)$ increases approximately by 1.7 times, and for the coefficient $b_2(N)$ decreases by approximately 2.7 times. And according to the data in Table 1, we can see that for the first period, the increase in migration flows is characteristic, and for the second, the decrease.

5 Conclusion

Thus, as the main result of the study, it can be noted that even a highly aggregated regression model (1) at a qualitative level reflects a change in trend from an increase to the reduction of migration flows to the CFD and N-WFD occurring in 2012–2013.

References

1. Tikhomirova, N.E.: Statistical study of population migration in the regions of the Russian Federation. Candidate's thesis, Gos. Economy. University of Samara (2008)
2. Batishcheva, G.A.: Research of factors of migratory exchange between federal districts of Russia. In: Regional Economy: Theory and Practice, vol. 30, pp. 65–73 (2009)
3. Vasilyeva, T.P., Myznikova, B.I., Rusakov, S.V.: Mathematical modeling of the process of grading: a probabilistic approach. In: Scientific and Technical Statements of the St. Petersburg State Polytechnic University, vol. 1, no. 140, pp. 73–79 (2012)

Creative Tools for Interaction of Brands with Consumers on the Digital Technologies Basis

Alexandra Ponomareva[1]([✉]), Maxim Ponomarev[2], and Alexandr Ponomarev[3]

[1] Department of Marketing and Advertising, Rostov State University of Economics and South Federal University, B. Sadovaya, 69, Rostov-on-Don 344002, Russia
alexandra22003@rambler.ru
[2] Department of Economic Theory and Entrepreneurship, Russian Academy of National Economy and Public Administration, Pushkinskaya Street, 70, Rostov-on-Don 344011, Russia
kardiogramma@rambler.ru
[3] Department of Taxation and Accounting, Russian Academy Om National Economy and Public Administration, Pushkinskaya Street, 70, Rostov-on-Don 344011, Russia
ponomarev@uriu.ranepa.ru

Abstract. Creative tools based on the use of digital technologies form a new type of "brand to consumer" communication, its special feature is the realization of the socio-psychological function of marketing. The brand's solution of social and psychological problems of target audiences using digital technologies allows building meaningful, interactive, partner, personalized brand relationships with consumers.

Keywords: Communication marketing · Digital marketing · Social marketing
Art-marketing · Partisan marketing · Flashmob · White paper
Entertainment-marketing · Provocative marketing · Aggressive marketing
Viral marketing · Buzz-marketing

1 Introduction

Nowadays, new creative and innovative tools and tools for communication marketing are being developed and developing along with traditional tools of marketing communications - advertising, Public Relations, Sales Promotion, Direct Marketing, Personal Selling. Digital technologies usage obtain an institutional basis for new types of communication marketing for their implementation in practice, methods and receptions.

Our hypothesis is that the new tools are completely different: the mechanism of their impact on consumers is not pressure, repetition, to attraction and manipulation techniques – but the brands attempt to be consumer-friendly, to give him new information, to solve socio-economic and/or psychological problems that are relevant to an

© Springer Nature Switzerland AG 2019
T. Antipova and A. Rocha (Eds.): DSIC 2018, AISC 850, pp. 40–47, 2019.
https://doi.org/10.1007/978-3-030-02351-5_6

individual consumer, target audience or society, in general. The basis for this communication type is digital technologies, the digital environment for brand-to-consumer interaction, which allows to precisely and subtly orient the brand's communications to the consumer taking into account his style and lifestyle, preferences, problems, values, stereotypes, aspirations, knowledge, principles of integration and omnichannel.

2   Methods and Methodology

To describe the specifics, the nature of the communication interaction of brands with consumers in the digital environment, we use general scientific empirical methods of observation, comparisons, descriptions and measurements, general scientific methods of theoretical knowledge - formalization, hypothetico-deductive method, general methods and methods of research - analysis and synthesis, induction and deduction, generalization, analogy, modeling, system approach, structural-functional method, the method of historical analysis. Disciplinary methods used in our research include marketing analysis methods (qualitative and quantitative, desk, event-analysis, modeling), and interdisciplinary methods - questioning, associative experiment, semantic analysis, etc. The study of the socio-psychological functions of creative tools and digital tools for their implementation was carried out on the basis of a structured-semantic and comparative analysis of materials collected by the authors themselves - more than 300 case studies, which are a description of new popular marketing tools, developed by creative, advertising and marketing- communication agencies, specialized in printed publications and on Internet sites. The object of the study were those case studies in which creative tools were used, the subject being the mechanics of realizing the socio-psychological function of creative tools and tools.

3   Discussion

The research interest in digital media, the nature of communication in the digital environment is growing every year. Thus, Hervás from Universidad Católica de Murcia writes about the Internet as a new interaction environment, and shows the commonality of traditional and digital media. The author shows how the digital environment changes the technology of communication, a special role is given to the visual identity and flexibility of communication - with the preservation of the main message [1].

On the contrary, Polo, Sánchez and Meroño believe that digital communications turn over the previous 5000 years of impersonal communication between people, and that it is digital media that can capture, hear and immediately satisfy the needs of customers. In addition, according to the author, this section of marketing should not be aimed at sales, but, first of all, to collect information: digital communication is called upon to improve itself [2].

Solarte-Vásquez and Nyman-Metcalf of Tallinn Law School considers digital media as the basis for a Smart contracting - a proactive proposal to operationalize the relational contract theory for the upgrade and improvement of legally relevant exchange. The author's vision shows the revolutionary character of communication in a digital environment [3].

Dr Ben Walmsley represents digital media as an opportunity for digital engagement for the purpose of audience enrichment. Digital platforms can promote a slower, deeper, more relational and more democratic engagement with audiences. The author dosen't discuss the mechanisms of digital engagement, but believes in their efficiency.

Generally, research in the field of digital media is focused on formation effective mechanisms and obtaining data.

4   General Characteristics of Creative Tools and Their Socially-Psychological Functions

In our opinion, the usage of marketing creative tools for communications is the basis for creating an enterprise communication marketing system. It is the study of the specifics of creative tools and means of marketing communications that allowed the authors to develop a concept of communication marketing.

What does unite all these different new tools and means of communication marketing? First of all, we can talk about a functional community. The functions of the new marketing communication tools include:

- strengthening the impact on the consumer due to the presence as an obligatory component of creativity;
- increased storage due to the novelty effect;
- building loyalty to the brand;
- building knowledge about the brand;
- increase in demand and increase in the number of sales;
- launching a new product on the market;
- Building brand recognition;
- creation of the effect of integrated impact on the consumer;
- Reduction in the frequency of paid contacts (sometimes up to 1) while preserving, and sometimes increasing the effectiveness of the action;
- Initiation of free communications that promote the product;
- decrease in the degree of rejection of marketing communications by consumers as a factor in increasing their effectiveness;
- creation of such property of marketing communications, as interactivity with the purpose of increase of their efficiency;

The creation of personal communication effects, which can also positively influence the effectiveness of the shares.

A socio-psychological function.

Let us reveal some details on socio-psychological functions, since it is the basis for creating a motivational base for the use of creative tools, and provides interest to consumers. It is based on the fact fact that the creative tool overcomes mistrust and rejection of consumers due to the fact that with its help the advertiser not only solves its marketing tasks, but also addresses important social and psychological problems of the individual consumer. And it doesn't only address, but often offer a way to solve them, and sometimes decides. This kind of mechanics, due to the implementation of this function, have an increased ability to involve consumers in the advertiser marketing communications, create an interactive communication effect that can positively influence their effectiveness. As a result of the analysis, we have identified those creative and innovative tools, in the use of which advertisers are trying to solve certain social and psychological problems of consumers (Fig. 1).

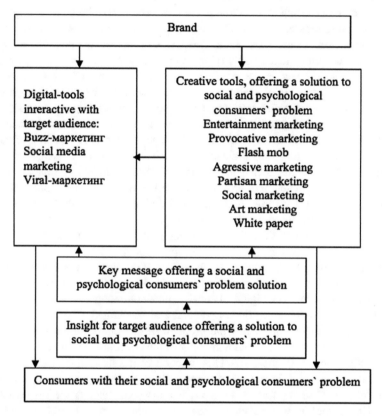

Fig. 1. Social and psychological function of creative tools for of brands to consumers interaction, based on the digital technologies basis

Let's explain these promotion tools.

Entertainment-marketing is a creative tool of communication marketing, consisting in the development and production of entertaining events online or offline, used in the market conditions for the purpose of promotion, during which the consumer comes into direct contact with the product, interacts with it. Interaction leads to the fact that the participant of the E-marketing event becomes a consumer of the product with a high degree of loyalty. The creativity of the event launches viral marketing and buzz communications in the digital environment, reinforcing its effect many times. The essence of E-marketing most accurately expresses the formula: creative interaction with the product for the purpose of consumer entertainment and sales increase based on digital technologies. The following action for a liquor brand is an example of E-marketing: a nightclub entertainment event, based on a famous musicians concert for the target audience; during the event, presenters are offered tasting an advancing beverage with the prize announcement. One of the wine glasses contains a large diamond; a diamond in a glass becomes a prize for the finder; the action has a significant noise-effect and SMM-effect.

Provocative marketing is a communication marketing tool based on a bright event, publication, presentation, etc. violating the moral, ethical, behavioral and other norms that exist in society, causing a scandal, which due to this has the potential energy that can "launch" the viral mechanisms of buzz-informing consumers about provocation and the promoted object, often refers to the teenage sense of protest against the foundations of society, family and school regulations. At the same time, the share mechanics takes into account two target audiences - the adolescents themselves, for whom the product is targeted and the communication is targeted, and the adult audience. Provocation is estimated positively by teenagers and causes rejection of the adult audience, launching buzz communications, which increases the coverage of the tool and additionally motivates teenagers to consume. Some modern companies, e.g., the Russian company Euroset, use provocative marketing as the main promotion tool, combining events with various mechanics and digital tools to enhance the effects of provocative marketing by increasing coverage.

Flash mob is a communication marketing tool based on a strange, not always understandable event, lasting from 1 to 5 min, initiating due to its strangeness and incomprehensibility into the buzz communication. It is calculated both for the primary audience - present at the event, and for the secondary audience, to which the information will reach both via digital channels, buzz-communications, and through the media. Initially flashmob as a non-commercial tool was aimed at solving an important social and psychological problem - the problem of loneliness. In the framework of a flash mob a person can feel himself a member of the collective, jointly carry out certain actions - albeit devoid of meaning. This function of the flash mob was partially preserved in its commercial form. One of the most famous flash mobs, which has already become "classical" (many tried to copy it), is a massive dance at the Liverpool street station. The promotion was developed for the mobile operator T-Mobile, and if you click on the link https://youtu.be/VQ3d3KigPQM, the video will have more than 41 million views (possibly more at the time of publication).

Aggressive marketing is a communication marketing tool that, as a basis for initiating viral marketing and buzz, uses aggressive behavior, an event, a publication, etc., whose elements are violence, cruelty, appeals to the psychological problem of aggression and irritation, as well as the social problem - the tension in society. When using the tool, the consumer gets the opportunity to "discharge", throw out the accumulated negative energy while contacting the product, which affects the economic performance of the stock performance evaluation. So, the hotel management in the center of Madrid provided an unusual marketing experiment, allowing the audience to completely crush its hotel decoration on one of the floors. The idea was prompted by the need to restructure the rooms. Before the bacchanalia, 40 decent citizens who suffered from stress were selected, and they did not miss their own: they broke TV sets on the floor and smashed mini-bars in the rooms. Free information support is impressive - more than 200 journalists with cameras, reporters and photographers.

Social marketing is directly designed to organize a solution of a socially significant problem while simultaneously implementing the marketing objectives of the enterprise. Social marketing is capable of simultaneously performing two important functions: opportunity of media information creating, such as blogging, the image of the product (and this is how social marketing is similar to PR), and the function of promotional product sales increasing (the main function of the sales promotion). It is the sharpening of this tool to solve acute social problems that makes it interactive, increases the involvement of consumers on the basis of their goodwill and willingness to help. Digital-channels at the same time perform both informational and selling functions.

A vivid example of this interpretation of social marketing is the campain of RED: famous fashion & lifestyle brands present their goods of red color for sale. Some of the proceeds from sales go to charitable purposes, namely: assistance to the peoples of Africa. Thus, at the expense of the mechanics of the action, it is possible to realize several goals at once - social, on the one hand, image and increasing the number of sales - on the other. Moreover, the goods in red colour are a design statement for the traditional brands.

Partisan (guerilla) marketing is a creative tool of communication marketing aimed at increasing sales. Its essence is an event that includes a personal contact of the consumer with the brand and communicator in the living space of target audiences, in which the consumer does not realize that they are being influenced for advertising purposes. Thus, the marketing problem of consumer's rejection of communications is solved and an interactive effect is created. Partisan marketing has been particularly developed in the digital environment. Unfortunately, at the present time it plays not only a positive role in the life of society, distorting the communication space in the political, social and economic spheres. Examples of offline guerrilla marketing are various kind of events, fake consumers, in the digital environment these are comments in social networks, participation in forums, online discussions, texts, illustrations, product representation by blogger, videos that cause brand associations that the consumer does not identifies as advertising.

Art marketing - a creative tool for communication marketing, when the consumer is engaged into creativity while simultaneously contacting the product or consuming it. Art marketing usually consists of the following stages: creativity of consumers (a demonstration of creativity of consumers), viral marketing and buzz-communication. Art marketing addresses the need to engage in creativity, create something new, which, on the basis of high involvement and the effect of interactivity, increases the effectiveness of actions aimed at promoting the product. The following action is an example. Consumers are distributed Red Bull drinks in a big pack, at the same time there is an offer to participate in the contest for the best design made from Red Bull cans. Demonstrations of art objects created by consumers are held in the best museums of Russian cities and are covered in the press. The event becomes the starting point for viral, SMM and buzz communications - both personal and through the media.

Viral marketing, unlike Buzz, uses an intermediary - a certain material object - games, flash games (Viral Game), SMS, MMS, video clips, music, electronic games, free soft, pictures, links, anecdotes. All that content is sent to each other by consumers, it is used by communication information means. Viral marketing relies on the socio-psychological need to share information and uses as a basis for creating additional motives for humor, shock effect, fake, joining trends, jokes, unexpected combinations, sex, celebrities, etc. The interactive viral video of the Tipp-ex brand in the active campaign period gained more than 46 million views from 217 countries via Youtube, 1 million reposts via Facebook, 220,000 tweets, 60,000 online article, sales results in Europe + 30%.

Buzz-marketing is a marketing communication tool in which information about an enterprise or product is distributed through personal communication of consumers. We use social media marketing to solve social and psychological problems of lack of communication - both qualitatively and quantitatively, and to create an interactive communication field of an advertiser with a consumer, so and consumers among themselves. An interesting solution was invented by experts f the advertising agency Design-Art (Rostov-on-Don, Russia). The generating base of buzz was the creative SP action with the name "Nostalgia". It was based on the mechanics that the Zhigu-levskoye beer was sold at a "Soviet" price - 37 kopecks. for a half-liter bottle (that is very cheap nowadays). Intensive buzz-communication of citizens both offline and online forced the media to address this topic, and absolutely free of charge. Thus, the creative decision turned the traditional sampling tool into an international event: information about it appeared not only in regional but also in national media, and even in international ones (the action was mentioned in one of the BBC broadcasts).

White paper – granting on a free of charge information with educational, developing character to the consumer. White paper helps consumers to solve the problems of self-development, self-improvement, personal growth, business development. The condition for access to important information, as a rule, is the transfer by the consumer of a certain amount of information about yourself, which in future allows the brand to organize accurate, point-to-point communications with target audiences based on digital technologies. An example are the free educational web-seminars, e-books conducted by a digital agency, offering a large amount of useful information for consumers of b2b services and simultaneously promoting a digital agency.

5   Conclusion

Thus, the brief overview of the implementation of the socio-psychological functions of the creative tools of communication marketing in the field of promotion allows us to conclude that increasing the efficiency of using these tools is based on creating an interactive effect through digital technologies based on the involvement of consumers in solving social and psychological problems, existing both in the individual representative of the target audience, and in the society as a whole.

The creative and innovative tools that allow to implement the socio-psychological function of consumer impact using the digital technologies are Entertainment-marketing, Provocative marketing, Flash mob, Aggressive marketing, Social marketing, Partisan (guerilla) marketing, Art marketing, Viral marketing, Buzz-marketing, White paper.

Research Limitations. The results of the research can not be used for most traditional tools of marketing communications - Sales Promotion, Direct Marketing, Public Relations, Personal Selling, all types of advertising (television, outdoor, newspaper, magazine, outdoor, souvenir, etc.), because they are mainly ignore the establishment of the emotional connection of the brand with consumers on the basis of addressing the significant socio-psychological problems of target audiences, continue to use traditional methods of influencing consumers based on repetition, planned frequency of contacts and audience coverage.

Scope of Future Research. The directions of future research are related to the study of the features of integrating the interests of brands and target audiences on the basis of digitalization of promotion, identifying specific integration of digital and offline promotion tools, differences in integration between mobile app, site and SMM-centered models of digital integration of brand communications based on the use of brands of creative tools and tools. One of the directions of future research is the study of the attitude of consumers to the creative tools of brand promotion, which solves social and psychological problems in society.

References

1. Hervás, D.S.: Marketing digital. Guía Básica Para Digitalizar tu Empresa. J. adComunica. Revista Cientí ca de Estrategias, Tendencias e Innovación en Comunicación **12**, 247–249 (2016)
2. Polo, J.M.M., Sánchez, J.T.M., Meroño, M.C.P.: Marketing Digital. Guía Básica Para Digitalizar tu Empresa. J. Revista de Comunicacion de la SEECI **40**, 171–172 (2016)
3. Solarte-Vásquez, M.C., Nyman-Metcalf, K.: Smart contracting: a multidisciplinary and proactive approach for the EU digital single market. J. Balt. J. Eur. Stud. Tallinn Univ. Technol. **7**(2(23)), 208–246 (2017)
4. Walmsley, B.: From arts marketing to audience enrichment: how digital engagement can deepen and democratize artistic exchange with audiences. J. Poet. **58**, 66–78 (2016)

The Problems of Digital Economy Development in Russia

Zhanna Mingaleva[1,2]($\boxtimes$) and Irina Mirskikh[1]

[1] Perm State National Research University, Perm 614000, Russia
mingal1@psu.ru
[2] Perm National Research Polytechnic University, Perm 614000, Russia

Abstract. Digital economy is a kind of economic activity where digital form of data becomes a factor of production. It stimulates creating information space taking into account needs of citizens and the society. At the same time the development of digital economy gives rise to a number of problems. Different threats of unfair use of information appeared. One of the most serious problems is the problem of intellectual property and data protection. Successful realization of innovation projects and implementation of innovations requires creating an effective mechanism of intellectual property protection and protection of information. The paper investigates the main problems of digital economy development in the Russian Federation the most serious of which is inadequate system of information security. The comparative analysis made it possible to indentify that economic security is closely connected with intellectual property protection and innovation activity within employment relations and contract relations. Unfair access to commercial confidential information can cause damage and loss of important data and lead to innovation program failure. The study found that in order to provide protection against unauthorized use of information by the third parties it is important to create a system of limited access to intellectual property, commercially valuable information, and the results of innovation activities.

Keywords: Digital economy · Intellectual property · Information security

1 Introduction

In Russia, most of the problems connected with the use of information and telecommunication technologies in various fields of activity have been resolved. However, the legal regulation in this sphere has a number of shortcomings, which can cause serious problems for creating new institutions of the digital economy, the development of information and telecommunication technologies and related economic activities.

The citizens of the Russian Federation have recognized the necessity to possess digital competence and skills, but the level of use of personal computers and the "Internet" network in Russia is still much lower than in Europe. And there is a serious gap in digital skills between different groups of the population [1].

A special international network readiness index was created to assess countries' readiness for the digital economy. This index is used to reveal readiness of countries'

T. Antipova and A. Rocha (Eds.): DSIC 2018, AISC 850, pp. 48–55, 2019.
https://doi.org/10.1007/978-3-030-02351-5_7

economies to use digital technologies to improve competitiveness and prosperity, and assesses the factors that influence upon the development of the digital economy.

The research showed that such countries as Finland, Switzerland, Sweden, Israel, Singapore, the Netherlands, the United States of America, Norway, Luxembourg and Germany are the leaders in the sphere of the digital economy. The Russian Federation takes only the 38th place [2].

Such a significant difference in the development of the digital economy can be explained by the shortcomings of legal regulation of the digital economy, the lack of a favorable environment for business and innovation and, as a result, the low level of use of digital technologies by business structures.

The multimedia revolution is affecting economics, science, and law; thereby involving in a global debate issues concerning fundamental freedoms and access to knowledge [3]. The development of digital economy caused a lot of problems in the sphere of intellectual property protection.

The study, carried out with the Science Applications International Corporation, found that cyber criminals are increasingly turning their attention towards corporate assets such as trade secrets, marketing plans, research and development findings and even source code [4].

Economic agents must protect secret information by using data loss prevention and other tools. Intellectual property is the most valuable but acceptable kind of property. Corporate intellectual property has become the target for criminals [5]. Organizations must protect know how and secret information from theft and abuse.

Raj Samani aVice President, Chief Technical Officer for Intel Security in Europe, the Middle East, and Africa pointed down that firms need to start by understanding the value of their data. Intellectual property is the lifeblood of any organization and failure to protect it could be the death knell of the business [6].

But organizations prefer not to publish the information about intellectual property theft done by the employees because it could influence upon their reputation.

Russia faces a very difficult situation connected with violation of intellectual property rights. Unlawful manufacture and distribution of counterfeit video and audio products, replication of the most popular computer programs are widespread [7].

Digital space is very attractive for criminals who commit crimes against intellectual property. The potential of Internet, high degree of anonymity, and lack of control on the transfer of information objects create favorable conditions for digital leakage and gives the possibility to use the Internet as a tool to commit offences in the field of copyright and intellectual property. The damage from copyright violations amounts millions of euro [8].

Copyright infringement takes place when the original creation is used without permission. Legal regulation in the sphere of intellectual property was created to protect the intellectual property of individuals for a limited amount of time, and to encourage creativity [9].

Digital law provides responsibility for actions and violations in the digital sphere [10].

2 Theoretical Background

Digital economy is a kind of economic activity where digital form of data becomes a factor of production. It stimulates creating information space taking into account needs of citizens and the society to receive reliable information development of information infrastructure of the Russian Federation and creating and use of Russian information-telecommunication technologies and forming the new technological base (background) for social and economic sphere.

Digital economy consists of 3 levels which influence upon the life of citizens and the society:

(1) market and branches of economy where interaction between economic agents such as suppliers and consumers take place;
(2) platforms and technologies which form capacities and tendencies for the development of markets and branches of economy (spheres of activity);
(3) environment which creates the conditions for the development of platform and technologies and stimulates an effective interaction between economic agents and branches of economy (spheres of activity). It embraces normative regulation, information, infrastructure, personnel and information (digital) security [11].

In order to stimulate the development of digital economy a special program of strategy of scientific and technological development was adopted in the Russian Federation.

The Programme takes into account the main objectives which are implemented within the framework of National technological initiative and documents of strategic planning, including such acts as scientific and technology development of the Russian Federation for the period of the time up to 2030; Strategy of Scientific and technological development (confirmed by Decree of the President of Russia 01.12.2016 № 642 "On Strategy of Scientific and technological development of the Russian Federation"; Strategy of development of the Information society in the Russian Federation for the period of 2017–2030 and priority project "Improvement of process of medical help organizing based on information technology implementation (25.10.2016) and documents and acts of the Euroasian economic society.

The information security doctrine of the Russian Federation, approved by the Decree of the President of the Russian Federation dated December 5, 2016 № 646 "On approval of information security Doctrine of the Russian Federation", formed the basis for creating the state policy and development of public relations in the field of information security, as well as for the development of measures to improve the information security system. The Russian Federation has traditionally paid much attention to improvement of information security of gas- and energy supply. However, the majority of Russian companies believe that the number of crimes in the digital sphere has increased by 75% over the last three years. It requires improvement of the information security system in all sectors of economy.

The Program defines the five basic directions of the development of the digital economy in the Russian Federation for the period up to 2024.

The basic areas include improvement of legal regulation and education, creating research competencies and technical achievements, development of information infrastructure and information security.

The main aim of the improvement of legal regulation in the sphere of digital economy is to create favorable legal regime for the development of modern technologies, as well as for the implementation of new economic activities associated with their use (digital economy) [12].

Internet access, in particular the ability to move data across borders, needs to remain free from unnecessary and restrictive rules [13].

On the basis of the «road map», a special plan of activities in the sphere of digital economy development was worked out [14]. The "road map" identifies the 3 main stages of development of the digital economy:

(1) Creating a concept of priority measures to improve legal regulation for the development of the digital economy. The aim is to remove key legal restrictions for the development of the digital economy, and identify the priority basic legal concepts and institutions necessary for the development of the digital economy - 2018.
(2) Implementation of the concept of medium-term measures to improve legal regulation for the development of the digital economy - 2020.
(3) Implementation of the concept of complex legal regulation of relations arising in connection with the development of the digital economy to provide a favorable legal regime for the emergence and development of modern technologies and economic activities associated with their use (digital economy) - 2024.

3   Informational Threats

Protection of innovations and commercial information can be defined as a system of legal, economic, organizational and technical measures providing information security of innovations. Information policy provides rational use of confidential information and economic security [15].

There are different threats of unfair use of information such as internal or external, active and passive, intentional and unintentional (see Fig. 1).

Passive threats are not aimed at the destruction or damage of information or information systems, computer programs, databases, or intellectual property. They are connected with the unauthorized use of the secret information by the third parties to obtain benefits. Active threats cause the intentional deliberate damage of information resources, software. Active threats include actions of hackers.

Threats can also be intentional and unintentional (accidental).

Unintentional threats are associated with accidental damage or access to information resources. They can occur as a result of force majeure (floods, fires, natural disasters) or as a result of an error (for example, sending information for the wrong e-mail addresses).

However, unintentional threats can cause the damage comparable with the damage from intentional threats.

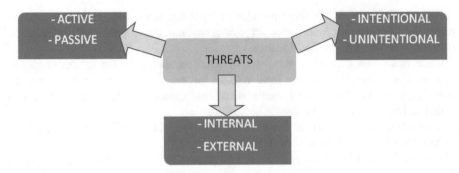

Fig. 1. Types of threats in the sphere of digital economy.

Intentional threats are specially created for the destruction of the information security system. Intentional threats can be divided into internal (occurring within the company) and external. Internal threats are often determined by social tensions and "heavy moral climate", the actions of employees. Figure 2 demonstrates the three main types of intellectual property theft done by the employees.

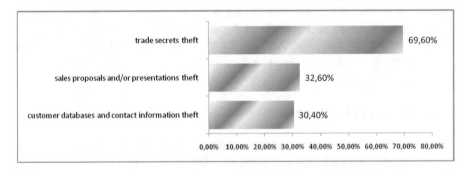

Fig. 2. Types of intellectual property theft.

Figure 2 demonstrates that trade secrets theft form more than 2/3 of all the crimes. About 1/3 of the crimes refer to sale proposals and presentations theft. And another 1/3 of the crimes were committed to steal customer databases and contact information.

The most valuable know-how form trade secrets and technological secrets. "It is possible to define the following features of Know-how. It is the information, technical knowledge, experience, connected with the development, launching the production, operation, service, repair, improvement and utilization of new hardware, technology and materials.

1. It has real or potential technical and commercial value;
2. It is being applied or can be applied.
3. It has confidential nature, it is not in the public domain;
4. It has no legal protection on the national, regional and local levels.

5. It is necessary to possess special knowledge and experience for using it.
6. The way of its fixation can be different (written form, oral or visual).
7. Without obtaining know-how it is impossible to improve technical objects, technology and materials, in which know-how is involved" [16].

Most of the thefts were committed by the employees leaving a job.

In 2003 US Department of Justice revealed the information about the two former Boeing managers who were charged in a plot to steal trade secrets (documents concerning a multi-billion dollar rocket program for the US Air Force) [8].

In Great Britain about 25% of companies suffered from intellectual property theft.

The results of intellectual activities of such employees as engineers, researchers, inventors, their inventions and achievements belong to the employer and the organization in which they work [17]. Activities of this group of employees provide innovations and guarantees competitive advantage. Disclosure of trade secrets and technical secrets, know how by this category of employees can cause a lot of problems. Sometimes they can do it unintentionally during presentations at congresses, seminars and conferences [18].

Temporary workers and employees who have access to commercial and trade secrets of the employer can disclose information also (Fig. 3).

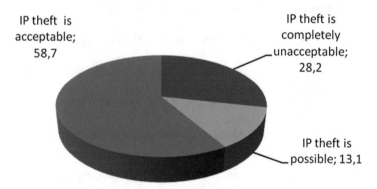

IP theft is acceptable; 58,7

IP theft is completely unacceptable; 28,2

IP theft is possible; 13,1

Fig. 3. Intellectual property theft (attitude of the employees)

More than 2/3 of the employees consider that intellectual property theft is acceptable. About 13% of respondents think that intellectual property theft is possible. And only 1/3 of the employees answered that intellectual property theft is completely unacceptable.

External threats can arise as a result of deliberate malicious actions of competitors. Most often, they are realized in the form of economic or industrial espionage [19], which refers to the illegal collection, assignment and transfer of commercially valuable information, know-how, results of intellectual activities to obtain benefits at the expense of the rightful holder of the information [6].

Economic and industrial espionage cause external threats which are aimed to collect and transmit commercial information for providing competitive advantage [20].

Today, new problems and threats influence upon the development of the digital economy of Russia. Such as the problem of protection of human rights in the digital world, including the identification (correlating a person with his digital image), the safety of digital data of the user [21]. Threats to the individual, business and state in the sphere of digital activities can be caused by virtualization, remote (cloud) data storage, communication technologies and devices.

4 Conclusions

The analysis of the digital economy development made it possible to reveal some serious problems in the sphere of intellectual property and data protection. These problems can be explained by the nature of information and potential of Internet which creates favorable conditions for digital leakage and gives the possibility to use the resources of Internet for committing intellectual property theft.

In order to solve the problems it is necessary to introduce differentiated legal regulation of any actions aimed at creation, storage, copying dissemination and use of information.

To achieve good results in this sphere it is important to:

(1) create an effective mechanism of management in the field of digital economy regulation;
(2) remove legal restrictions and create a special legal regulation of the digital economy;
(3) accept a number of measures aimed at promoting economic activities related to the use of modern technologies,
(4) provide the policy of the development of the digital economy on the territory of the Eurasian Economic Union,
(5) work out methodological basis for the development of competencies in the field of digital economy regulation.

It is also important to adopt special legal acts regulating rights and duties of participants of information interaction in social networks and other social communication.

The state must install sufficient legal regulation to ensure safe access to cloud operators in the processing of personal data. The requirements for the identification of users of communication and other services of participants of information interaction must be worked out. Special mechanisms must be created to prevent the emergence of illegal information in the Russian segment of the Internet, including mechanisms for its removal.

References

1. McKinsey: Nine questions to help you get your digital transformation right. https://www.mckinsey.it/idee/nine-questions-to-help-you-get-your-digital-transformation-right. Accessed 21 June 2018
2. Kajtazi, M.: Information asymmetry in the digital economy. In: 2010 International Conference on Information Society, pp. 135–142. I-Society (2010)

3. Lucchi, N.: Access to network services and protection of constitutional rights: recognizing the essential role of internet access for the freedom of expression. Cardozo J. Int. Comp. Law (JICL), **19**(3), 2011. http://papers.ssrn.com/sol3/papers.cfm?abstract_id = 1756243. Accessed 13 May 2018
4. Benedek, W.: Internet Governance and human rights. In: Benedek, W., Bauer, V., Kettemann, M.C. (eds.) Internet Governance and the Information Society: Global Perspectives and European Dimensions, pp. 31–49. Eleven, Utrecht (2008)
5. Kravets, D.: U.N Report Declares Internet Access a Human Right. Wired, San Francisco (2011)
6. Houlding, D., Samani, R., Vuckovic N.: Healthcare Security: User Experience, Compliance, and Risk. https://www.intel.ru/content/dam/www/public/us/en/documents/solution-briefs/healthcare-mcafee-security-brief.pdf. Accessed 18 May 2018
7. Mirskikh, I.Yu., Mingaleva, ZhA: Legal regulation and protection of intellectual property rights in russia and abroad. Vestnik Permskogo Universiteta-Naucnyj Zurnal **3**, 62–70 (2015)
8. Intellectual Property Theft: How Can You Prevent It? https://www.locklizard.com/intellectual_property_theft. Accessed 11 June 2018
9. Digital Law: Purdue University Copyright Office (2009). http://cunedigitalcitizenship.wikispaces.com/Digital%20Law. Accessed 16 May 2018
10. Lee, S.J., Watters, P.A.: Gathering Intelligence on High-Risk Advertising and Film Piracy: A Study of the Digital Underground. Automating Open Source Intelligence: Algorithms for OSINT, pp. 89–102. Syngress Publishing, Maryland Heights (2011)
11. Ronzano, A.: Data processing: The French Competition Authority publishes a joint paper with the Bundeskartellamt on data and its implications for Competition Law and announces the launch of a sector inquiry on data, commercial strategies and competition in the digital economy. Concurrences, **3**, 222 (2016)
12. Holt, T.J., Copes, H.: Transferring subcultural knowledge on-line: practices and beliefs of persistent digital pirates. Deviant Behav. **31**(7), 625–654 (2010)
13. Meltzer, Joshua P.: A New Digital Trade Agenda. E15 Initiative. International Centre for Trade and Sustainable Development (ICTSD) and World Economic Forum, Geneva (2015)
14. The program "Digital Economy of the Russian Federation", approved by the order of the Government of the Russian Federation on July 28, 2017, No. 1632-p (2017)
15. Tapscott, D.: The Digital Economy: Promise and Peril in the Age of Networked Intelligence. McGraw-Hill, New York (1997)
16. Mingaleva, Zh., Mirskikh, I.: On innovation and knowledge economy in Russia. World Acad. Sci. Eng. Technol. **66**, 1032–1041 (2010)
17. Mendis, D.: Digital economy Act 2010: fighting a losing battle? Why the 'three strikes' law is not the answer to copyright law's latest challenge. Int. Rev. Law Comput. Technol. **27**(1–2), 60–84 (2013)
18. Filyak, P.Yu.: Information security in the context of the digital economy. In: CEUR Workshop Proceedings, vol. 2109, pp. 25–30 (2018)
19. Holt, T.J., Brown, S.C.: Contextualising digital piracy. In: Digital Piracy: A Global, Multidisciplinary Account, pp. 3–12 (2018)
20. The digital advantage: how digital leaders outperform their peers in every industry, 24 p. Capgemini Consulting (2012)
21. Klang, M., Murray, A.: Human Rights in the Digital Age. Routledge, Abingdon (2005)

Economic Safety Level Increase of the Electric Power Industry Enterprises

Kseniy Lapshina[✉] and Zhanna Mingaleva

Perm National Research Polytechnic University, Perm 614000, Russia
lapshina.km@yandex.ru

Abstract. One of the priority directions of development for the advanced countries today is the digitization of the economy. The article deals with the most important aspects of this process in the one of the largest sectors of the Russian economy - electric power industry. Russian power industry is characterized by a technical lag, so the introduction of modern digital technologies and the formation of an innovative development strategy should become the defining vectors of the electric power enterprises development. The aim of this research work is to analyze the opportunities of Russian power industry in the conditions of digitization of the economy. The methodology of the work includes accumulation of information, their correlation with each other and comparison, analysis and assessment of actual information. The conclusion of the importance of the power industry transformation and development is necessary to drive digital technological innovations. Selection of the main directions of the industry development in the conditions of digitization of the economy and the system of indicators for the assessment of the level of the power industry enterprises economic security.

Keywords: Digitization · Electric power industry · Digital business models
Innovations · Innovative technologies

1 Introduction

Globalization and digitization of the economy is one of the priority directions of development for the Russian state. Ensuring the effective implementation of modern business processes is carried out through the digital economy, the main tools of which are new "breakthrough" technologies, innovative services, digital business models. The value of digital solutions is growing today, and it requires organizational and technological modernization of enterprises.

Since the beginning of the XXI century, the world has been actively using information technologies in electrical networks, including the development and implementation of "smart" networks and the required infrastructure for them. The need for new developments escalated with the active development of electricity production based on renewable energy and industrial generation distribution. Connection of such generation facilities requires the latest technologies for their activation, ensuring reliability and efficiency in the transmission and distribution of electricity.

© Springer Nature Switzerland AG 2019
T. Antipova and A. Rocha (Eds.): DSIC 2018, AISC 850, pp. 56–63, 2019.
https://doi.org/10.1007/978-3-030-02351-5_8

The most active "digitization" of the electricity grid infrastructure is carried out in Europe, where the percentage of renewable sources in electricity production is already about 15%, and it should reach at least 27% by 2030. In order to support the development of intelligent networks, the EU aims to replace at least 80% of electricity meters with smart meters by 2020 [1]. According to the European Commission, the use of smart counters and networks can reduce the annual primary energy consumption in the EU by 9% by 2020, as well as reduce harmful emissions. Financial support for research and development and projects for the implementation of such networks are also used to achieve these goals. The EU provides funding for about 30% of digital network projects in Europe [2].

According to the report 2016 of the joint research center of the European Commission Directorate General, the EU countries are realizing more than 308 new projects on the implementation of "smart" networks with a total cost of about 2.15 billion euros. At the end of 2016, more than 642 projects were implemented (2.82 billion euros). In total, more than 950 projects to create a "smart" infrastructure have been implemented and are being implemented. The largest number of networks being implemented is in Germany (140) and Denmark (105) [3].

2 Theory and Methods

The term "digital economy" was introduced by Don Tapscot in 1997 [4]. In addition, such terms as Internet economy, web economy, e-economy, New economy are used. The term "digital economy" was used first also by Alexander Kuntsman as "a modern type of economy characterized by the predominant role of information and knowledge as the determining resources in the production of material products and services, as well as the active use of digital technologies for storage, processing and transmission of information» [5].

Today the digital economy is a system of economic, social and cultural relations based on the use of digital information and communication technologies. Thus, the usual relationship is now replaced by an electronic analog. These changes are accompanied by the need to transform classical public institutions in order to ensure the level of trust in the electronic sphere.

Russia has also created the Digital economy programme, which is aimed at organizing systemic development and introducing digital technologies in all spheres of life: in the economy, in business, in public administration, in the social sphere and in the municipal economy [6]. According to the McKinsey report, Russia's goal is to increase the share of the digital economy by 3 times, because today digitization level of the domestic economy is insufficient [7]. The process of digitization in Russia is slower than in the whole world. First, this process threatens to change culture. Russia is a multi-ethnic country, where culture has great importance for people, those features that are inherent in Russian culture can't be digitized [8]. For example, according to Moody's, digitization in Russia doesn't mean growth, which will soon be at the level of 1.5% per year in Russia because of the lack of the digital environment development in the country and structural constraints [9].

For the effective development of the digital economy in Russia it is also perspective to apply the cluster approach.

The methodology of the work includes accumulation of information, their correlation with each other and comparison, analysis and assessment of actual information.

3   The Problems of Energy Industry in Russia

The energy industry of Russia needs to be modernized with smart systems to improve efficiency and reduce capital and operating costs. In General, the process of digitization in the power industry is at an early stage. Despite the fact that the trend of transition to digital technologies in the systems of data collection and processing, control and automation of substations was outlined more than 15 years ago, the world's first digital substation was launched only in 2006 [10]. However, today almost all the leading companies-producers of the electric power industry are actively working in this direction, and also telecommunications are the basis of monitoring and control systems in any other sector of the economy.

The problem of developing "reliable and flexible networks" has become one of the tasks of the national technology initiative EnergyNet (along with the development of distributed generation and consumer services in the power industry), which aims to create competitive technological solutions in the world by 2035. According to the high-cost map of the national technological initiative EnergyNet, the concept of digital networks includes the development of intelligent metering systems of energy currents, distributed automation systems, systems for monitoring the operational status of equipment and quality of energy supply, the formation of digital models for optimal control of the power system operation and development [10].

Digitization of the electric power industry implies both modernization of its technological processes and introduction of innovative methods of enterprises economic activity assessment. Power industry enterprises activity, which is based on outdated principles of monitoring and management, are not able to maintain a high level of economic security, and the industry as a whole can pose a threat to the state.

Innovative technological development of the electric power industry today is characterized by the unification of the power grid and information infrastructure in the network nodes—digital substations [11]. Digital substation is an element of active—adaptive (intellectual) power supply network with control, protection and management system based on information transmission in digital format. This technology allows to reduce the cost of construction of substations, reduce their size, improve reliability and, ultimately, improve the quality of power supply to the consumer without increasing the cost. This, in turn, gives an increase in noise immunity, reducing the number of equipment, secondary switching circuits and saving space. Digital substation can be built faster and it is easier to develop model projects for replication. Currently, there are hundreds of such substations installed in China, USA, Canada and other countries [9]. The transition to the new automation and control systems was possible with the occurrence of new standards and technologies of digital substations, which, first of all, includes a specially developed standard. All information links at the substations are digital and form a single process base. This opens up the possibility of rapid and direct exchange of information between devices, which ultimately allows to abandon the mass of copper cable connections, individual devices, and to achieve a more compact location. So, the main feature of the digital substation is that all its secondary circuits are digital data channels forming a single information network.

Thus, it is expected that the digital substation can improve the security of electric power facilities, to obtain a significant metal consumption reduction, reduce the number of elements in the control and monitoring systems with their efficiency increasing, improve the level of reliability and minimize the cost of engineering and adjustment. The successful implementation of such substations contributes the transition from the hardware market to the software market in the future. The first ultra-high voltage class power facility, which implemented digital technology was introduced in Russia only in 2018 (substation "Tobol" in the Tyumen region). It is planned to build more than 30 such facilities in the country by 2025 [10].

Pilot projects in Kaliningrad and Sevastopol have been developed and are being implemented for the testing of intelligent technologies within Energy Net. Other pilot projects of networks digitization are also being implemented, for example, in Ufa and St. Petersburg, pilot projects are being implemented by network companies together with «Siemens». It is expected that the pilot projects should pay off by reducing the loss of electricity in the network, reducing operating costs and the need for investment to upgrade the networks. It is also expected that the new technologies will improve the reliability of power supply, including reducing the number of de-energized consumers in case of emergency situations [10]. Such technological innovations, together with organizational novelties, can qualitatively transform the electric power industry and ensure the proper level of economic security [12].

4 Digitization and Economic Security of Electric Power Industry

In conditions of the economy innovation and digitization for an adequate assessment of both the economic activity of the organization as a whole, and the level of economic security and competitiveness, it is necessary to introduce into the existing system of indicators the new list of indicators, including indicators of the enterprise's innovative activity and their thresholds to determine the level of economic security and sustainable development [13, 14]. The system of indicators for the assessment of the level of the power industry enterprises economic security, which is based on several approaches of scientists and experts competent in the field of innovation and is created into account the specifics of the electric power industry, is presented (see Table 1).

The indicators given in the table reflect the level of economic security of power enterprises in terms of implementation and use of innovations. For the quick calculation of these indicators, it is necessary to develop and implement appropriate computer programs in the management of enterprises [15]. Such programs will allow for an accurate assessment of the level of economic security in an automatic mode without high labor costs [16]. The requirements to the results of projects are presented:

- to provide the possibility of obtaining exact and current data of economic processes;

- minimize the costs of collecting indicators from the enterprise;
- minimize errors and probability of manifestations of the "human factor" in the calculation of indicators.

Within the enterprise, there should be a special department - the economic security department, the work of which must meet the requirements of the modern economy, striving for digitization. This means that the economic security department also needs to implement the necessary technology to be more effective. For example, providing access to electronic programs of all departments of the enterprise, providing access to closed databases, replacing old computer equipment with new and more progressive technologies, avoiding a large amount of paper work.

From theoretical and practical points of view, virtual organizations with the highest level of virtualization management and assets have great prospects in the conditions of digital economy [17]. Blockchain is an information technology that allows to carry out transactions (such as data transfer, funds transfer, conclusion of a contract, etc.) between equal participants of a single network (P2P-network) without intermediaries and to store information about all transactions carried out in the registers [18]. The most famous example of the use of the blockchain technology is bitcoin, the cryptocurrency, launched in 2009. So far, it has only become widespread, but interest in blockchain is growing outside the financial sector (see Fig. 1) [18]. Decentralization is considered as the most perspective direction of blockchain technology implementation in the energy sector. Existing centralized multi-level energy systems are complex and costly,

Table 1. The system of indicators for the assessment of the level of the power industry enterprises economic security

Group of indicators	Indicators
Financial indicators (costs)	Unit costs of scientific research and development in sales, characterize the indicator of the knowledge intensity of the company's products
	The level of innovation activity (the volume of investment in innovation, The cost of purchasing innovative products)
	Availability of funds for the development and implementation of innovative developments
	Specific costs for the licenses acquirement, patents, know-how
Financial indicators (profit)	Profit obtained through the use of innovations in production
	Profit from the sale of innovative services and products
Renewability indicators	Number of developments or implementations of innovations-products and innovations-processes
	Volume of innovative services and products
	Number of transferred new technologies (technical achievements)
Structural indicators	Number of researchers, developers and other scientific and technical structures
	Number and structure of employees involved in research and development
	Structure of employees involved in research and development (number, age, qualification)

blockchain can simplify the interaction, connecting directly producers and consumers of energy (primarily electricity). It is assumed that in such systems the electricity produced at small distributed generation facilities will be delivered to the end users via micro networks [19]. The volumes of produced and consumed electricity will fix the «smart meters», and trade operations and payments, will fix «smart contracts». The participation of brokers or electricity companies will not be required.

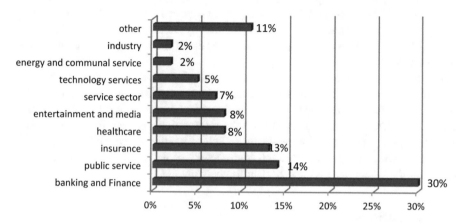

Fig. 1. Use of blockchain technologies in the sectors of economy in 2017.

According to the consulting company Indigo Advisory, the total number of cases of blockchain technology application in the energy sector has already exceeded 100. Most of them are in Europe, and more specifically in Germany, which has declared an "energy turn" towards clean energy. In August 2016, the German Energy Agency conducted a survey among 70 managers in the energy sector about the perspective of blockchain implementation. More than half of them indicated that they were implementing or planning pilot projects using blockchain. Sreaking abouy Russia, in November 2016, the initiative for the implementation of block chain technology for the accounting of electricity transactions was made by the Russian Qiwi group together with "Tavrida Elektrik" [10].

There are several difficulties, that prevent the blockchain development in the electricity sector:

- the technology has not yet reached maturity, "smart contracts" are not sufficiently developed, and there are practically no user-friendly applications for consumers;
- energy is rather inertial: changing business models of producers and consumer behavior takes time;
- blockchain is not the only alternative to promote decentralized energy, because the other technological solutions exist. It is attractive in terms of reducing operating costs and risks, but the costs of hardware and electricity (for transaction verification) are significant;
- blockchain technology diffusion requires legislative changes, including the development of common rules and standards [18].

In order to overcome the obstacles to the development of digital technologies and the introduction of blockchain in the energy sector, power energy enterprises should join forces with the state, research centers and other organizations, whose activities are aimed at accelerating the commercial development of innovative technologies in the energy sector [20]. The assessments show that the resource provision of innovation developments requires state support and regulation, and at the federal level.

Today, the list of energy companies, that use the blockchain technology, includes Consolidated Edison (USA), Enercity and the German RWE AG, the Austrian Wein Energie, Finnish Fortum OYJ, Japanese E-net Systems and the Australian Living Room of Satoshi. As we see from the list, very few of these companies are the leaders of the national energy markets [21].

In Russia the blockchain will allow to manage the power system, to conclude "smart" contracts and to conduct financial calculations, but the technology will be efficient only in the presence of intelligent commercial accounting devices, combined through the Internet and allowing to use the information obtained in real time. Since July next year, Russia plans to gradually switch to intelligent metering devices, while blockchain technology can ensure full transparency of payments in the housing sector and limit the electricity loss and heat in the networks [22].

5 Conclusions

The electric power industry is one of the largest sectors of the Russian economy, which is characterized by a technical lag, so the introduction of modern digital technologies and the formation of an innovative development strategy should become the defining vectors of the electric power enterprises development. Moreover, the use of modern technological developments and computer programs helps to increase the level of the economic security of the power industry enterprises. The key areas of the electric power industry digitization include tele- control, development of telemetry (remote control of the network from the unified network operation centers through digital communication channels), digital accounting systems that automatically send data to the head office. These activities will provide the best quality of data analysis, speed of response to faults, reducing maintenance costs and improving reliability of electricity supply.

Acknowledgment. The work is carried out based on the task on fulfilment of government contractual work in the field of scientific activities as a part of base portion of the state task of the Ministry of Education and Science of the Russian Federation to Perm National Research Polytechnic University (topic # 26.6884.2017/8.9 "Sustainable development of urban areas and the improvement of the human environment").

References

1. Eurostat. http://ec.europa.eu/eurostat/statistics-explained/index.php/Renewable_energy_statistics. Accessed 13 May 2018
2. Research Centre of the Directorate General of the European Commission (2016). https://ses.jrc.ec.europa.eu/smart-grids-observatory. Accessed 07 May 2018

3. European Commission. http://ec.europa.eu/eurostat/statistics-explained/index.php/Renewable_energy_statistics. Accessed 12 May 2018
4. Tapscott, D.: The Digital Economy: Promise and Peril in the Age of Networked Intelligence. McGraw-Hill, New York (1997)
5. Kuntsman, A.A.: Transformation of internal and external business environment in the conditions of digital economy. Econ. Syst. Manag. Electron. Sci. J. **11**(93), 1 (2016)
6. Leonova, K.S.: Necessity and possible consequences of digitization of the Russian economy. Econ. Bus. Theory Pract. **12**, 103–105 (2017)
7. McKinsey: Nine questions to help you get your digital transformation right. https://www.mckinsey.it/idee/nine-questions-to-help-you-get-your-digital-transformation-right. Accessed 10 Mar 2018
8. Golovinsky, V.V., Dorofeev, R.A.: The digital economy is our new everything? Obosnik. http://www.oboznik.ru/?p=56149. Accessed 08 Apr 2018
9. Moody's: Digitization of the Russian economy will not ensure its growth. TASS Official website. http://tass.ru/ekonomika/4680682. Accessed 28 May 2018
10. Grigoriev, L.: Digital technologies in the network complex. Energy Bull. **53** (2017)
11. Institute of Energy. https://energy.hse.ru/digitization. Accessed 11 May 2018
12. Mingaleva, Zh, Balkova, K.: Problems of innovative economy: forming of "Innovative society" and innovative receptivity. World Acad. Sci. Eng. Technol. **59**, 838–843 (2009)
13. Guertin, J.D.: What is sustainability? Tunn. Undergr. Space Technol. **11**(4), 373–375 (1996)
14. Sheehan, B.: What is sustainability? Control Eng. **56**(9) (2009). http://www.controleng.com/article/print/339879-What_is_sustainability_php. Accessed 11 May 2018
15. Lane, N.: Advancing the digital economy into the 21st century. Inf. Syst. Front. **1**, 317–320 (1999)
16. Marz, N.: Large data. Principles and practice of building scalable data processing systems in real time. M. Williams (2016)
17. Gumerova, G.I.: Virtual organization of the digital economy as an object of the study of the study of management theory. Azimuth Sci. Res. Econ. Adm. **7**(1/22), 107–110 (2018)
18. Blockchain - new opportunities for producers and consumers of electricity. PwC (2016). https://www.pwc.ru/ru/publications/blockchain.html. Accessed 18 June 2018
19. Mingaleva, Z., Shpak, N.: Possibilities of solar energy application in Russian cities. Therm. Sci. **19**(2), S457–S466 (2015)
20. Donichev, O.A., Fraimovich, D.Y., Grachev, S.A.: Regional system of economic and social factors in the formation of innovation development resources. Econ. Soc. Chang. Facts Trends Forecast **11**(3), 84–99 (2018)
21. Mengelkamp, E., Gärttner, J., Rock, K., Kessler, S., Orsini, L., Weinhardt, C.: Designing microgrid energy markets: a case study—the Brooklyn Microgrid. Appl. Energy **210**, 870–880 (2018)
22. Vershinina, P.: Blockchain in the Russian energy sector (2017). http://bellona.ru/2017/05/29/blockchain. Accessed 18 June 2018

Environmental Accounting in Digital Economy

Tatiana Sheshukova[1(✉)] and Evgeniia Mukhina[2(✉)]

[1] Perm State National Research University, Bukireva street 15,
Perm 614990, Russia
sheshukova@psu.ru
[2] Perm National Research Polytechnic University, Komsomolsky prospekt, 29,
Perm 6145990, Russia
scancens@rambler.ru

Abstract. The market economy is characterized by high uncertainty of the business environment, the rapidity of the development. It leads to the need to adapt enterprises to a dynamic market environment; the development of mechanisms for managing the competitiveness of an economic entity comes to the fore. Innovative technologies are among the most important factors of competitiveness. Our task is reflection of these innovations at accounting. The paper is devoted to the study of key features of eco-innovations and identification of basic problems at environmental economy. The aim of the study is to develop the theoretical provisions of environmental accounting, the objects of the system (ecological innovations) and determine the range of key problems in research field. The results of the authors' study are useful in practice, because in contributes to a more justified adoption of managerial decisions at industrial enterprises.

Keywords: Ecological accounting · Eco-innovation · Information
Green economy · Competitiveness · Environmental accounting
Development · Innovative economy · Digital economy · Artificial intelligence

1 Introduction

Formation and development of innovative economy is impossible without the production of high-tech competitive products, use advanced technologies, minimizing environmental damage. However, innovative development of the economy is hampered precisely by the environmental factor, covering almost all types of natural resources and technological systems based on their use. Environmental safety is directly related to the rational use of nature. Natural resources are depleted. It indicates the need for their rational use and the creation of a system capable of recording objects of innovation accounting in the environment. This system should provide timely and transparent information on the costs of environmental measures, on eco-innovations, and on the economic efficiency of introducing eco-innovations for the management team. All these factors actualize the development of accounting and analytical support for the management of innovation in the environment.

© Springer Nature Switzerland AG 2019
T. Antipova and A. Rocha (Eds.): DSIC 2018, AISC 850, pp. 64–70, 2019.
https://doi.org/10.1007/978-3-030-02351-5_9

2   Method

A significant contribution to the development of the theory of the accounting and ecological process in Russia was made by Russian researchers, such us Alimov S.A., Barkhatov A.P., Belousov A.I., Vinokurov Yu.S., Demina T.A., Karelov A.M., Konstantinov V.A., Korostelkina I.A.., Melnik M.V., Mironova O.A., Nikitina S.A. The problems of environmental accounting and audit are studied in the works of Alborov R.A., Baryshnikova N.P., Bogatoy I.N., Bogatyreva M.A., Grishanova S.V., Grishina N.N., Popova L.V., Tatarinova M.N., Saenko K.S., Suvorova S.P., etc. Among foreign economists engaged in the problem of ecologization in the economy, it should be noted Ashford [1], Arundel [2], Sarkar [3], James [4], Rennings [5], Spaargaren [6] and others.

The problems of the evaluation of innovation activity have been studied in the works of domestic and foreign scientists, including Anchishkin A.I., Arkhipova M.Y., Bagrinovsky K.A., Varshavsky A.E., Glazyev S.Yu., Golichenko O.G., Efimova M.R., Ivanova N.I., Larionova E.I., Nesterov L.I., Schumpeter J. et al.

The problems of development of the system of management accounting of innovation activity are presented in the works of Basova A.V., Krasnova N.A., Molchanov S.S., Proniaeva L.I., Sayenko K.S., Sidorenko A.Yu., Trubnikova L.S., Fayzrakhmanova G.R., Fedotenkova O.A. and other researchers.

The problems of development of digital economy (as economy the main characteristic feature of which is maximum satisfaction needs of all its participants through the use of information) are presented in the works of Andieva E.Yu, Ivanov V.V., Malinetskiy G.G., Keshelava A.V. etc.

The directions of introduction of the digital economy for Russia are development of artificial intelligence and robotics. The authors of the study note that the Russian market of artificial intelligence and machine training is only beginning to develop, showing a backwardness from foreign markets.

So far, there is no unanimity in the scientific community regarding the terminology field of categories associated with the research topic: "managerial accounting of innovations", "environmental accounting", "eco innovation", "competitiveness". That is caused by the failure of theoretical, methodological and methodological developments in the analyzed field. It requires the improvement of existing ones, as well as the creation of new conceptual representations, the development of methodological tools.

So, Volodin O.N. considers environmental accounting as a system providing users with reliable information about the organization's environmental performance and forecasting negative events and risks; information model about the environmental activity of the enterprise, which includes the processes of: registration, accounting, collection, monitoring of costs, raw materials, materials and research results [7].

Gubaidullina, Ishmeeva wrote: "Environmental accounting is a system that can be used to identify, organize, regulate and report environmental information in natural and value terms [8]."

Zhelbunova characterizes environmental accounting as an integrated accounting system, a unique set of indicators that reflect the process of consumption of natural resources and their protection at the microeconomic level [9].

Ilicheva considers environmental accounting as a complex system that combines various components: definition of goals, forecasting, planning, accounting, analysis, development of management solutions [10].

According to Ishmeeva, environmental accounting is a segmental area of accounting, scientifically based system of continuous and continuous monitoring, evaluation, systematization and generalization of information on economic and environmental processes arising from the activities of an economic entity [11].

Mishigdorj interprets environmental accounting as accounting aimed at determining the costs of protecting the environment during normal production activities and creating a numerical profit measure taking into account environmental costs [12].

Rubanova wrote: "Environmental accounting is the result of the activities of economic entities for the purpose of managing and determining the environmental potential of the enterprise; it is the process of collecting, recording, summarizing and reflecting in the system environmental costs, natural assets, environmental funds, reserves and liabilities, as well as the results of activities of economic entities for the purpose of managing and determining the environmental potential of the enterprise" [13].

Famous Russian scientists Khmelev and professor Suglobov consider environmental accounting as a system for collecting, recording and summarizing information that will enable the identification, assessment, planning and forecasting, monitoring and analysis of environmental costs and environmental liabilities [14].

According to Sergeeva, environmental accounting should be considered in the context of methods of internal (management accounting) financial accounting for the purposes of external reporting, as well as from the perspective of strategic accounting based on the inclusion of environmental protection parameters in the modeling of enterprise value [15].

The problem field of environmental accounting includes the following issues:

- the lack of a unified approach among scientists to the content of the concept of "environmental accounting";
- the lack of s no explanation of the accounting terminology;
- the absence of requirements for disclosure of environmental information in financial statements;
- the ack of a common understanding and classification of environmental costs;
- the need to formulate the principles of corporate responsibility on which environmental accounting should be based;
- the need to establish standards for environmental accounting in Russia;
- the need to identify environmental accounting objects;

Thus, the system of regulation of accounting and financial reporting is not sufficient to reflect the environmental aspects of enterprises.

3   Results

One of the key long-term goals of the company is maintaining competitive positions in order to continue to stay on the market. In this connection, there is a need to study the term "enterprise competitiveness". The analysis of the main definitions allowed the authors to identify a number of directions in the characterization of the term:

- Competitiveness as an opportunity to carry out effective economic activity, determined by the release and sale of competitive goods to the market.
- Competitiveness as a sequence of implementation of functions (from the position of the functional approach).
- Competitiveness as a comparative characteristic of an object from the point of view of the subject.
- Competitiveness as a measure of reaction to market demands [16].

The essence of the concept of "competitiveness" from the point of view of Russian researchers to a large extent characterize such features as advantage, efficiency, differences, competitive market.

The competitiveness of an enterprise is inextricably linked with the competitiveness of a product, service or technology, that is, with the features of the represented object. Since consumers, in our opinion, determine the distinctive features of the produced product or technology the competitiveness of the economic entity should be characterized, first of all, by using such categories as "subject" and "object" in direct correlation with the efficiency of the business entity.

Among the factors providing (sometimes restraining) the formation of competitive advantages of the subject, is the environmental factor. In modern conditions, innovation and ecology are regarded by researchers as key factors in the strategic development of the enterprise. The main purpose of innovation is to reduce production costs, as well as reduce the anthropogenic load on nature. This can be achieved by reducing production waste and more efficient utilization of waste products. Thus, we can talk about the effective consumption of natural resources, or about the rational use of natural resources.

The desire to maintain stability and strengthen competitiveness inevitably leads to the issue of organization of environmental accounting aimed at studying the influence of environmental factors, taking into account environmental innovations, providing timely information required by responsible persons. Environmental accounting is an important step in the development of an innovative development system aimed at increasing the level of competitiveness.

4   Discussion

While appreciating the contribution of scientists, it should be noted that the issues of accounting for eco-innovations as an object of environmental accounting are insufficiently studied. In our opinion, these aspects require complementation, deepening.

We can illustrate the relationship between environmental problems arising from the use of obsolete technologies, society and the economy using the scheme (Fig. 1).

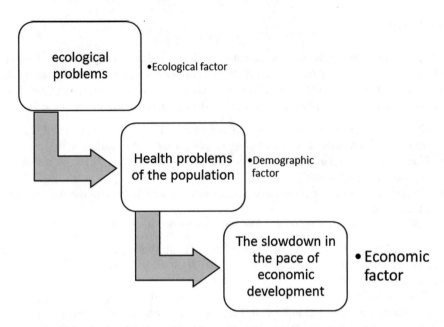

Fig. 1. Interrelation of problems of ecology, economy and society

Thus, the system of environmental accounting is an important step in the development of an innovative development system aimed at increasing the level of competitiveness of the organization as a whole, which once again emphasizes the importance of issues of setting environmental accounting and the priority of ecooriented production in the business entity. The task of the subject is to establish the relationship between economic, environmental and social indicators; development of key indicators for assessing the environmental impact of production and improving control over the environmental performance of production activities; application of economic efficiency indicators that meet the principles of green economy.

The use of modern, equipped with the newest systems of quality control, safety, equipment that minimizes environmental damage to the environment, awareness of the social responsibility of organizations for both the state of nature and the health of an individual [17], the production of competitive products satisfying the requirements of the consumer can be regarded as the most important prerequisites for the transition to an innovative way of economic development. It is widely known that the pace of growth in innovative development of the economy has been slowed by the impact of an environmental factor, the requirement of taking into account it determines the importance of developing a system for collecting, recording and systematizing information in various spheres of the ecological and economic space [18, 19].

5   Conclusions

The objects of environmental accounting are primarily environmental innovations and environmental obligations.

In order to determine the problems of accounting for eco-innovations, it is necessary to address the key aspects of financial accounting as a whole.

The key issues solved by the financial accounting system are the following: "When did the operation take place?", "What is the value of the accounting object?", "How is the operation recorded in the account?". This raises the problem of recognition of the accounting object, the problem of object evaluation, the problem of the qualification of the operation. We can achieve the solution of these issues by reflecting each operation by a certain date, assigning a certain amount and fixing the accounting entries. For a long time, the innovative activity as a whole and, consequently, its results were not considered as independent objects of financial accounting in general. In the context of transition to an innovative economy, it is necessary to clearly understand the importance of allocating innovations in general and eco-innovations in particular to individual accounting objects, which points to the problem of developing normative and methodological support in this area.

First of all, the issues of ensuring the totality of the elements of the resource support of the accounting and analytical system (procedures, tools, methods) that provide accounting and analytical support for the process of making managerial decisions in the sphere of eco-innovations require development.

Improvement of the current regulatory and legal framework, development of a single standard for recording the results of innovation activities, guidelines and recommendations for the incorporation of eco-innovations will contribute to the development of an environmental accounting system, as well as the creation of environmentally-oriented production. The key tasks of environmentally-oriented production are establishment a correlation between the system of economic, environmental and social indicators; increasing control over the environmental friendliness of production activities; using in practice of a set of economic efficiency indicators, corresponding to the basic principles of green economy.

References

1. Ashford, N.: Understanding Technological Responses of Industrial Firms to Environmental Problems: Implications for Government Policy, Environmental Strategies for Industry: International Per-spectives on Research Needs and Policy Implications. Island Press, Washington, DC (1993)
2. Arundel, A., Kemp, R.: Measuring eco-innovation. Working paper series. United Nations University, UNU-MERIT, vol. 017 (2009)
3. Sarkar, A.: promoting ecoinnovations to leverage sustainable development of ecoindustry and green growth. Eur. J. Sustain. Dev. 2(1), 171–224 (2013)
4. James, P.: The sustainability circle: a new tool for product development and design. J. Sustain. Prod. Des. 2, 52–57 (1997)

5. Rennings, K., Zwick, T.: Employment Impacts of Cleaner Production. Physika-Verlag, New York (2002)
6. Spaargaren, G., Mol, A.P.J.: Sociology, environment and modernity. ecological modernization as a theory of social change. Soc. Nat. Res. **5**(1), 323–344 (1992)
7. Volodin, O.N.: Environmental accounting in the enterprise. Contemporary Society and science: the socio-economic problems in the research of the university teachers. In: Proceedings of International Scientific Conference, pp. 73–79 (2015)
8. Gubaidullina, I.N., Ishmeeva, A.S.: Environmental accounting as a factor of economic security business entity. J. Mod. Probl. Sci. Educ. **1**(1), 657 (2015)
9. Zhelbunova, L.I.: Problems of formation and accounting of environmental accounting border. Russian Economy in the XXI century. In: Proceedings of XI International Scientific Conference on "Economics and applied research: fundamental problems of modernization of the Russian economy", pp. 68–70 (2014)
10. Ilicheva, E.V.: Environmental Accounting in the Conditions of Realization of Ecological Balance Policy: Abstract Dissertation Dr. Degree, Eagle, GOU VPO. Orel State Technical University, p. 28 (2010)
11. Ishmeeva, A.S.: Balance environmental accounting system in management information sources. The rule of law: the problem of understanding and realization. In: Proceedings of International Scientific Conference, pp. 225–229 (2015)
12. Myagmar, M.: Environmental accounting in the modern economy. Theory and practice of economic development at the international, national, regional levels. In: Materials International Scientific Conference, pp. 174–177 (2014)
13. Rubanova, N.N.: Environmental accounting in the building materials enterprises: Abstract Dissertation Ph.D. in Economics, Eagle, GOU VPO. Stavropol State University, p. 23 (2005)
14. Khmelev, S.A., Suglobov, A.E.: Methodological aspects ecologic accounting and auditing in order to ensure the economic security of industrial enterprises. J. Vector Sci. Togliatti State Univ. **3**, 95–101 (2011)
15. Sergeeva, E.A.: On the eco-oriented statements in the analytical assessment of the sustainability of enterprises. J. Her. Taganrog Inst. Manag. Econ. **1**, 50–53 (2011)
16. Mukhina, E.R.: Analysis of the main trends in the definition of the term "competitiveness". J. Vector Econ. 1 (2016). http://www.vectoreconomy.ru/images/publications/2016/1/EconomicsAndManagement/Muhina.pdf
17. Sudova, T.L.: Individual social responsibility. J. Vector Econ. 2 (2017). http://www.vectoreconomy.ru/images/publications/2017/2/economictheory/Sudova.pdf
18. Sheshukova, T.G., Mukhina, E.R.: Towards a theory of the development of environmental accounting at an industrial enterprise in the conditions of the formation of an innovative economy. J. Her. PNIIP Socio-Econ. Sci. **2**, 141–151 (2017)
19. Rudskaya, E.N., Voronina, A.K.: "Green economy" in the system of ecoinnovations: practice and problems. J. Vector Econ. 9 (2017). http://vectoreconomy.ru/images/publications/2017/9/innovationmanagement/Rudskaya_Voronina.pdf

The Corporate Reporting Development in the Digital Economy

Olga Efimova$^{(\boxtimes)}$ and Olga Rozhnova

Financial University, Moscow, 125993 Moscow, Russia
oefimova@fa.ru

Abstract. The paper examines the effects of data and technology on the reporting development—the data collection; analysis, interpretation and use in the preparation and disclosure of connecting financial and non-financial reports. Next, we study the use of technology in the delivery of financial data to interested parties in the information supply chain and in the consumption of financial and related non-financial information for decision making process. That is, we investigate how data, data analytics, and technology may potentially transform the financial reporting process to make it more effective, resulting in greater transparency for capital providers. Finally, we outline current trends in structured data using to increase efficiency and transparency for all users.

Keywords: Financial reporting · Disclosure · Non-financial information
Reporting development · Digitalization

1 Introduction

The economic resources of the reporting entity and its claims may change for reasons not related to financial results[1]. That is why disclosures of material non-financial information provided in connection with financial statements are essential to an investor's understanding and analysis of the business underlying the information in financial statements. Information on these types of changing and their consequences is necessary for interested users, primarily investors and analysts to develop a more reasonable forecast of future cash flows needed to assess the business fair value and make investment decisions (Graham et al. 2005). Resent numerous studies had evidenced that the main issue is the harmonization of financial statements and non-financial information provided by the reporting companies (Khan et al. 2015; Amel-Zadeh and Serafeim 2017). The issues of corporate reporting based on comprehensive disclosure both financial and non-financial information prepared in friendly to investors manner are actively discussed at numerous scientific congresses and academic publications that examine various aspects of the reliable content, completeness and relevance as well as the prospects for further reporting development (Amel-Zadeh and Serafeim 2017, Wenxiang (Lucy) Lu et al. 2015).

---

[1] Conceptual Framework for Financial Reporting' (2018).

© Springer Nature Switzerland AG 2019
T. Antipova and A. Rocha (Eds.): DSIC 2018, AISC 850, pp. 71–80, 2019.
https://doi.org/10.1007/978-3-030-02351-5_10

Over the past ten years the number of voluntary non-financial reporting require-ments around the world had increased from under 10 to 182^2. At the same time, the process of legislative regulation and non-financial reporting standardization had been actively developing. For example, the EU's Non-Financial Reporting Directive 2014/95, which determined the requirements for ESG information disclosure has now been incorporated into the national legislation of many EU countries. Among inter-national guidelines for non-financial reporting the most well-known and widely used are the Global Reporting Initiative (GRI), Integrated Reporting standards, Sustain-ability Accounting Standards Board (SASB).

Despite that all these standards and requirements are oriented at presenting of material non-financial information to interested users, they have significant differences concerning the target audience, requirements for structure and content, recommenda-tions for selecting performance indicators and criteria for their evaluation, industry specific features and other important aspects. This undoubtedly creates difficulties for interested users as well as for reporting companies (Amel-Zadeh and Serafeim 2017).

2 Literature Review

Recent studies (CFA Survey 2017; Amel-Zadeh and Serafeim 2017) investigated current trends in financial reporting disclosure in content of investor perspectives on transparency and evidenced that investors are looking for better not less disclosure.

The study (Barton and Wiseman 2014) indicated that investors are considering strategies that take into account environmental, social and corporate governance (ESG) criteria. The ESG information is associated with numerous economically meaningful effects for investors. Specifically, ESG disclosures are associated with lower cost of capital (Cheng et al. 2014; Dhaliwal et al. 2011), stock price movements (Graham et al. 2005) and better financial performance. There is data indicating that companies adopting sustainability measures may perform better than those that do not (Khan et al. 2015).

The research provided by Financial Reporting LAB^3 classified the information that were determined as more reliable for investors like financial metrics and wider metrics. The financial metrics were divided to GAAP (or IFRS metrics) and related to GAAP (IFRS) metrics and other additional disclosing indicators (for example operating profit, free cash flow, total shareholder return and others). Wider metrics include non-financial indicators that can be preparing according to non-financial standards, for example, GRI or SASB and company specific metrics, developed on voluntary basis (Financial Reporting LAB, 2018). The numerous surveys of interested users outlined that infor-mation of corporate reports disclosed on a voluntary basis is often incommensurable, which makes companies' comparative analysis difficult.

Other relatively new areas of concern to investors are cyber security and climate risks. The past few years have seen massive and unprecedented cyberattacks against

^{2} www.reportingexchange.com.

^{3} www.frc.org.uk/Lab.

some of largest companies in the world. Russian companies lost about 116 billion rubles due to cyberattacks in 2017[4]. Investors need information relating to how companies are addressing these risks and how might them affect the performance of a company in the future? (COSO & WBCSD 2018).

As numerous studies shown the interviewed respondents emphasized the insufficient level of the reporting disclosure usefulness for financial and investment decision process (CFA survey 2017, Khan et al. 2015). This makes it necessary, on the one hand, to improve the information disclosure quality, and, on the other hand, to develop a mechanism for involving stakeholders in the preparation, discussion and public verification of non-financial reports.

The providing research had allowed to identify the main difficulties that hinder the solution of this problem (Amel-Zadeh and Serafeim 2017; join research COSO & WBCSD 2018):

- Comparability of non-financial reporting data of companies in the same industry, as well as one company for different periods of time;
- Co-ordination of financial and non-financial reporting requirements;
- Timeliness of obtaining information by interested users;
- Reliability and completeness of material information disclosure;
- Difficulty quantifying non-financial issues;
- Longer time horizons for non-financial risks and opportunities;
- Lack of practical examples for the formation of "best practices" for the preparation of non-financial reporting.

The surveys (Eccles et al. 2013, CFA 2017) found out, that the respondents asked to identify key difficulties of ESG integration emphasized lack of comparability across firms (44.8%) was the most frequently identified impediment, followed by lack of reporting standards (43.2%), cost (40.5%), data usefulness (39.4%), lack of quantifiability (37.8%) and lack of comparability over time (34.8%).

All these obstacles create barriers for investors and financial analysts to timely integrate financial and non-financial information, which is necessary for making decisions. Fundamental investing depends on the analyst's ability to make comparative analyses among companies and across time periods. Decisions about which stocks to purchase are based on the observations and insights gained from this analysis. Investors want structured quantitative data which tends to be both quantitative and qualitative for the reason that greater transparency results in decrease in the cost of equity capital, public debt, private debt capital and thus increases equity values (Graham et al. 2005).

At present, there is no longer any doubt that these problems cannot be solved by local improvements in the current system of corporate reporting without digital transformation. Digital transformation, also known as digitalization, refers to a business model driven by "the changes associated with the application of digital technology in all aspects of human society" (Stolterman and Fors 2004). It can be implemented through digitization, i.e. the "ability to turn existing products or services into digital variants, and thus offer advantages over tangible product" (Gassmann et al. 2014;

---

[4] https://www.rbc.ru/technology_and_media/19/12/2017/5a38f3749a794710aa15581b?from=main.

Amin). CFA Institute research (CFA Survey 2018) showed that the use of structured data and technology can result in a more effective and efficient overall financial reporting process in which investors and analysts receive more transparent, better-quality information on a timely basis.

3 The Main Body

3.1 Approach

We focused on the financial and non-financial reporting usefulness for investors and other capital providers and investigated the financial reporting process from the beginning to end to evaluate the effectiveness of the current system and the ways that data and technology may potentially improve or when necessary to transform that process.

We analyzed 168 public corporate reports of major Russian companies (all available reports) registered in the National Register of corporate reports to assess the quality of non-financial information disclosure from the investor's needs. The analysis had outlined similar problems to international ones. Despite the companies' majority use GRI standards, in practice, the information is disclosed in such a way that users have to spend significant time searching for the comparative data (sometimes without positive results). Another problem is related to the availability and timeliness of obtaining non-financial reports by interested users.

As a part of the study we used the companies' interviews, investors surveys and Russian companies' financial and non-financial reports' analysis to assess their relationship and interactions. We also used a systematic literature review method to accumulate reporting development research results.

The paper examined the ways that data, data analytics, and technology may potentially improve or even transform corporate reporting system to fully meet stakeholder information requests. New technologies for the collection, processing and disclosure of accounting information and reporting were formed in accordance with the development of accounting requirements (Kuter et al. 2017). We analyzed the possibilities of digitalization in the delivery on-line data to all parties in the information supply chain and in the consumption of financial and non-financial information by all users (Amin et al. 2018; CFA 2016; Stolterman and Fors 2004). At last we assessed how digitalization may potentially transform the financial and non-financial reporting production, delivery and consumption processes to make them more effective and transparent for investors.

3.2 Statement of Basic Materials

Information which is not regulated by International financial reporting standards should disclose the features of financial results and financial position; a description of the uncertainties the features of financial results and financial position; a description of the uncertainties; dividend policy; and company's target leverage; provide reports on environmental protection, etc.[5]. Many companies report such information, but it is not comparable between entities, reports' structure is too specific, often there is no direct connection between information in non-financial and financial reports. Frequently the information in non-financial reporting is sketchy, some statements about future performance are too lengthy and contain a lot of "air" - redundant information and daring declarations.

There are no interactive cross-links between reports, at only basic links at best.

These additional reports are often either too short or too long and although include tables, charts, figures and are presented in a certain manner, but, in fact, are poorly structured in contrast to the financial reporting. As a rule, reports, which also include financial reporting, give an impression of a complex and intricate documents, which cannot be read fully and comprehend by anyone. We believe that Comprehensive reporting, which includes all forms – financial, non-financial, management, should be built on "matryoshka" principle: on opening of a report, a user first sees an enlarged picture, where each item can be expanded by clicking on it and further down. This principle should also be applied to the text itself, the user should be able to choose what to see first – the text, the tables, or the graphs and then be able to switch from one option to another... It is advisable that at user's request any section of the text could be opened and relevant prompts should appear, for example, to open previous periods, or to go to additional available information. Thus, reporting should be based on client servicing principles - attract and serve users in the best way as shown in Fig. 1.

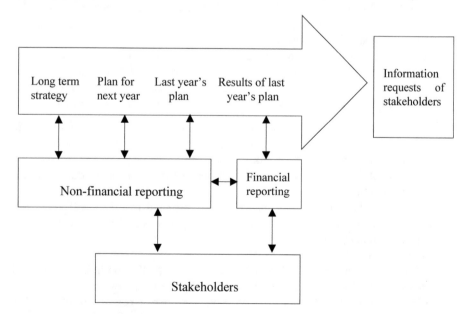

Fig. 1. Interaction of financial and non-financial reporting in the conditions of digitalization in order to meet information requests of stakeholders

---

[5] https://www.ifrs.org/issued-standards/.

Preparation of such reports is only possible with digital technologies, allowing user instant access to any reporting point with the most details, explanations of the causes of events, factors affecting them, and proactive actions that the company takes or will take in response to various situations. Artificial intelligence can help process users' fuzzy requests - understand what information the user is looking for and offer the most appropriate response. So, a user may be interested in how a certain professional judgment was made for reporting purposes. For example, for revenue recognition under IFRS 15, it may be necessary to determine which method of estimating the variable part of the remuneration is the most relevant, and for different components of the variable remuneration, different methods may be applicable, for example, for one part - expected value, for another - the likely cost. In addition to method selection, IFRS requires determination of probabilities. In financial reporting it is not possible to give detailed descriptions on these choices for each contract, but it may contain the important information for users. Interactive digital report can provide users with explanations for each professional judgement applied. It is important that in response to, for example, a request what forecasts were made by the company for reporting purposes, the user is provided with a list of information, including information about the essential contracts for which revenue with a variable component of remuneration was recognized.

As a result, reporting will no longer be perceived as a heavy multi-page document, but as a convenient tool for building responses to any information request.

The Report should be cross-linked, not only to the company information, but also to external data, including market data, reports on any industry enterprise/country/ world, competitors, legal framework.

The same approach should be applied to internal users, for whom, in addition to open sources, closed internal information of the enterprise will be available.

To improve financial reporting in non-financial reports, we propose the following. *To improve the usefulness of the SFP (the statement of financial position):*

- Determine risky resources (by projects, processes) of an enterprise, referencing to non-financial reports, which describe significant risks and disclose enterprise's preventive risk-minimization actions. This approach may be less relevant to financial assets because they are subject to detailed risk disclosures directly in the financial reporting.
- Disclosure of detailed risk information in non-financial reporting will allow risk comparison with competitors, overseeing risk dynamics and the effectiveness of company's counter actions.
- Disclose company's digitalization costs.
- Disclose investments comparable in importance to digitalization in other advanced technologies critical to the future of the company, by nature of activity.
- Distinguish these investments between capitalized and expensed, allowing refer-encing to the financial reporting in order to assess the extent to which fixed assets correspond to the advanced technology used by the company.

Suggestions for developing the non-financial reporting information to improve the perception of financial statements are shown in Table 1.

Table 1. Information to be presented in the non-financial reporting to improve the perception of the financial reporting.

Statement items	Determine objects (in terms of value and provide description of the main characteristics), which are used for
Fixed assets	Advanced technologies and digital transformation; digital technologies; controlled by artificial intelligence; supporting and improving the environment technologies; remotely controlled assets
Non-recognized Intangible assets (under IFRS)	Goodwill composition, including portions related to digital technologies; artificial intelligence; reputation of an environmentally friendly company; workers engaged in intellectual tasks, including those associated with digital technologies; number of employees with both professional knowledge and knowledge of digital technologies; the section of company's strategy, which focuses on the application of new technologies
Recognized Intangible assets (under IFRS)	Applied digital and other advanced technologies
Financial instruments	The forecast of financial markets development included in the sphere of the company's interests, the use of digital technologies for transactions with financial instruments
Estimated liabilities, commitments and obligations	Submit a detailed description of the associated risks (risk area, actions to mitigate these risks, including risks associated with digital transformation)

To increase the usefulness of SCI (the statement of comprehensive income):

- Disclose development costs (for non-recognized intangible assets): in digital technology, advanced technology, to protect environment; employee development.
- Disclose total development costs (capitalized and expensed);
- Describe research costs, purpose; connection with digital technologies; connection with advanced technologies, with strategic partnership; with retraining of personnel or personnel cost involved with digital technologies.
- Classify impairment losses (natural for current level of digitalization and current market volatility, occurring from insufficient digitalization (of the company or its partners) and high level of uncertainty and variability in the economy; related to the remaining subjectivity due to the incomplete transition to artificial intelligence, ignored by management signals provided by digital technologies (accurate forecast by digital technologies should have resulted in appropriate actions, e.g. sale, purchase, improvement, reorganization); related to the digital technologies themselves (hacker attacks, failures in advanced technologies); related to corruption, force majeure, wars, natural disasters, etc.

To improve the overall usefulness of financial reporting

Disclose information on considered, but not chosen options in professional judgment, so that user can better assess the situation. This approach will increase user's confidence in reporting.

On one hand, the digital economy creates a great potential for IFRS to achieve a new quality level as a system of financial reporting, overcoming existing limitation for accounting of reality, but on the other hand digital economy by itself introduces new risks, primarily for business but also for IFRS reporting, management reporting, non-financial reporting.

The risks introduced by digital technologies should be considered as natural, because any progress carries a downside. However, the power of digital technologies is so great that associated risks will be reduced by further improvement of digital technologies and artificial intelligence.

According to IFRS financial reporting must meet the two fundamental characteristics – relevance and reliability. The financial reporting defines three components of reliability: completeness, absence of errors, neutrality – all of this can be significantly improved by digital technologies. The development period of digital technologies is inevitably associated with unauthorized access to the company's website, but any new progressive phenomenon inherently carries growth diseases. It should be noted, that for the second fundamental characteristic, - predictability of reporting information - digital technologies create ideal conditions. The accuracy of the forecasts can be assured only by digital technology and artificial intelligence. The problem of relevance is easily solved with the help of demonstrated above reporting tool. The problem of reliability in financial and non-financial reporting is also close to resolution with the help of digital technologies:

From the position of the reporting entity -

- Big data allows more accurate predictions, due to significant reduction of subjectivity.
- Online availability of information allows instant access to any information necessary for professional judgment, such as technical, legal, geographical, social, political, economic, etc.
- Calculations accuracy is guaranteed.
- The choices become more warranted, because any option can be considered and calculated, confirmed by experts, any advice instantly obtained.

From the position of reporting entity, auditors and users - the verifiability of the reporting information increases many times, the subjective factor is excluded from verification; the control function becomes non-materialized, verification can be carried out on-line non-stop or at a specified interval (the auditor also has online access to all information); there is a possibility of auditing all related parties at the simultaneously. From the user's perspective - any significant information should be extracted from company's report in desired format to provide interested users on data for forecast preparation, performance evaluation and comparative analysis of products, markets, share price fluctuations, etc.

The issue is to adapt financial and non-financial reporting to digital technology using structured data (XBRL - e**X**tensible **B**usiness **R**eporting **L**anguage) or to wait for the full digitalization of the entire economy including reporting. In our opinion, each company needs to strive to digitize its reporting, because only such an approach can ensure competitiveness and trust of stakeholders.

4 Conclusion

We outline the main advantages of structured data for interested users including better financial statement accuracy, systematic risk management and appraisal, ensuring the interconnection of financial reporting and related non-financial information, improving the information disclosure quality, increasing opportunity for higher returns.

There is a possibility and necessity to develop interrelated approach to preparing financial and non-financial reporting. This requires wide utilization of the digital technology opportunities. The standards of financial and non-financial reporting can no longer be developed separately, it is necessary to create a unified concept for the formation of corporate reporting (including financial and non-financial reporting) in the face of the digital technologies' risks. Some significant directions and ways of integration were highlighted in the work.

We expect that the considering issues will stimulate further discussion within the impact of new digital technologies on corporate reporting and future studies developed the subject.

References

Amel-Zadeh, A., Serafeim, G.: Why and how investors use ESG information: evidence from a global survey. Harvard Business School. Working Paper (2017). http://www.hbs.edu/faculty/

Barton, D., Wiseman, M.: Big investors have an obligation to end the plague of short termism. Harvard Bus. Rev. 48–55 (2014)

CFA: ESG Survey 'Global Perceptions of Environmental, Social and Governance Investing. https://www.cfainstitute.org/en/research/survey-reports/esg-survey2017

CFA: Data and technology: transforming the financial information landscape. Investor Perspectives (2016). https://www.cfainstitute.org/en/advocacy/policy-positions/data-and-technology-transforming-the-financial

Cheng, B., Ioannou, I., Serafeim, G.: Corporate social responsibility and access to finance. Strateg. Manag. J. **35**, 1–23 (2014)

COSO, WBCSD: Enterprise risk management: applying enterprise risk management to environmental, social and governance-related risks by the Committee of Sponsoring Organizations of the Treadway Commission (COSO) and the World Business Council for Sustainable Development (WBCSD) (2018). https://www.wbcsd.org/Projects/Non-financial-Measurement-and-Valuation/Resources/Applying-enterprise-risk-management

Graham, J., Harvey, C., Rajgopal, S.: The economic implications of corporate financial reporting. J. Account. Econ. **40**(1/3), 3–73 (2005)

Dhaliwal, D., Zhen, L., Tsang, A., Yang, Y.: Voluntary nonfinancial disclosure and the cost of equity capital: the initiation of corporate social responsibility reporting. Account. Rev. **86**(1), 59–100 (2011)

Eccles, R.G., Ioannou, I., Serafeim, G.: The impact of corporate sustainability on organizational processes and performance. Harvard Business School Working Paper Series. Harvard Business School (2013). https://www.hbs.edu/faculty/Publication%20Files/SSRN-id1964011_6791edac-7daa-4603-a220-4a0c6c7a3f7a.pdf

Khan, M., Serafeim, G., Yoon, A.: Corporate sustainability: first evidence on materiality. Account. Rev. **91**(6), 1697–1724 (2015)

Kuter, M., Gurskaya, M., Andreenkova, A., Bagdasaryan, R.: The early practices of financial statements formation in Medieval Italy. Account. Hist. J. **44**(2), 17–25 (2017)

Stolterman, E., Fors, A.: Information technology and the good life. In: Information Systems Research, pp. 687–692 (2004)

Lu, W., Taylor, M.E.: Which factors moderate the relationship between sustainability performance and financial performance? A meta-analysis study. J. Int. Account. Res. **15**(1), 1–15 (2015)

Self-adaptive Intelligent System for Mass Evaluation of Real Estate Market in Cities

Alexander O. Alexeev[1], Irina E. Alexeeva[1], Leonid N. Yasnitsky[2(✉)], and Vitaliy L. Yasnitsky[2]

[1] Department of Applied Mathematics and Informatics, Perm State University, Bukirev Street 15, 614600 Perm, Russia
[2] Department of Building Engineering and Materials Science, Perm National Research Polytechnic University, Komsomolsky Prospect 29, 614990 Perm, Russia
yasn@psu.ru

Abstract. This article is devoted to the method of creating an intelligent neural network system. Unlike existing similar systems, the proposed system does not require frequent updates, because it is able to adapt itself to the constantly changing state of the economy and to the peculiarities of a particular region. Besides, the proposed system allows performing scenario forecasting of regional real estate markets depending on virtually changing economic parameters such as the dollar rate, the market price of oil, gross domestic product and gross regional product, the volume of housing construction in the region, the parameters of the state's credit policy, etc.

Keywords: Real estate market · Neural network · Scenario forecasting
Economic situation

1 Introduction

As analysis of literature [1–4, etc.] shows, there is a large number of works that note the importance and urgency of creating high-precision techniques for mass valuation of real estate. To create such techniques, both regression and neural network technologies are actively used [5–10, etc.]. The authors of [11] made an attempt to develop a methodology that does not have these drawbacks. This paper is a continuation of the investigations begun in [11].

2 Formulization of a Mathematical Model and Its Testing

The following factors that characterize the static construction and performance factors were included as input parameters in creation of the model of mass appraisal of residential property in city Yekaterinburg: total area of the apartment, number of rooms, floor, number of floors, house type, walls type, availability of a balcony/loggia, district, distance to the city center, as well as a number of macroeconomic indicators:

© Springer Nature Switzerland AG 2019
T. Antipova and A. Rocha (Eds.): DSIC 2018, AISC 850, pp. 81–87, 2019.
https://doi.org/10.1007/978-3-030-02351-5_11

GDP, RTS quotes, Brent crude oil price, dollar rate, new housing supply, housing loans issued.

The output variable of the model corresponds to the declared price of the city apartment.

Many examples for training and testing the neural network were formed on the basis of statistical data of the real estate market of Yekaterinburg over the last 10 years: from 2006 to 2016. Selling prices of apartments were taken from open sources. Thus, many instances included the data during economically calm times for Russia (2006), period of economic growth (2007–mid-2008), crisis and turning point of the Russian and world economy (2008–early 2010), period of recovery after the crisis (2010–2012), growth retardation (2013–early 2014), strong fall in the background of Russian foreign policy, imposition of western sanctions, sharp drop in oil prices and ruble rate relative to dollar and euro, financial blockade and closure of access to international capital (2014–2016). During this decade, the RTS quotes have varied from 625 to 1,733, the price of Brent oil – from \$40.11 to \$126.90, the US dollar exchange rate –from 23.45 to 66.49 rubles, housing construction in Sverdlovsk region – from 1284.2 to 2483.7 thous. sq.m., issued mortgage loans – from 4,369 to 59,829 mln rubles, GDP – from 26,916 to 80,4125 bln rubles.

Overall, data on 2360 city apartments was collected and processed. This set was divided into training set, which contained 2,160 examples, and testing set, which contained 200 examples. The optimal structure of the neural network was a perceptron with one hidden layer of three sigmoid neurons.

After removing erroneous examples, the training error amounted to 6.2%, and testing error – to 6.5%. Moreover, additional checks on the quality of the network using multi-fold cross-validation method did not show any significant increase in training and testing errors.

The coefficient of determination R2 on the test set (between the predicted and observed values) was 0.87, which suggests that the constructed approximating model describes the market by explaining the input variables by 87%.

3 Computational Experiments and Discussion of Results

After the work of the neural network has been checked in the test cases, and thus the adequacy of the mathematical model of neural network has been proved, we can begin to investigate it. The trained neural network model responds to changes in input variables and behaves in the same way as the subject field itself. Therefore, the dependence of the predicted values on the input parameters of the model can be explored using the neural network of the model.

The first question that can be answered using the models is to determine the degree of influence of its input parameters on the simulation result – the value of apartments in Yekaterinburg. The objective assessment of this influence can be obtained, for example, by the technique using the same neural network by alternate exclusion of input parameters and observation of the error of its testing. The higher the testing error is, the more significant the excluded parameter is. The histogram constructed in this way is shown in Fig. 1. The height of the columns corresponds to the testing error obtained with the excluded parameter marked under the column. Moreover, the values of the column heights are scaled so that their sum totals 100%. The height of the columns is interpreted as the value of the parameter, which corresponds to the column. As can be seen from the figure, the following parameters were the most significant:

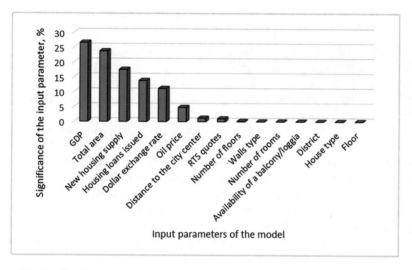

Fig. 1. Significance of the input parameters computed using neural networks.

- Gdp: 26.5%,
- Total area: 23.7%,
- New housing supply: 17.5%,
- Issued housing loans: 13.8%,
- Dollar exchange rate: 11.1%,
- Oil price: 4.8%,
- Distance to the city center: 1.2%.

As noted above, the neural network model is adequate to the modeled topical area, so it can be used to study the laws of the market of real estate in Yekaterinburg. This can be done, for example, through computations using the trained neural network along with a gradual change of either one of the input parameters and observation of computational results. Four apartments, which differ in their technical characteristics and district, were selected to perform computer experiments:

- One-room apartment with total area of 33 m², located on the 8th floor of a 10-storeyed panel building with improved layout, with a loggia, the building is located in "Elmash" district of Yekaterinburg at a distance of 9.6 km from the city center.
- Two-room apartment with total area of 59 m², located on the 6th floor of a 9-storeyed panel building with the full-length layout, with a loggia, the building is located in "Uralmash" district at a distance of 9.9 km from the city center.

- Three-room apartment with total area of 67.3 m², located on the 6th floor of a 13-storeyed building of "gray panel" type, "Monolith" walls type, with a balcony, the building is located in "Avtovokzal" district at a distance of 3.7 km from the city center.
- Four-room apartment with total area of 118 m², located on the 7th floor of a 16-storeyed panel building with improved layout, with a loggia, the building is located in "Uralmash" district at a distance of 8.4 km from the city center.

The apartments were appraised at the time of the market condition in the 1st quarter of 2016, when the macroeconomic indicators had the following meanings: RTS quotes were 876; oil price was 42.93 US dollars; US dollar exchange rate was 66.49 rubles; new housing supply in Sverdlovsk region was 2,483.7 thous. sq.m.; issued housing loans were 40,822 mln rubles; GDP was 80,412.5 bln rubles.

Figure 2 shows the results of the virtual computer experiments performed in order to study the value of apartments depending on their location in the city. As can be seen from the figure, the value of all four apartments uniformly decreases with their virtual distancing from the city center. Moreover, the patterns are different in nature: the curve related to the 4-room apartment has a negative second derivative at all points, whereas the curves corresponding to one-, two- and three-room apartments have a positive one. This means that the rate of decrease in the prices of the 4-room apartment increases when distancing from the city center, whereas for the other considered apartments it decreases.

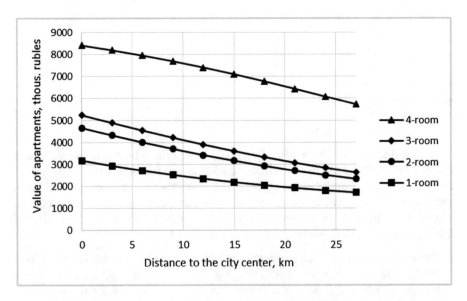

Fig. 2. Dependence of the value of apartments on their location in the city.

The next series of experiments is devoted to the study of the impact of the lending program implemented by banks on the residential real estate market in Yekaterinburg.

This time, the computer experiments on neural network mathematical model were carried out through a virtual change of the input parameter "Issued housing loans", all other input parameters remain unchanged. Figure 3 shows the value of apartments in 2016 corresponding to the volume of loans issued in 2015, which amounted to 40,822 mln rubles (marked with an incremental marker). As the figure shows, the simulation results predict an increase in the value of all four apartments with an increase in housing lending. In particular, for example, if the banks increase the volume of existing housing lending from 40,822 to 41,000 mln rubles, the value of the one-room apartment will increase from 2,249,000 to 2,550,000 rubles, i.e. by 2.4%, while the value of the four-room apartment will increase from 7,745,000 to 7,833,000, i.e. by 1.1%. Thus, it can be concluded that an increase in housing lending in Yekaterinburg will lead to price of luxury apartments with a larger area growing about 2.2 times faster than the cheaper apartments with smaller area.

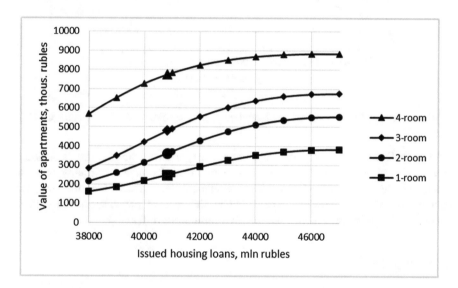

Fig. 3. Dependence of the value of apartments on the volumes of mortgage lending.

Figure 4 shows the dependences of the value of the apartments under study on the volume of housing construction, obtained in a similar manner. The results are obtained through neural network computing by gradually changing the input parameter "New housing supply" and the preservation of all other input parameters unchanged. As before, the enlarged marker shows the value of apartments corresponding to the condition existing by the I quarter of 2016: level of new housing supply in Sverdlovsk (Yekaterinburg) region amounted to 2,483.7 thous. sq.m. As can be seen from the figure, with an increase in housing construction by about 2,550 thous. sq.m., there is an increase in the value of all four apartments under study. This is explained by the fact that the apartments in new buildings, as a rule, are more expensive than in older buildings. However, as follows from the figure, with an increase in new housing supply above the mentioned figure, the prices cease to grow, and then their decline begins. Thus, the results of mathematical modeling predict the saturation of the housing market in Yekaterinburg, which will occur if the volume of housing construction exceeds the mark of 2,550 thous. sq.m.

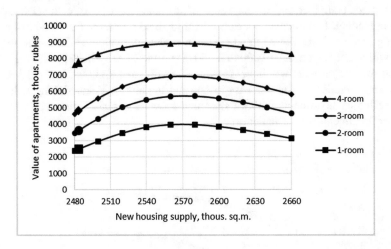

Fig. 4. Dependence of the value of apartments on the volume of housing construction.

We shall note that this series of computer experiments was performed using the "freezing" technique – the volume of housing construction was virtually increased, while all the other macroeconomic parameters, and hence incomes, remained unchanged. We can therefore expect that in the case of non-compliance with this condition, the prediction results would have turned out different.

4   Conclusion

Thus, an integrated economic and mathematical model of mass appraisal of residential real estate in city Yekaterinburg was created, taking into account both construction and performance parameters of apartments and the evolving economic situation in the country and the world. In contrast to the static economic and mathematical models that take into account only construction and performance parameters, the developed model does not require frequent updating and is also suitable for medium-term forecasting of the behavior of the real estate market in order to extract useful knowledge.

The developed integrated model has allowed us to conduct research of the residential real estate market in Yekaterinburg, identify patterns and perform some forecasts, the most interesting of which are the onset of market saturation effect with the increasing housing construction (Fig. 4) and the effects of the increase in mortgage lending volumes (Fig. 3).

In conclusion, we shall note that the proposed technique was demonstrated by the example of appraisal and prediction of the residential real estate market of

Yekaterinburg, which refers to the developed cluster of Russian cities with the high incomes and relatively high prices on the housing market. Similar studies and forecasts using the proposed technique can be made for other countries and cities.

References

1. Hefferan, M.J., Boyd, T.: Property taxation and mass appraisal valuations in Australia adapting to a new environment. Prop. Manag. **28**(3), 149–162 (2010)
2. Davis, P., McCluskey, W., Grissom, T.V., McCord, M.: An empirical analysis of simplified valuation approaches for residential property tax purposes. Prop. Manag. **30**(3), 232–254 (2012)
3. Kilpatrick, J.: Expert systems and mass appraisal. J. Prop. Invest. Finan. **29**(4), 529–550 (2011)
4. Manganelli, B., Pontrandolfi, P., Azzato, A., Murgante, B.: Using geographically weighted regression for housing market segmentation. Int. J. Bus. Intell. Data Min. **9**(2), 161–177 (2014)
5. Curry, B., Morgan, P., Silver, M.: Neural networks and non-linear statistical methods: an application to the modelling of price-quality relationships. Comput. Oper. Res. **29**(8), 951–969 (2002)
6. Gonzalez, M.A.S., Formoso, C.T.: Mass appraisal with genetic fuzzy rule-based systems. Prop. Manag. **24**(1), 20–30 (2006)
7. Guan, J., Shi, D., Zurada, J.M., Levitan, A.S.: Analyzing massive data sets: an adaptive fuzzy neural approach for prediction, with a real estate illustration. J. Organ. Comput. Electron. Commer. **24**(1), 94–112 (2014)
8. Kontrimas, V., Verikas, A.: The mass appraisal of the real estate by computational intelligence. Appl. Soft Comput. J. **11**(1), 443–448 (2011)
9. Mao, Y.H., Zhang, M.B., Yao, N.B.: Hangzhou housing demand forecasting model based on BP neural network of genetic algorithm optimization. In: Applied Mechanics and Materials, vol. 587–589, pp. 37–41 (2014)
10. Zhang, H., Gao, S., Seiler, M.J., Zhang, Y.: Identification of real estate cycles in China based on artificial neural networks. J. Real Estate Lit. **23**(1), 67–83 (2015)
11. Yasnitsky, L.N., Yasnitsky, V.L.: Technique of design of integrated economic and mathematical model of mass appraisal of real estate property by the example of Yekaterinburg housing market. J. Appl. Econ. Sci. **11**(8), 1519–1530 (2016)

Comparability of Financial Reporting Under Different Tax Regimes

Svetlana Grishkina⬤, Vera Sidneva⁽✉⁾⬤, Yulia Shcherbinina⬤,
and Galina Dubinina⬤

Financial University Under the Government of the Russian Federation,
Leningrad Prospect, 49, 125167 Moscow, Russia
v.sidneva@gmail.com

Abstract. The article analyzes the procedure for the formation of financial statements when various tax regimes are applied, reveals differences in the methods for the formation of individual elements of financial statements, depending on the taxation systems used. The practice of applying special tax regimes in the Russian Federation is investigated, their advantages and disadvantages for certain economic entities are analyzed and the appropriateness of accounting for tax factors in the creation of an expedient accounting system in every organization is substantiated. The presented material indicates that the balance of economic benefits from improving the quality of reporting information by ensuring comparability of indicators and costs for accounting can be achieved through additional disclosures in the financial statements.

Keywords: Financial reporting · Simplified taxation system
Unified tax on imputed income · Unified agricultural tax

1 Introduction

Under a market economy, the successful performance of economic entities is possible only as long as financing economic activity is sufficient, and in this regard access to the resources involved is of paramount importance for organizations with any form of ownership. This is especially relevant for small and medium-sized businesses, whose economic activity provides the country as a whole with a healthy economic climate.

Making a decision on providing funds the investor is primarily interested in the possibility of an adequate risk assessment and efficiency of investment. Of course, to a certain extent risks and efficiency are dependent on general macroeconomic and other external to the subject of investment factors, but in the first place, the financial condition of the potential funds recipient should be subject to assessment. The main source of information for business activity economic analysis by external users is the accounting (financial) statements of the organization – the result of summarizing the current accounting data.

T. Antipova and A. Rocha (Eds.): DSIC 2018, AISC 850, pp. 88–93, 2019.
https://doi.org/10.1007/978-3-030-02351-5_12

2 Review of the Current State of the Issues Under Research

In recent decades, the Russian Federation has been reforming the accounting and reporting system, in order to shift the priorities in formation and disclosure of information from fiscal goals to managerial decision-making. A guide conducted changes serve as international financial reporting standards (IFRS). Particular attention in the reform process focuses on the comparability of information, without which it is impossible to talk about the usefulness of financial statements for management decision-making.

International financial reporting standards (IFRS) are the benchmark for the changes. Particular attention in the reform process focuses on the comparability of information, without which the usefulness of financial statements for managerial decision-making is hardly worth talking about. Two aspects of the comparability of accounting information can be identified: the comparability of consecutive time periods within a single economic entity and the comparability of reporting data from different organizations.

Comparability of the economic entity's accounting information is ensured by the Accounting Regulation "Accounting Policy" (PBU 1/2008) in the accounting continuity assumption [1, item 5].

Comparability of reporting data from various organizations is in the first place determined by the uniform requirements to accounting and accounting regulation mechanisms prescribed by in the Federal law "On accounting" [2]. In practice, this is implemented through the use of Federal accounting standards by economic entities. However, both IFRS and the current Federal standards lay down alternative methods of accounting for individual objects, the choice of which is determined by the accounting policy of the organization.

Therefore, much attention in the system of regulatory accounting is paid to the disclosure of the methods used for accounting in the financial statements of the organization. PBU 1/2008 imposes a general obligation to disclose methods of accounting [1, p. 17], not knowing which the interested users of accounting statements won't be able to reliably evaluate the economic entity's financial position and results of operations Besides [3–5] certain Accounting Regulations contain more specific requirements for the disclosure of relevant aspects of accounting, such as how to accrue depreciation for individual groups of fixed assets [6, p. 32]. Problems of comparability of financial statements are subject to active discussions on the pages of Russian and foreign publications [7–14].

At present, the accounting indicators are influenced not only by the selected accounting methods within the current system of accounting Regulations (Federal Standards), but also by the tax system chosen by the economic entity [15, 16].

3 Statement of Basic Materials

In the Russian Federation, the current system of taxes and charges presupposes a general taxation system and special tax regimes established in part one of the Tax Code (hereinafter referred to as the Tax Code of the Russian Federation) [15, art.18].

Special tax regimes include:

(1) the system of taxation for agricultural producers (single agricultural tax);
(2) simplified system of taxation;
(3) the system of taxation in the form of a single tax on imputed earnings for certain types of activities;
(4) taxation system for production sharing agreements;
(5) patent system of taxation.

In the study of special tax regimes operating in the territory of the Russian Federation, it was revealed that the largest share in the total amount of revenues to the state budget from special tax regimes used as tax preferences for economic entities of small and medium-sized businesses belongs to those taxed under the simplified taxation system (Table 1). In 2017 the share of tax with the simplified taxation system was 78.6%.

Analysis of the dynamics of revenues shows that the amount of taxes within the special tax regimes is constantly growing. So, in 2017, compared to 2013, the increase in the total amount of taxes amounted to 146 186 million rubles, or 49.9%.

At the same time, there is an increase in taxes for all special tax regimes, with the exception of receipts for a single tax on imputed income, which in 2017, as compared to 2013, decreased by 3,840 million rubles, or by 5.2%.

In 2017 the amount of tax in the simplified taxation system increased by 132,929 million rubles, or by 62.6%.

Table 1. Analysis of the dynamics of tax payments under different tax regimes

						Mln rub.
Years tax regimes	2013	2014	2015	2016	2017	2017 to 2013%
Tax amounts for a simplified taxation system	212 287	229 316	254 164	287 068	345 216	162.6
Unified tax on imputed income for certain types of activities	74 471	76 631	78 507	74 327	70 631	94.8
Unified agricultural tax	4 041	4 713	7 431	11 438	11 891	294.3
Tax amounts for the patent taxation system	1 947	3 433	5 285	7 559	11 194	574.9
Total income taxes in the form of income from special tax regimes	292 746	314 093	345 387	380 392	438 932	149.9

Source: Compiled by the authors on the basis of data from the Federal Tax Service [17]

The greatest increase in relative terms is characteristic for the single agricultural tax and the patent taxation system (194.3% and 474.9%, respectively).

However, it should be noted that the patent system of taxation is not applied to legal entities, so its application does not affect the reporting of organizations.

In the case of small and medium-sized businesses that are non-agricultural producers, the choice is between the basic tax system and simplified system of taxation. Features of simplified system of taxation application are given in Chapter 26.2 of the Tax Code of the Russian Federation [16].

In compliance with the legally established criteria, the transfer to the simplified taxation system provides for exemption from the obligation to pay corporate profits tax, corporate property tax (except for the tax paid on the basis of cadastral value). Likewise, economic entities that have transferred to the simplified taxation system are not payers of the value-added tax.

The procedure for settlements with the budget for the value-added tax is disclosed in Chapter 21 of the Tax Code of the Russian Federation [16]. In general, the amount of VAT payable to the budget is defined as the difference between the amount of VAT accrued to the payment and the amount of VAT required by taxpayers to be deducted in accordance with the established procedure.

The tax period for VAT is the quarter, on the basis of which VAT Declaration is formed and submitted to the tax authorities, no later than the 25th day of the month following the end of the reporting quarter. The credit mechanism for the formation of the VAT tax base determines the fact that the amount of value added tax paid to suppliers when purchasing goods, works and services are not included in the cost of assets and expenses for ordinary activitiesи but are accumulated on a separate account for subsequent deduction.

Organizations that have transferred to the simplified tax system, exempt from the obligation to pay VAT to the budget, at the same time lose the right to tax deductions. As a result of this, the VAT amounts claimed by suppliers of goods, works and services are included in the cost of assets/expenses for ordinary activities.

As a result, the accounting statements of organizations applying the simplified tax system, compared to the reporting of organizations that are on the *General tax system*, will have the following differences:

(a) the book value of fixed assets of the simplified tax system payers will be higher by the amount of input VAT;
(b) the value of current assets will also be higher by the amount of input VAT;
(c) material costs for which a separate line of VAT was allocated, will be higher under the simplified tax system for the amount of input VAT.

These indicators are taken into account when determining the structure of assets, the value of its own working capital, profitability and turnover. Therefore, users of financial statements should take into account the tax regime used to assess its impact on the financial and economic indicators reflected in the financial statements. At the same time, the widespread use of the simplified tax system implies that organizations that apply different taxation systems do not always accompany all the acquisitions of goods, works and services by the presentation of input VAT. Therefore, it is not worthwhile introducing the average arithmetic correction factors.

The most objective way of recording the specifics of the value of assets/expenses formation, in our view, is the disclosure in the financial statements of not only the applied tax system, but also the following additional information:

(a) the amounts of value-added tax included in expenses for ordinary activities;
(b) the amounts of value added tax included in the cost of acquisition of fixed assets and intangible assets;
(c) the amount of value-added tax included in inventories.

In addition to accounting for value added tax, another distinctive feature of the simplified tax system is the cash method of recognition of income and expenses: the date of income receipt is the day of crediting funds to the accounts in banks and (or) to the cash desk, and expenses are recognized only after their actual payment [16, article 346.17]. In this case, the following ratios of accounting and tax accounting indicators are possible.

If the organization applying the simplified tax system falls under criteria of the subjects having the right to apply the simplified methods of accounting and compiling financial statements, it has the right also to establish in accounting the order of recognition of revenue and expenses on the cash method [3, item 12; 4, item 18], having made the corresponding disclosures in the financial statements of the organization. This will ensure consistency of accounting and tax accounting data and will allow a fairly transparent assessment of the final financial result. If the organization adheres to the traditional method of recognizing the income and expenses, most of the automation in accounting is focused on, then it is impossible to adequately link the information in the Report on financial results about income and expenses and the amount of tax calculated by the simplified methods of accounting. To a greater extent, the indicators of the Cash Flow Statement will conform with the tax accounting, but this form is hardly ever not included in the interim financial statements.

Accounting Regulations on Accounting for Corporate Income Taxes (PBU 18/02) dwell upon the issues of reconciling accounting and tax accounting data for organizations that use the basic tax system and are payers of corporate income tax [5]. The same mechanism for economic entities applying the simplified system of taxation does not exist.

4 Conclusion

Special tax regimes, including the simplified system of taxation, are intended primarily for small businesses, and are designed to reduce not only the tax burden, but also administrative costs for maintaining accounting and tax records. However, their application leads to the formation of incomparable indicators of financial reporting (income, expenditure, value of acquired values and production costs). The balance between the economic benefits of improving the quality of accounting information by ensuring its comparability and the costs of increasing complexity of accounting can be achieved by additional disclosure of information on special tax regimes.

References

1. Regulation on Accounting "Accounting Policies of the Organization" (PBU 1/2008). Approved by Order of the Ministry of Finance of the Russian Federation No. 106n from 06 Oct 2008 (Ed. from 28 April 2017)
2. Federal law "On accounting" Dated 06 Dec 2011 N 402-FZ (As amended on 31 Dec 2017)

3. Regulation on Accounting "Incomes of the Organization PBU 9/99". Approved by Order of the Ministry of Finance of the Russian Federation No. 32n from 06 May 1999 (Ed. from 06 April 2015)

4. Regulation on Accounting "Expenses of the Organization PBU 10/99". Approved by Order of the Ministry of Finance of the Russian Federation No. 33n from 06 May 1999 (Ed. from 06 April 2015)

5. Regulation on Accounting "Accounting of Calculations for Corporate Income Tax "PBU 18/02". Approved by Order of the Ministry of Finance of the Russian Federation No. 144н from 19 Nov 2002 G (Ed. from 06 April 2015)

6. Regulation on Accounting "Accounting of Fixed Assets PBU 6/01". Approved by Order of the Ministry of Finance of the Russian Federation No. 26n from 30 March 2001 (Ed. from 16 May 2016)

7. Grishkina, S.N., Rodionova, O.A.: The indicators of financial statements of agricultural enterprises and their comparability under different tax regimes. Int. Res. J. **11**(42), 32–35 (2015)

8. Pantic, B.: Comparability of financial reports: a literature review of most recent Studies. Working Paper # 17 (E)–2016. Graduate School of Management, St. Petersburg State University: SPb (2016)

9. De Franco, G., Kothari, S.P., Verdi, R.S.: The benefits of financial statement comparability. J. Account. Res. **49**, 895–931 (2011)

10. Yip, R.W.Y., Young, D.: Does mandatory IFRS adoption improve information comparability? Account. Rev. **87**, 1767–1789 (2012)

11. Bhojraj, S., Lee, C.M.C.: Who is my peer? A valuation-based approach to the selection of comparable firms. J. Account. Res. **40**, 407–439 (2002)

12. Cascino, S., Gassen, J.: What drives the comparability effect of mandatory IFRS adoption? Rev. Acc. Stud. **20**, 242–282 (2015)

13. Kim, S., Kraft, P., Ryan, S.: Financial statement comparability and credit risk. Rev. Acc. Stud. **18**, 783–823 (2013)

14. Kim, J.B., Li, L., Lu, L.Y., Yu, Y.: Financial statement comparability and expected crash risk. J. Account. Econ. **61**(2–3), 294–312 (2016)

15. "The Tax Code of the Russian Federation (Part One)" of 31.07.1998 N 146-FZ (Ed. of 19 Feb 2018)

16. "The Tax Code of the Russian Federation (Part Two)" of 05.08.2000 N 117-FZ (Ed. of 04 June 2018)

17. Data on Forms of Statistical Tax Reporting. https://www.nalog.ru/rn48/related_activities/statistics_and_analytics/forms/. (Ed. of 04 June 2018)

Methodology of Construction Management Models of Actors of Nature

Kirill Litvinsky[(✉)]

Economy Department, Kuban State University, Stavropol'skaya street, 149,
350040 Krasnodar, Russia
litvinsky@econ.kubsu.ru

Abstract. The article examines the management of simulation ecological-economic systems. In the research introduced the concept of a locally closed ecological and economic system as a system consisting of production and economic system, environmental ecological system population and the center (coordinator). Authors introduced a scheme of coordination of management processes in the locally closed eco-economic system, defines the types of impacts, which shows impossibility of constructing classifier assess interactions structural subsystems of the locally closed ecological and economic system, and identified ways to coordinate their application. It is proved in the study that coordination office should be combined with a reflexive control. Authors identified ways of reflexive control, motivating and stimulating factors, ways to find a compromise.

Keywords: Natural resources · Ecological and economic system
Reflexive management
Motivating and stimulating factors and ways to search compromise

1 Introduction

We define the management of the ecological and economic system as coordinating management – a way of structuring management in hierarchic multilevel economic systems featured by [7]:

(a) the main goal of coordinating is to find and realize such a management solutions which, on the one hand serve to the individual interests of natural resources subjects (human-nature) and on the other hand correspond with the objective function of the whole community (multicriteria optimization);

(b) coordination supposes management specialization, that is differentiating the general management function on particular functions imposing them on different organizational components;

(c) the coordination problem arises when natural resources subjects who have contradictions but not antagonism are quite self-sufficient in choosing management solutions;

© Springer Nature Switzerland AG 2019
T. Antipova and A. Rocha (Eds.): DSIC 2018, AISC 850, pp. 94–102, 2019.
https://doi.org/10.1007/978-3-030-02351-5_13

(d) in coordinating the superior body has the right to interfere with market subjects work without imposing on them a line of behavior and helps to solve appearing problems.

Not only exclusion but any infringement of independence of natural resources subjects in choosing their management destroys the integrity of an ecological and economic system. At the same time the freedom of choice of their own behavior leads to contradicting goals of natural resources subjects not coinciding with the whole community interests. Conflict of interests "private-private" and "private-general" arises. That is why, in contrast to usual management, coordination supposes conflict situations analyses and search of ways of conflict resolution due to reconciliation of private interests of natural resources subjects with pursuing global goals of the whole ecological and economic system [7].

2 Review of Prior Literature

An extensive theoretical literature [2, 3, 5, 6, 13] is devoted to the management of simulation ecological-economic system. However, it should be noted, that there is no one common long-held viewpoint on the majority of methodological problems in the sphere of working out criteria and principles of the simulation up to the present time. There is not observed one common treatment of fundamental concepts such as "environmental management and protection", "ecological and economic system", "natural resources", etc. in the scientific literature.

Many scientists define this system as the least unit of the noosphere featured with integrity, and moreover, an ecological and economic system should include three subsystems: ecological, economic and social. So, for example, Lipenkov A. defines an ecological and economic system as "...limited with certain territory part of noosphere where ecological, social and industrial structures and processes are connected with intersupporting streams of material, energy and information. An ecological and economic system is relatively closed system, which inside material streams are significantly bigger than the streams through its boarders. All the noosphere represents an interconnected system of regional ecological and economic systems."

3 Statement of Basic Materials

Let ecological and economic system S (Fig. 1) consist of management system MS and controlled process CP, that is S ={MS, CP, u, p, f(t)}, where u is managing actions; p is information about current state of controlled process; f(t) is disturbing effects.

Let us take management system MS function in choosing of managing actions u leading to the least deviation δ_P of controlled process CP from the given goal state in the period of time [t, t + T]. Then optimal are such managing actions u^*, that

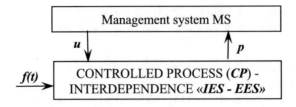

Fig. 1. Ecological and economic system management model S

$$\arg\left(\min_{u \in U} \delta_p[u^*, \mathrm{p}, \mathrm{f}(\mathrm{t})]\Big|_{\mathrm{t,t+1}}\right). \tag{1}$$

where U is the sphere of acceptable management, and T is observation time.

Equation (1) is a classic well-studied task of optimal management. Mathematical programming methods can be used to solve the equation. With the practical point of view difficulties in solving these tasks (1) start from the researching the structures and mechanisms of managing actions forming.

Let us define an ecological and economic system S as stochastic locally closed ecological and economic system (LCEES) which can be understood as a totality of industrial undertakings (let call them IES – industrial-economic systems), implementing on some territory their industrial-economic activities in certain ecological conditions and having mutual environmental influence, that is population and ecological [3].

From the aforesaid LCEES structure consists of two subsystems IES and EES: LCEES = {IES$_1$, IES$_2$, ..., IES$_n$, EES}. Then the object of research is relationship of varieties of IES and EES, more exactly managing (relations search) of the relationship (Fig. 2).

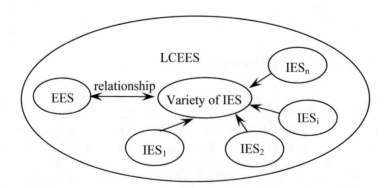

Fig. 2. Locally closed ecological and economic system structure.

Industrial-economic systems have a common functioning goal (profit-making), their principal feature is certain freedom in choosing of their behavior, understood as, for example, possibility of managing working-out u according to their own point of view, that is basing on the information which often does not correspond to the information of LCEES elements relations character. Besides, they can independently form their behavior goals and choose criteria of local management solutions, which may not to coincide with the LCEES goal and even contradict it. Anyway, managing subsystem division on elements is equal to giving the elements incongruous functions, which is the main factor, arising the problem of coordination. This factor can be removed but in that case the Centre will be responsible for all the process managing functions (management body of federal, regional and other ranks). This usually happens when planned centralized economics is preferred and it is based upon the Centre being possible to cope with all the managing functions and growing information flow.

However, in the context of market economy in natural resources sphere a reverse situation is typical when the central body is supposed not to be able to manage LCEES and has to delegate their functions to the natural resources subjects (IES). So, the coordination problem is as a peculiar charge for decentralization of LCEES management or as a reaction of the whole on its breakdown.

Figure 3 shows the scheme of the LCEES coordination office. The scheme symbols mean: K_1, K_2 – coordinating effects; m – effects of IES on EES; P_{IES}, P_{EES} is information about IES and EES state accordingly; z is information about mismatching of needed quality of cooperation V of LCEES structure subsystems; $f(t)$ are disturbing effects.

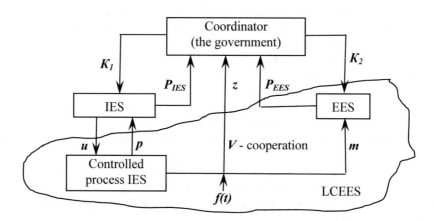

Fig. 3. Locally closed ecological and economic system coordination office.

This model is functioning the following way. The coordinator (the government), getting information z about current mismatching of needed quality of cooperation V of LCEES structure subsystems, tries to minimize the mismatching of controlled process CP from the given state. It should be underlined that this way it bases not on the whole information about the process state p but only on the part reflecting appearing mismatches among the components of the controlled process. Moreover, coordinator does not influence directly the LCEES processing but manages it indirectly, through the coordinating effects K_1 and K_2 on IES and EES respectively.

It should be noted that the coordinator has a wide specter of coordinating effects: administrative, economical, legal, etc. It is almost impossible to offer a simple general classifier of rules and mechanisms of coordinating effects managing solutions on the assumption of multiplicity and uniqueness of LCEES, their ongoing processes and consequences with the human factor strong impact.

Coordination problems' solution precedes the choice of coordination mode, which is the rule regulating relations between the coordination bodies (the Center) and coordinated objects (natural resources subjects). We distinguish five main modes of coordination office [8], which, in this case, can be interpreted the following way:

I mode is coordination through contradictions forecasting when the coordinator, based on the current LCEES situation analyses forecasts the character and development tendency and informs subjects about possible contradictions and their ways of development and subjects act basing on the information.

II mode is coordination through direct regulation of contradictory relations between natural resources subjects where the coordinator commands subjects completely excluding any uncertainty of their actions and the subjects strictly obey the commands.

III mode is coordination through "provoking" contradictions, when the coordinator does not interfere with contradictory relations of the natural resources subjects, allowing them to solve the problems, giving tasks, and checking the results.

IV mode is coordination through giving responsibilities, when the coordinator differentiates natural resources subjects' competency in solving arising contradictions and the last ones act by themselves in the frames of their competency.

V mode is coordination through making coalitions, when the coordinator unites natural resources subjects in groups basing on some characteristics, for example, common interests, and allows them to act independently in the group but having the right to correct the group behavior.

The above-mentioned coordination modes [8] have verbal scale of appliance depending on the state of the current process: 1 mode when the controlled process is stable; 2 mode is when the process is disorganized; 3 mode is when the process is consistently stable; 4 mode is when the process is very unstable; 5 mode is when the process is unstable.

According to the above-said, the general trend is the closer LCEES is to the disorganization the higher is the centralization degree, and on the contrary the more stable are the LCEES processes the less centralized should be the managing structure of the process functioning.

In real LCEES management, mentioned coordination modes can be in different combinations and change over one another. Besides, specific modifications, differentiating not formally but informatively, are possible within each coordination mode. Particularly, there can be distinguished target, resource, period, space coordination and coordination of objects of effects and used ways of actions. Combining and matching, these modes make practically unlimited number of possible ways of coordination office. It makes us use more detailed models considering individual features of her coordinator as well as of coordinated subjects [1, 4].

Now let us dwell on the questions of human factor in making decisions in LCEES, because in coordination office we should consider how the coordinator and performers interests coincide. Three situations are possible here: the coordinator interests dominate the performers' interests; the performers' interests dominate the coordinators interests; their interests are equal.

In accordance with the target goals of the government and taking into account ecologization of all the society spheres we are more interested in the first situation (the coordinator interests dominate the performers' interests). This way coordination office should coincide with reflexive management.

According to [9, 11] in the context of system management, reflection is meant a behavior process based on mutual simulation of state and possible ways of behavior of opponents. The office meaning is in part A (here it can be the government, the center, and the coordinator) passing through, for example by coordinating effects, to the opponents, for example, part B (here it is IES) information making them choose the strategy efficient to part A. In this case it is said [9, 10]: part A motivates part B. Within it part A should:

1. Find out the needs and interests of part B, i.e. understand their motives, defining their solutions, acts and behavior lines;
2. Find out the possible ways of actions of part B (usually through complex marketing procedures), their particular goals and intentions, ways of its achievements, resource and communication possibilities and external limiting factors;
3. Make a decision about their own behavior and based on it to define self-efficient behavior strategy for part B;
4. Define and give part B such information about themselves and their intentions which will lead them to choose behavior strategy efficient to part A.

Therefore, reflexive management is systematic mutual process of information transfer to the opponents, motivating them to make decisions and manage in the way, efficient to the part making reflexive management.

All the variety of stimuluses making performers compromise with the coordinator can be limited with five motivating factors [2]:

1. SGP – Belief in the system goals priority over each performer goals (criterion weight – w_1);
2. PE – Promotion expectation, the performer will get for the system goals achievement (criterion weight – w_2);
3. PP – Performer punishment, in case the system goals will not be achieved (criterion weight – w_3);
4. PC – Performer consciousness, according to which the performer consciously gives the priority to the system goals (criterion weight – w_4);
5. PA – Personal attitude, sympathies and antipathies of performers and coordinators (criterion weight – w_5).

Figures 4 depicts a hierarchical scheme of decomposition of the category "the management of the locally closed ecological and economic system".

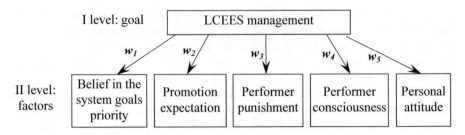

Fig. 4. Hierarchical scheme of LCEES management.

It is reasonable that LCEES management task needs further decomposition of target criteria and eventually showing concrete indexes defining variability of these categories assessment. Basing on the decomposition scheme in Fig. 4 we should stick to the following demands:

- *Representativeness* – in the given list all the main indexes of the category should be represented, i.e. the most significant characteristics of ecological and economic interests of the coordinator and the performer should be listed for the highest level;
- *information availability* – all the analyzed indexes and private criteria should be available for usage;
- *Information reliability* – used statistic data and private criteria should adequately reflect the state of the analyzed aspect of LCEES management.

Let us see an example of assessment of the LCEES management quality. Let us define the priorities and alternatives vector. Basing on the expert questioning of performers and coordinators (interested representatives of business-sector and the government) we made a preferences matrix of motivating factors of the coordinator and the performer cooperation by Saaty technology [12].

For the following analyses we can introduce quality category "goals compromise" K_{on}, which can be defined as a weighted total of above-mentioned factors which are quality categories. Quantity representation of quality criteria K_{on} allows us to assess cooperation level of the coordinator and the performer. Not changing the presentments generality, we can suppose $K_{on} \in [0,1]$, and besides if K_{on} is equal to zero, then in the functioning process performers do not compromise with the coordinator taking into account only their own goals. If K_{on} is equal to one there is also no compromise because performers take into account system goals. In any other case, there is a compromise. The mentioned approach allows analyzing coordinators and performers relations.

The obtained vector of criteria priorities $w = (0,4379; 0,0878; 0,1459; 0,2189; 0,1095)^{\mathrm{T}}$ is given in the last column of the Table 1.

So, the analytical record, considering only the first level of the hierarchy, of the integral index of LCEES management can be shown as:

$$LCEES\,management = 0,4379\,SGP + 0,0878\,PE + 0,1459\,PP + 0,2189\,PC + 0,1095\,PA$$

Table 1. Preferences matrix of motivating factors

LCEES management	Which motivating factor is more important?					w_i
	Belief in the system goals priority	Promotion expectation	Performer punishment	Performer consciousness	Personal attitude	
Belief in the system goals priority	1	5	3	2	4	0,4379
Promotion expectation	1/5	1	3/5	2/5	4/5	0,0878
Performer punishment	1/3	5/3	1	2/3	4/3	0,1459
Performer consciousness	1/2	5/2	3/2	1	2	0,2189
Personal attitude	1/4	5/4	3/4	1/2	1	0,1095

$\lambda_{max} = 5$; $CR = 0$

4   Conclusion

The most important and priority criteria having the most specific weight is belief in the system goal priority, the second one is performer consciousness. This criteria ranking is ideal for the whole system and does not contradict the government and society interests.

Quantities of the found out priority vector components w have clear interpretation they show the average result change (LCEES management) with the one change of the criteria. The problem of quantity representation of quality categories of the second level is the direction for the following interdisciplined research, which is the topic of the next article.

References

1. Ankoff, R.: Planning for the Corporation. Economika, Moscow (1985). Акофф Р.: Планирование будущего корпорации. Экономика, Москва (1985)
2. Arzhakova, N.V., Degtiareva, O.N., Demin, B.E.: Search and evaluation of compromises in the relationship «center-to-business». Control Syst. Inf. Technol. **2**(24), 59–64 (2006). Аржакова, Н.В., Дегтярева, О.Н., Демин, Б.Е.: Поиск и оценка компромиссов во взаимоотношениях «центр-предприятие». Системы управления и информационные технологии **2**(24), 59–64 (2006)
3. Velichko, S.V., Serbulov, Y.S.: Statistical analysis of locally-closed ecological systems. Econ. Math. Methods **2**, 41–47 (2004). Величко, С.В., Сербулов, Ю.С.: Статистический анализ локально-замкнутых экологических систем. Экономика и математические методы. **2**, 41–47 (2004)

4. Kondrat'ev, V.V.: The task of harmonization, coordination, optimization in the active systems. Autom. Telemech. **5**, 3–28 (1987). Кондратьев, В.В.: Задачи согласования, координации, оптимизации в активных системах. Автоматика и телемеханика. **5**, 3–28 (1987)

5. Litvinsky, K.O.: The forming of market trends reproduction of natural resources. Nat. Interests Prior. Secur. **6**(243), 9–14 (2014). Литвинский, К.О.: Современные тенденции формирования рынка воспроизводства природных ресурсов. Национальные интересы: приоритеты и безопасность. **6**(243), 9–14 (2014)

6. Litvinsky, K.O., Shevchenko, I.V.: Conditions and priorities of eco-economic system: laws, regulations, specifications, indicators. Reg. Econ. Theory Pract. **21**(78), 2–9 (2008). Литвинский, К.О., Шевченко, И.В.: Условия и приоритеты развития эколого-экономической системы: законы, правила, характеристики, показатели. Региональная экономика: теория и практика. **21**(78), 2–9 (2008)

7. Mesarovich, M., Marko, D., Takakhara, I.: Theory of Hierarchical and Multilevel Systems. Mir, Moscow (1973). Месарович, М., Мако, Д., Такахара, И.: Теория иерархических и многоуровневых систем/Пер. с англ. Мир, Москва (1973)

8. Mesarovich, M., Takakhara, I.: The General Systems Theory: Mathematical Foundations. Mir, Moscow (1978). Месарович, М., Такахара, Я.: Общая теория систем: математические основы/Пер. с англ. Мир, Москва (1978)

9. Novikov, D.A.: The Theory of Management of Organizational Systems. Fizmatlit, Moscow (2007). Новиков, Д.А.: Теория управления организационными системами. Физматлит, Москва (2007)

10. Novikov, D.A., Tsvetkov, A.V.: Mechanisms of the Functioning Organizational Systems with Distributed Control. IPU RAN, Moscow (2001). Новиков, Д.А., Цветков, А.В.: Механизмы функционирования организационных систем с распределенным контролем. ИПУ РАН, Москва (2001)

11. Novikov, D.A., Chkhartishvili, A.G.: The Reflexive Games. SINTEG, Moscow (2003). Новиков, Д.А., Чхартишвили, А.Г.: Рефлексивные игры. СИНТЕГ, Москва (2003)

12. Saaty, T.L.: Mathematical Principles of Decision Making (Principia Mathematica Decernendi). RWS Publishing, Houston (2010). Саати Т.Л. Математические принципы принятия решений. РВС, Москва (2010)

13. Usik, N.I.: The transformation of coordinating principle of competition. Reg. Econ. Theory Pract. **6**(285), 2–7 (2013). Усик, Н.И.: Преобразование координационного принципа конкуренции. Региональная экономика: теория и практика. **6**(285), 2–7 (2013)

Digital Innovations in the Global Exhibition Industry – Synergy of New Digital Technologies and Live Communication Measures

Igor Shevchenko[✉] and Anna Tololina

Kuban State University, Stavropolskaya street 149, 350040 Krasnodar, Russia
dean@econ.kubsu.ru

Abstract. In this paper, we examine whether exhibitions industry players who are using the new information and communication technologies show higher performance. We also present the information about digital economy as a new phenomenon, which has to be determined.

Keywords: Digital economy · Exhibition industry
Information and communication technology · UFI · AUMA

1 Introduction

Speaking about digitization of the economy and the application of new information technologies, the exhibition industry does not appear on the list of pioneers in this field. Moreover, for a long time it seemed quite difficult to bring together two completely different ways of communication: the classical format of live communication in the exhibition and the new digital one. The first involves the exchange of information between exhibitors and visitors, suggesting a multi-sensory impact on the target audience through vision, hearing, touch, smell, taste, which allows customers to have a strong impact and is a key tool for trust marketing. Digital communication is a method of instant data exchange, when the information immediately reaches its consumer after placement. Recently, having completed many various successful projects, the exhibitors have concluded that exhibitions and fairs, as a tool for live communication, and digital technologies do not contradict each other. It is rather the opposite: digitization of the processes and inclusion of new digital instruments in the exhibitor's program increases the efficiency of participation and return on investment.

The goal of this research is to identify the best practices in digital innovations implemented in the exhibition industry and to ensure that the combination of live and digital communication measures are the key factors for providing added value. For achieving this, it is necessary to evaluate the results of using digital innovations complimentary to live communication in relation to the previous experience of classic exhibition communication. The integration of digital tools and innovations for all parties - visitors and exhibitors, exhibitions companies and venues - can bring synergy effects and generate the value that make the exhibition more efficient.

© Springer Nature Switzerland AG 2019
T. Antipova and A. Rocha (Eds.): DSIC 2018, AISC 850, pp. 103–110, 2019.
https://doi.org/10.1007/978-3-030-02351-5_14

The first part of current research covers one of the most increasing trend of digital economy. We analyzed how to measure what is "digital economy". We also attempted to consider how big the impact of the information and communication technology on economic performance is. Last part of this paper serves this task by gathering best practice cases and putting the cases in a form that allows identifying the fields where research evidence makes the most difference. The structure of each case contains the description of the starting position and the initiator company, as well as objectives and actions for implementing the digital innovation.

The research issues of this paper are mostly based on the statistics and data from UFI's and AUMA's reports and surveys. UFI is a global association of the world's tradeshow organizers and exhibition center operators, as well as the major national and international exhibition associations, and selected partners of the exhibition industry. The information provided by UFI shows up-to-date development of tradeshows across different business sectors around the world. As German trade fair industry's umbrella organization, the Association of the German Trade Fair Industry AUMA represents the interests of exhibitors, organizers and visitors of trade fairs in Germany and serves to exchange experiences between the members, make recommendations and prepare decisions for other committees. The institute of the German trade fair industry as a part of AUMA plays its role as a platform for accumulating the knowledge about the exhibitions in Germany and around the world and for providing research findings. Both databases allows getting the wide range of information to undertake this study.

2 Transformation of Economic Fundamentals Through Digitization

More than any other technical innovation, the explosion of information and communication usage in our everyday lives is contributing to a radical change in the economy and society. Over the next decade, ICTs will be able to bring a paradigm shift in society and production systems, enabling greater growth and prosperity through efficiency gains, new products and services, and smarter public services.

The trend of digitization of products and workflows is increasing every year. It reveals an enormous and exponentially growing quantity of data presented in various forms; examples include users' web-activity and machine-to-machine interactions. This increased data availability has made it possible to measure and analyze the phenomena of digitization to an extent never achieved before. This, in turn, makes it easier to realize controlled studies to measure the success of them with great precision getting verification of different hypothesis.

As it is said above, before the availability of necessary amounts of data it seemed quite difficult to measure what is "digital economy". Nevertheless, there were some attempts to make a research in this field of study and some findings were published in different papers.

Provided by the European Commission, the Digital Agenda for Europe in 2010, information and communication technology sector represents 4.8 per cent of the EU economy; generates 25 per cent of total business R&D; and ICT sector and investment in ICT are responsible for 50 per cent of productivity growth [9].

McKinsey has made a study of data from the G8 and five other countries (Brazil, China, India, South Korea, and Sweden) to understand how big is the impact of the internet on economic performance. They calculated that the internet accounted for 3.4 per cent of GDP, and had fuelled 21 per cent of GDP growth in the preceding five years. Internet usage by small and middle enterprises was estimated to create a 10 per cent rise in their productivity.

Boston Consulting Group according to the report [8] estimated that by 2016 the Internet economy in the G-20 the economy will cost 4.2 trillion. (compared to $ 2.3 trillion in 2010) and that the Internet contributes 8 percent and over 12 percent of GDP in South Korea and the United Kingdom, respectively. The study notes that while economic growth is generally slow in most G-20 countries, the Internet economy will grow at an annual rate of 8 percent, far outpacing growth in the more "traditional" sectors.

Because of the fast-paced development of technologies and the widespread diffusion of the digital economy within the all sectors of economy, it can be described only as a part of overall changes. However, it can be characterized by a set of key functions: mobility, data using and network effects.

The digital economy increases mobility in many different dimensions. For example, an intangible property is one of the features of the digital economy. Data as a source of value is a key feature of the digital economy. Data are collected from several market participants and activities. The increasing ability to collect, store and process massive data flows has led to the concept of "big data" that could benefit both in private (marketing) and in public (government) activities.

Network effects are widespread in the digital economy. The advancement and diffusion of networks effects has made possible the creation of private value. In so-called multi-sided business models, several groups of persons interact through a platform, getting as a result positive or negative output (e.g. payment card system, operating system, media industry).

Future innovations and technological trends, which are given in the European agendas, are Cloud technology, Big Data, Internet of Things, Advanced Robotics, Autonomous Vehicles, 3D Printing, and Automation of Knowledge Work.

In his research paper, Brynjolfsson used different approaches for calculating the impact of information technology investments. The most commonly used factor is multifactor productivity. Other alternative methods of measuring firm performance is to relate an accounting measure of profitability to the construct of interest and other control variables and the total market value of the firm. However, these approaches can be used to estimate the impact in the firms working with Data-Driven Decision-making [4].

For the exhibition industry with its specific objectives and goals, the entry to the ICT era was not easy but very efficient for those who worked on it. In the next part of the paper, we will present the examples of successful implementation of innovations for different groups of exhibition and fair industry.

3   Digital Innovations in the Global Exhibition Industry

The basis of the fair, the exhibition and the opportunity to meet and contact, remains, the core is stable. However, this implementation is being reinterpreted, digitization and constant socio-economic change call for changes. A fusion of formats from all sorts of communication areas can be regarded as a relevant model for fairs: Everything goes; the main thing is to give visitors an experience. After all, in the knowledge society, a conference or convention or fair is always an exhibition of knowledge for the purpose of gaining customers through the power of interpretation. Experience and this is what these mergers are all about, is supported or initiated by a new narrative: that of participation, interaction, immersion, adventure. For the near future, the progression of digitization, and thus, above all, the individualization in communication (the speech before, during and after) and the immersion techniques in the staging will be seen (such as virtual reality).

The marketing mix of a company consists of the product design, the price and condition policy as well as the measures for the distribution and communication; it serves to fulfill the goals of the company in the target market. The goals of participation in a trade fair are derived from the corporate goals, because a trade fair participation should represent not just a single product solution, but also the entire company. In addition, participation in the trade fair includes all elements of the marketing mix and offers great potential for impact on the end customers. Communication is a central function of trade fairs and exhibitions, so participation in a trade fair has a stable position in the communication portfolio of a company. Personal communication has high priority in marketing policy and will continue to increase in importance.

Five hundred German companies took part in the survey about the position of exhibition and fairs in their marketing-mix portfolio. The Fig. 1 shows that the companies who regularly participate exhibitions and fairs put the participation in trade shows and fairs to the second place after webpage of a company.

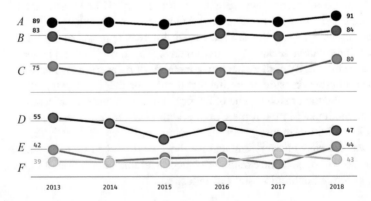

Fig. 1. Marketing-mix of German enterprises who regularly participate the exhibitions and fairs, in focus on B2B communication. Estimation of tools in % where A is Company Homepage, B – Exhibitions and Fairs, C – External Sales, D – Direct Mailing, E – Events, F – Advertising in the press. (Source: AUMA MesseTrend 2018)

Three instruments have been the leading group for many years. In the first place - and thus belonging to the standard - stands the own homepage (91%), followed by fairs and field service. This is followed by direct mailings, followed by advertising in trade journals and events.

The importance of the trade fair in the marketing mix has been at a very high level for years: currently, trade fair participations are important to very important for 84% of companies.

The new Internet medium is used today in almost all companies, but companies will not refrain from trade fairs and personal sales because they allow face-to-face contact and customer proximity, which is very important especially for new customer contact and customer loyalty, as well as for the launch and development of new markets.

Digitization in all its facets is also in the trade fair industry for years one of the dominant themes, without being very clear what is meant by that. In the AUMA MesseTrend study it is examined on the one hand, whether digitization has advantages for exhibitors by making trade fair participation more efficient and on the other hand, whether virtualization threatens the real trade fairs. The use of digital offers before and after the fair seems to pay off for the exhibitors. 26% of them say that the participation in the fair has definitely become more efficient by digitization and another 45% think that the participation becomes more efficient. Only 28% see no effects on efficiency through digitization. In this point, it has no significant difference between larger or smaller companies and even by sector. For seven out of ten companies digitization has positive effects positive effects on trade fairs. The often-described competition for real fairs by so-called virtual fairs does not seem to be real in the near future according to the available results. Only 6% of the surveyed companies have ever presented themselves on these virtual platforms. Of these, not a single company has canceled real fairs in favor of virtual fairs. The vast majority of respondents (91%) say that a presentation on a virtual fair has no influence on the participation in a real fair. This also runs through all industries and size classes.

In conclusion, it can be said that virtual metering in B2B marketing and sales is currently not an issue for companies, even though the technology has become quite advanced. The manifold central functions of a real fair, above all the real meeting and communication between humans, can not be replaced by virtual offers at the moment and are therefore no real alternative for the enterprises [3].

Since 2008, UFI organizes a best-practice competition for different members of exhibition industry - world's tradeshow organizers and exhibition center operators, as well as the major national and international exhibition associations, and selected partners of the exhibition industry.

The national exhibition associations and unions are also using the advantage of new technologies for promoting national exhibitions and trade fairs. AUMA offers a useful instrument for decision-making MesseNutzenCheck on its website [2]. This software for planning, calculation and evaluation of participation for exhibitors is available free of charge in four languages. It is useful for quantifying the benefits of trade fair participation and compare them with the costs.

The AUMA represents the interests of the trade fair industry vis-à-vis the legislature and executive at federal and state level, in particular in the areas of sales and export promotion, sustainability, but also in tax, construction and employment law topics. It maintains contacts with foreign and international trade fair organizations. AUMA is a member of the world trade fair association UFI.

Germany is the world's number one in international trade fairs. Around two-thirds of the world's leading trade fairs take place in Germany. Every year around 160 to 180 international and national trade fairs are held, with up to 180,000 exhibitors and around 10 million visitors [1].

As a leading exhibition country, Germany has very professional approach to the promotion of national exhibition and fairs and the AUMA offers very profound knowledge and activity platform for this. Recently AUMA launched the MyFairs App, which gives the access to all important data and facts for the preparation for trade fair participation or your visit. MyFairs links the user to the AUMA trade fair database with the trade fairs, addresses and dates relevant to your planning [17].

The AUMA informs about dates, offers, numbers of exhibitors and visitors of domestic and foreign fairs, to make it easier for exhibitors and visitors from all over the world to decide on participation in a trade fair. For this purpose, AUMA offers data on more than 5,000 events in its worldwide four-language trade fair database at its webpage. He publishes exhibition dates and tips for the planning and implementation of trade fairs. With its rules on market transparency, it helps to balance the interests of exhibitors, visitors and organizers. FKM is the objective authority for the certification and publication of exhibition data in Germany [12].

The certification and gathering fair statistics are one of the most important tasks for the exhibition industry worldwide. The m + a ExpoDataBase offers current information and trade fair dates in its trade fair planner with more than 10,000 trade fairs and events worldwide. The Russian trade fair association RUEF and its member Russcom IT Systems are responsible for gathering the data and statistics about Russian exhibition companies. Since 2005, the independent testing organization and UFI member RussCom IT Systems have inspected the Russian trade fairs [19]. The transparent and up-to-date information about exhibitions allows exhibitors to make the right choice of trade fair.

Not only the associations but also the venues and exhibitors do their best to be successful participating the trade fair. UFI organizes the best practices competition of using new ICT solutions for the exhibition sector where the impact of the winner's cases to the success of participation in exhibition is shown.

In 2008, the ICT award category of UFI was for the best web application for visitors - to address the best method of improving visitor experience and numbers at events. Until 2013, there have been yearly a competition with the following topics: Best innovative web-based applications for exhibitors, Customer relationship management for exhibitions, Best practice of social media within the exhibition industry, Which new services did your mobile app provide to your exhibition customers? From data to success – best overall solutions for Exhibitor and visitor data management, What have you done to fit your it team to the future? What will be our tools for the future? What will be our solution for the future?

All the information above ensures that the new solutions in ICT is the key factor to success.

References

1. AUMA BranchenKennzahlen 2013–2017. http://www.auma.de/de/Messemarkt/Branchenke nnzahlen/InternationaleMessen/Seiten/Entwicklung2013–2017.aspx. Accessed 21 June 2018
2. AUMA MesseNutzenCheck. http://www.toolbox.auma.de/index.php?id=32. Accessed 23 June 2018
3. AUMA MesseTrend 2018. http://www.auma.de/en/DownloadsPublications/Seiten/AUMA-MesseTrend.aspx. Accessed 21 June 2018
4. Brynjolfsson, E., Hitt, L.M., Kim, H.H.: Strength in numbers: how does data-driven decision making affect firm performance? SSRN Electron. J. (2011)
5. Brühe, C.: Messen als instrument der Live communication. In: Kirchgeorg, M., Dornscheidt, M., Giese, W., Stoeck, N. (eds.) Handbuch Messemanagement, pp. 75–86. Gabler, Wiesbaden (2003)
6. Clausen, E.: Mehr Erfolg auf Messen. Verlag Moderne Industrie, Landsberg/Lech (2000)
7. Davidson, R.: Business Travel: Conferences, Incentive Travel, Exhibitions, Corporate Hospitality, and Corporate Travel. Pearson Education Limited, Harlow (2003)
8. Dean, D., DiGrande, S., Field, D., Lundmark, A., O'Day, J., Pineda, J., Zwillenberg, P.: The internet economy in the G-20 the $4.2 trillion growth opportunity. BCG (2012)
9. Die digitale Agenda für Europa – digitale Impulse für das Wachstum in Europa: Mitteilung der Kommission an das Europäische Parlament, den Rat, den Europäischen Wirtschafts- und Sozialausschuss und den Ausschuss der Regionen. Europäische Kommission, Brussels (2012)
10. Erfolg durch Messen: AUMA. http://www.erfolgdurchmessen.de/index.php. Accessed 23 June 2018
11. ExpoDataBase: m + a Messeplaner. http://www.expodatabase.de/. Accessed 21 June 2018
12. FKM: Authority for the certification and publication of exhibition data in Germany. http://www.fkm.de/fkm/. Accessed 27 June 2018
13. Ladkin, A., Spiller, J.: The Meetings, Incentives, Conferences and Exhibitions Industry. Travel & Tourism Intelligence, London (2000)
14. Ladkin, A., Weber, K.: Trends affecting the convention industry in the 21st century. J. Conv. Event Tour. 6, 47–63 (2004)
15. Leask, A., Whitfield, J.E.: UK conference venues: past, present and future. J. Conv. Exhib. Manag. 4(1), 29–54 (2002)
16. Manyika, J., Chui, M., Bughin, J., Dobbs, R., Bisson, P., Marrs, A.: Disruptive Technologies: Advances that Will Transform Life, Business, and the Global Economy. McKinsey Global Institute, New York (2003)
17. MyFairsApp from AUMA. http://www.myfairs.auma.de/index.html. Accessed 21 June 2018
18. Overall Economic Relevance of Exhibitions in Germany 2018. http://www.auma.de/en/DownloadsPublications/PublicationDownloads/AUMA-Dokumentation-5.pdf. Accessed 27 June 2018
19. Official website of RussCom IT Systems. http://www.auditexpo.ru/en/. Accessed 25 June 2018
20. The UFI Report on Best Practices in Digital Innovation 2018. http://www.ufi.org/wp-content/uploads/2018/02/UFI_Best_Practices_Digital_Innovation.pdf. Accessed 21 June 2018
21. Von Georgi, R., Wünsch, U.: Dimensionen des Erlebens, ihre Wahrnehmung und Hinweise zu ihrer Inszenierung. Institut für Publikumsforschung in der SRH Hochschule der populären Künste, Berlin (2018)

22. Weber, K., Ladkin, A.: Trends affecting the convention industry in the 21st century. J. Conv. Event Tour. **6**, 47–63 (2005)
23. Weber, K., Chon, K.S.: Convention Tourism: International Research and Industry Perspectives. Routledge, New York (2003)
24. Working Paper: Digital Economy—Facts & Figures: European commission directorate—General taxation and customs union, Brussels (2014)

Expert Systems of Real Time as Key Tendency of Artificial Intelligence in Tax Administration

Alexander Biryukov[✉] and Natalya Antonova

Sterlitamak Branch of the Bashkir State University, 49 Lenin Avenue,
453103 Sterlitamak, Bashkortostan, Russia
guzsa@ufamts.ru, antonov1891@mail.ru

Abstract. The problem of neural network mathematical modeling of economic objects and systems, including the objects of the tax control and taxation, attracts scientists. In practice, the economic systems operate under conditions of uncertainty, which makes the results of strict mathematical calculation ineffective for solving the task. Despite numerous developments in the field of neural network modeling for stochastic objects with very noisy data, in particular, tax control facilities, the methods and principles of adequate and sufficiently precise neural network modeling have not been fully developed. Our experience shows that the "frontal" building of effective neural network modeling on the basis of standard neuropackage is impossible without the development of the fundamental concepts and the use of preliminary data processing.

Keywords: Artificial intelligence · Economic advice-giving system
Expert system · Logical model · Frame model · Semantic network
Production model · Neural network model · Fuzzy logic model

1 Introduction

In difficult conditions of modeling, classical modeling techniques are ineffective or even completely unacceptable.

This is due to the fact that it is impossible to describe the reality adequately with the help of a small number of model parameters, because the calculation of a model requires too much time and computing resources, and, most importantly, conditions of these methods application are not fulfilled and, therefore, it becomes impossible to use appropriate statistical criteria to evaluate the adequacy of the produced models.

Because of the afore-named drawbacks of traditional methods analytical systems of a new type have been actively developing of the last 10 years. They are based on the artificial intelligence technologies that mimic natural processes, such as the activity of neurons in the brain or the natural selection process.

According to modern presentation, artificial intelligence (AI) is defined "as a scientific discipline whose goal is to develop hardware and software, allowing the user - non-programmer to set and solve the tasks that are traditionally considered intellectual, communicating with computers on a limited subset of a natural language." [1, 6].

When creating artificial intellectual systems a large number of information technologies (ES, advise-giving systems, decision-making support systems, execution of

© Springer Nature Switzerland AG 2019
T. Antipova and A. Rocha (Eds.): DSIC 2018, AISC 850, pp. 111–118, 2019.
https://doi.org/10.1007/978-3-030-02351-5_15

decisions) are used. The common feature of these techniques is the use of some form of human knowledge. If we highlight the technology aimed to solve economic problems, this class of systems will be called "economic advice-giving systems" (EAS).

Neural network models are widely used in the ECs to represent and accumulate knowledge [2, 3, 5]. Note that there are three reasons for the rapid development of methods for neural network modeling in general and particularly in the sphere of EAS:

(1) In the neural network models parallel calculation method is implemented, i.e. several steps for computing operations are carried out at the same time. Due to this, the speed of a neurocomputer (electronic structure or neyroemulyator) increases sharply.
(2) Neural network model does not require prerequisites of classical regression analysis, which is particularly important for the study of economic systems, where these prerequisites may not be fulfilled.
(3) Though neural network models are parametric, they do not require pre-guessing of the form (structure) for the model.

The use of neural networks provides the following useful properties for the model:

(1) Non-linearity. Artificial neurons can be linear or nonlinear. Neural networks which are constructed from compounds of non-linear neurons are nonlinear themselves.
(2) Display of the input information into the output information. One of the most popular paradigms of learning in neural network system is training with the teacher. This implies a change of synaptic weights based on a set of labeled training examples. From many examples one is randomly selected, and the neural network modifies the synaptic weights to minimize the differences between the desired output signal, and the signal, formed by a network in accordance with the selected statistical test.
(3) Adaptability. Neural networks have the ability to adapt their synaptic weights and the very structure of the model to the changes in the environment. In particular, a network trained to operate in a particular environment can easily be retrained to operate in the conditions of minor oscillations of the medium parameters in the environment [6].

An important restrictive feature of the neural networks used in EAS for representation knowledge is that unlike other models reproducing determined connections, clearly articulated by the expert, the neural network is not able to explain its results. Therefore in instrumental ES where it is possible to use several models of representation of knowledge, the neural network should be complemented by logic or production models.

Consider the advantages and disadvantages of models for representation of knowledge in the EAS in Table 1.

To sum up comments on Table 1, we note that the EAS can include a combination of different types of models for knowledge representation. This must generate the synergistic effect (the emergence) by strengthening the advantages of the basic model and reducing its negative characteristics and limitations. For example, when there is a combination of fuzzy and neural network models of knowledge representation in the EAS, fuzzy inference systems allow giving clear interpretation of the performed actions, but they cannot be taught, i.e. to perform automatic setting of parameters for

Table 1. Comparative characteristics of models for representation of knowledge in EAS

Models	Benefits	Restrictions	Degree of study
1	2	3	4
Logic models	A clear formal semantics, the use of advanced mechanisms of inference, based on mathematical logic	Lack of mechanisms for the critical evaluation of knowledge, revealing the contradictions; lack of automatic detection of patterns and using them to predict; lack of mechanisms for extraction of new knowledge. Knowledge Source is an expert, i.e., in the original database there is no output mechanism - deduction	High degree of scrutiny
Frame models	Nested frames (the principle of "Matryoshka") to describe the most significant relationship between the attributes of the object, the ability to quick search for the inference based on the principle of inheritance; knowledge organization preserving information about the structure of the object	Knowledge Source - expert, ES is a "passive assistant (adviser)"; inability to manipulate knowledge, simulating the inference process; the lack of mechanisms for critical evaluation of the knowledge gained from the expert and extraction of new knowledge; output mechanism is deduction	High degree of scrutiny
Semantic networks	Visibility (semantic visualization of information); variety of funds for representation of different relationships between the basic concepts; the ability to create rules for the knowledge base	Knowledge Source is an expert, i.e., there is no initial knowledge base; output method is deduction	High degree of scrutiny

(continued)

Table 1. (*continued*)

Models	Benefits	Restrictions	Degree of study
1	2	3	4
Production models	Ability to playback the way and style of human thinking; modularity; compatibility with other forms of knowledge representation; natural parallelism inherited in the production system, which is convenient for implementation of new architectures on a computer, such as neurocomputers	The output mechanism is deduction; knowledge source is an expert; limitations are the same as these for the above three models; lack of mechanism for adequacy evaluation of the knowledge representation model	High degree of scrutiny
Neural network models	Ability to extract new knowledge about the laws of the object, the process, the situation out of the data; compact form of knowledge representation; the possibility to manipulate with knowledge in quantitative form (e.g., finding the optimal control actions); the ability to solve a wide range of tasks (approximation, clustering, optimization, forecasting); ability to work in view of the of "NON-factors" triad; output method is induction, i.e., from separate facts to general laws; neural network KB is able to detect contradictions in the KB, to predict new facts, and to assess/evaluate its own adequacy. Remarkable features of neural network techniques	Neural network does not explain its results and therefore requires a superstructure in the form of the other models when creating the EAS; for high-quality training and testing the network requires a fairly representative database, which dimension N depends on n-dimensional vector of explanatory variables $\vec{X}$; there is no theoretical apparatus to assess the adequacy of neural network model similar to, for example, an apparatus for the regression models obtained by OLS (for neural networks the substantiation of their adequacy is possible in principle, but it requires additional research every time. It is impossible to add	Insufficient scrutiny, some problems have not been investigated adequately, in particular: the stability of neural network model according to Hadamard; regularization methods using different approaches (regularization theory by Tikhonov, Bayesian approach); ensuring consistency for network regularization; optimization of the choice for the paradigm and architecture of network; a comprehensive assessment of the adequacy of the neural network model in terms of manifestations of the "NON-factors" triad; and others

(continued)

Table 1. (*continued*)

Models	Benefits	Restrictions	Degree of study
1	2	3	4
	(NNT) (approximations of system function on the basis of the final set of observations) are their internal regularizing properties, allowing getting small generalization error. The usefulness of these characteristics manifests itself in situations where the data about the system contain internal redundancy, i.e. a lot of data. This allows us to submit a set of the data as a model that contains fewer parameters than available data. Thus, NNT squeeze the experimental data, weakening noise components and emphasizing the smooth components	any a priori information (expert knowledge) to accelerate neural network learning process: it is necessary to re-build the neural network taking into account a priori information, re-select the system of set parameters and train the neural network	
Fuzzy logic models	Fuzzy logic uses the notions of everyday speech to determine the behavior of the system and makes it possible to build robust, fault-tolerant systems; takes into account the large number of parameters to be analyzed (estimated), a large number of control actions, strong disturbance and nonlinearity; factors of "NON" triad, the use of technical knowledge such as «know-how»	Residual uncertainty remains in the formalization of fuzzy model when designing functions of membership and choosing terms of linguistic variables; fuzzy model is not capable of learning to determine the parameters of the membership function on the basis of familiar information	Coarsening of data in the database because of their fuzzy representations increases the resistance of the neural-fuzzy model to a random variation of the input data, but how much - this issue has not been studied, particularly for large organizational economic systems

membership functions on the basis of known information. In contrast, the neural network can customize their parameters (weights), but the functions, which they realize, can't be clearly interpreted. The most effective way for hybridization of fuzzy logic and artificial neural networks, is a neuro-fuzzy system (more precisely - "neural inference system"), which, on the one hand, can be regarded as a fuzzy inference system (and thus, to interpret clearly the obtained results), and on the other hand - as an artificial neural network, that contain a special type of neurons and therefore, it can be trained.

2   Results and Practice

Russian scientist [4] describes the ES, which is used by the Canadian Internal Revenue Service for verification of a company income tax and VAT. To develop the expertise rules for the selection of taxpayers in Canada, a group, consisting of 30 most qualified tax inspectors was created. These inspectors told the experts in artificial intelligence (AI), why some declarations seem suspicious to them, what things should be paid priority attention during the checking, and what amount of additional charges should be expected. To perform field tax audits, all these rules were introduced into the system of the computerized selection of taxpayers. The following sources of data are used: the data from tax returns, the data from previous field tax audits, the data on the structure of earnings in the area where the taxpayer lives. This ES allows to look through tax returns in automatic mode and to classify them into two classes: Class 1 – "field tax audits should be performed"; Class 2 – "it is not necessary to inspect". The value of the expected additional charges is estimated in Class 1.

The US Federal Revenue Service uses the following combined (hybrid) model for selection of taxpayers for performing field tax audits [4]. A special data base of individuals and small and medium-sized businesses is collected according to the results of their special scrutiny, conducted in the framework of the Program "Measuring of taxpayers' law-abiding." Randomly selected tax returns are classified by the main source of income. A discriminatory function is the amount of income (or gross income of the company). The classes' labels are known from the results of previous checks. Then, using statistical methods specialists estimate the probability of additional accruals after the documentary check of the declaration from this class. This statistical model links the simulated core indicator with others, which the taxpayer shows in his declaration. This model is constructed as follows. Take a sample of declarations for the class, where the results of additional charges are known from previous audits and all these declarations are divided into two subclasses: 1 - "should have been checked"; 2 - "should not have been checked" (0 or 1).

The belonging of the declaration to one of these two categories is the modeled variable Y.

The simplest ES, which takes into account the uncertainty factor, is as follows. For simple regression:

$$\hat{Y} = \varphi(x_1, x_2, \ldots x_j, \ldots, x_n) \tag{1}$$

A destructive version of multi-stage OLS is applied, i.e., first, a large number of order factors (150) is included in (1), then their number is reduced by an order and more, i.e. only the most significant factors for the explanation of the simulated variable are retained. Thus, the model of knowledge representation in the EC is a linear or non-linear regression Eq. (1), supplemented by the following production rule:

$$\text{If } \hat{y} \in [y_a; y_b], \text{ then } Y = 1, \text{ and } Y = 0 \text{ otherwise.} \tag{2}$$

The required value interval $[y_a; y_b]$ is given by the expert.

In this regression equation, non-absolute values of these factors $\{x_j\}$ from the declaration are used, but dimensionless complexes, formed from these factors.

After receiving the model (1), this simple regression formula is applied to all selected declarations of this class. The selected declarations are submitted to a highly qualified tax inspector (called a "classifier" in the US) for censorship. The classifier scans all the selected declarations and gives his verdict about each one: whether it should be checked or not, and what points should be especially focused on during the verification. In the United States, approximately half of all field inspections of taxes on personal and corporations income is organized using this ES selection.

According to Chernik's study, tax services of many countries process tax returns using statistical methods such as regression and discriminant analysis, and according to the results of this analysis they build formulas, which allow to draw conclusions about new tax declarations: whether the check of a taxpayer promises large accrual or not. The documentary audit of taxpayers selected in such a way, is given priority.

3 Conclusions

Despite the intensive development of the theory and practice of neural network modeling in relation to the difficult conditions of economic systems modeling (hard formalization of the processes and interactions in the system), very noisy data (deliberate distortion of the tax base) because of the unknown laws of noise distribution, many problematic issues of neural network modeling of these systems have been either poorly researched or not researched at all:

1. The budgetary system of any level faces the problem of the insufficient amount of own funds for the projects focused on the end result and in the broader sense, for the budget system functioning, i.e. providing timely and quality services for budget spending units, ensuring the preservation of the financial and social stability and the development of the territories. So, one of the urgent problems is forecasting of budgets' filling, particularly, in municipalities, taking into account the risk of going beyond the confidence limits of the forecast based on hybrid neural network models.

2. Lack of computer techniques for multi-criteria ranking of budgetary institutions and organizations, which would allow to assess objectively the results of the organization' activities at the current time and a forecast period. Such techniques would allow to distribute transfers more equitably and efficiently in terms of governance

and to determine the directions for the rational development of budgetary organizations.

3. The taxpayers' activity is characterized by uncertain external and internal environment. The result of these trends is the spread of the output parameters for the organizations' economic activity, which in many cases determines the high risk for inefficiency of tax audits. The objectives of the tax control are creative in nature; they require specialists with extensive knowledge, experience and developed intuition. Therefore, the transition to the mathematical formalization of the decision-making stages faces a number of difficulties associated with the problem of modeling for poorly formalized systems.

The Federal Tax Service of the Russian Federation use information technology for desk audits, selection of taxpayers for on-site inspections, These technologies are generally aimed to automate the monitoring of declared reporting data, their analysis for the logical consistency of the interrogation mode, where each subject of taxation is analyzed in turn.

The main disadvantage of the existing methods of tax control is as follows: all the technology of tax audits planning is subjective.

In such circumstances, there is need for new computer technologies for taxpayers' selection.

So, we can conclude, that the level of development of theoretical and methodological foundations for neural network modeling in intelligent ECS for economic systems, does not meet the requirements of the practice, due to the ongoing process of reforming in the budget and tax system of the Russian Federation.

Solving this problem is aimed at solving algorithmically complex problems, as well as accumulation of scientifically based knowledge about the object, i.e., it is designed to maintain the existing system of economic models of the object of study and to add it with missing models and objectives. In applied economic aspect the problem is focused on improving the efficiency of state administration in the area of fiscal systems at all levels.

References

1. Biryukov, A.N.: The Theoretical Basis for the Development of Neural Network Models in the System of Tax Administration. RB Academy of Sciences, Guillem, Ufa (2011)
2. Miroliubova, T., Biryukov, A.: Spatial differentiation and market potential of the regions: the case of Russia. Asian Soc. Sci. J. Can. **11**(9), 96–117 (2015)
3. Haykin, S.: Neural Networks: A Comprehensive Foundation. Publishing House "Williams", Moscow (2018)
4. Chernik, D.G., Shmelev, Y.: The Theory and History of Taxation [Teorija i istorija nalogooblozhenija]. Publishing House Yuraytm, Moscow (2014)
5. Yasnitsky, L.N.: Intelligent systems [Electronic resource]. El Republic. ed. – Knowledge Laboratory, Moscow (2016)
6. Yasnitsky, L.N.: Neural networks - tool for getting new II is known: successes, problems, prospects. Neurocomput. Dev. Appl. **5**, 48–56 (2015)

Digital Education

Fuzzy Logic Model for the Selection of Applicants to University Study Programs According to Enrollment Profile

Nemias Saboya$^{(\boxtimes)}$, Omar L. Loaiza, Juan J. Soria,
and Jose Bustamante

Facultad de Ingeniería y Arquitectura, Universidad Peruana Unión,
Km 19 Carretera Central, Ñaña, Lurigancho, Lima 15, Peru
{saboya,omarlj,jesussoria,joseb}@upeu.edu.pe

Abstract. This study consisted in elaborating a diffuse model that allows for the selection of applicants to university study programs considering their enrollment profile, based on the need to reduce errors of precision in the selection procedures and to minimize the human limitations of the evaluators who are faced with a wide variety of differentiated evaluation criteria for the study programs. This fuzzy model is composed of three input linguistic variables referring to knowledge competencies and an output variable regarding the enrollment profile level. It is also made up of 125 fuzzy rules developed by the Mamdani and centroid methods and validated by experts. The results of the application show its effectiveness with $p < 0.05$. The model is flexible and useful to ideally select as many new entrants as possible, and thus it offers advantages to comply with any institutional regulatory context and external quality assurance entities such as the SINEACE or others.

Keywords: Enrollment profile · Fuzzy logic · Fuzzy model
University study programs

1 Introduction

1.1 Context for the Research Study

One of the pillars of socio-economic development and the competitiveness of a country lies in education; mainly, in university-level education that must support its development [1] with the advancement of knowledge and technology that solve problems in its environment. The OCDE [2] in a report on the global strategy for the achievement of competencies, points out that these are the pillar and the compass of professionals who are adequately trained to bring together the supply and demand of the labor market in a given constituency.

In order to train students in the competencies of their profession, they must be conducted through a training process in a study program. In this regard, the University Law 30220 of Peru, in article 98 defines that the universities must hold an open call for their admission processes, where it must be proven that the applicants meet minimum entry competencies. The norm establishes that the admission process must have at least

© Springer Nature Switzerland AG 2019
T. Antipova and A. Rocha (Eds.): DSIC 2018, AISC 850, pp. 121–133, 2019.
https://doi.org/10.1007/978-3-030-02351-5_16

one knowledge evaluation and optionally an aptitudes (abilities) evaluation and another for attitudes. In a complementary way, the SINEACE [3] considers that the process of admission to study programs should hinge on an enrollment profile.

The process of admission to universities is generally based on a knowledge test, which should not be general [4] and the evaluation items should be differentiated for different study programs that in turn are linked to the development of study profiles. This form of evaluation does not allow the university to objectively know to what extent the applicant meets a profile. On the other hand, although there is a tendency to combine different types of instruments [4], such as: interview guides, psychometric aptitude and attitude tests, among others, their use still focuses on the evaluation of profiles in the evaluation processes.

The traditional style with methods and instruments [4] is employed, but it restricts an objective evaluation to identify how the applicants fulfill the competencies because it is based on the sum of the scores of answers to questions in objective tests. The fact that it does not show to what level or degree an applicant is close to or far from a profile, leads to the deficiencies of each applicant being hidden throughout the process of general and specialized training, so they are all treated as if they had the same capabilities such that there are no personalized tutoring programs, potentially causing problems such as higher risk of dropout, low grades, low identification with their careers in the short term and, in the long term, job displacement and under employment [2].

At present there are different models such as [5]: multiple regression, neural networks and fuzzy logic, among others, that help to evaluate characteristics that the applicants possess which need to be analyzed based on a profile.

Fuzzy logic is a computational intelligence tool used initially in the field of industry [6], medicine and security, among others [7]; it has proved useful when working with ambiguous and complex information in labor contexts for evaluation of positions in organizations and evaluations of income, formative progress and graduation of students in training centers [8]; as well as when evaluating ambiguous or imprecise results to define their proximity or distance from a competitor with respect to a profile.

In the case of this research project, a fuzzy logic model based on a method is developed, which considers the articulation of differentiated instruments and criteria that make it possible to more objectively demonstrate the results of the evaluation of the applicant according to their enrollment profile regarding their knowledge, skills and attitudes; the present research study is focused on the knowledge dimension.

2 Literature Review

2.1 Enrollment Profile

A profile is a set of reference parameters, with their rules and measurements, acting together as a pattern [9] against which other values are contrasted to make decisions; so mathematically [10] a profile can be defined as $P = (a, b, c)$ in that order and not $P = (a, c, b)$, where a, b y c are characteristic values of a profile. Profiles are constructed based on criteria of what is intended to be evaluated and they are relevant in their social or organizational context. In the labor and educational field, one of the primary sources is the competencies [11].

Competencies are complex constructs that are composed of qualitative and some quantitative characteristics in three dimensions [6, 12], cognitive, procedural (skills) and behavioral, which the subjects of evaluation demonstrate with a certain degree of skill, several of them are scored on Likert scales or another scale. Competencies have to do with the capacity of a person facing a difficulty to resolve it. These competencies can be as diverse as the fields of application in human performance; where not only the sphere of knowledge but also the resolution of problems is valued, hence they are complex and there is no systematized knowledge [10] which can be applied in a comprehensive way.

In this way organizations seek to close the gaps between the characteristics required for a position and those of an applicant [6] which are products of the ambiguity of the valuation in the characteristics of the competencies. The selection of the most suitable applicants ensures optimal performance [13] for a collaborator to achieve organizational goals; the higher education field is no different, because it seeks that those applicants who come closest to a predetermined pattern can study curricula with better possibilities [2].

Therefore, an enrollment profile is a set of competencies established in an educational model or curriculum for a study program that should be used as a basis to evaluate the applicants in the most objective way possible.

2.2   Fuzzy Logic

The theory about fuzzy sets, in several areas, is a tool to support decision making [14], where the constraints of a real-world situation constitute the elements of modeling the fuzzy sets. This theory has been used in the industrial sector, in the selection of personnel and even in the evaluation of cognitive processes [15].

A fuzzy set [16] contains elements that have different degrees of membership that range from zero to one, the elements of a fuzzy set can also be contained in another fuzzy set in the same universe. If an element x of the universe is a member of the fuzzy set $\tilde{A}$, then the mapping is given by the images of the characteristic function included between [0; 1], where $\tilde{A}$ is a fuzzy set of elements zero and one and $\mu_{\tilde{A}}(x)$ is the range which is [0; 1].

Among two or more fuzzy sets $(\tilde{A}, \tilde{B}, y \tilde{C})$ [17] there are certain operations, such that the universe X is defined for a given element x, considering the union, intersection and complement of fuzzy sets as shown in the Eq. (1), defined by:

$$\left.\begin{array}{ll} \text{Union}: & \mu_{\tilde{A}\cup\tilde{B}}(x) = M\acute{a}x\left[\mu_{\tilde{A}}(x); \mu_{\tilde{B}}(x)\right] = \mu_{\tilde{A}}(x) \vee \mu_{\tilde{B}}(x). \\ \text{Intersection}: & \mu_{\tilde{A}\cap\tilde{B}}(x) = Min\left[\mu_{\tilde{A}}(x); \mu_{\tilde{B}}(x)\right] = \mu_{\tilde{A}}(x) \wedge \mu_{\tilde{B}}(x). \\ \text{Complement}: & \mu_{\tilde{A}^c}(x) = 1 - \mu_{\tilde{A}}(x). \end{array}\right\} \quad (1)$$

Insufficient clarity in a fuzzy set is characterized by the shape of its membership functions [16], which classify an element of the set, whether it be discrete or continuous, as there are certain restrictions with respect to the way in which they are used. Its properties are defined by: core, support and boundary [18] as shown in Fig. 1. The core is characterized by having a complete membership in the fuzzy set $\tilde{A}$, that is, the elements that have the membership function with their image $\mu_{\tilde{A}}(x) = 1$. The support is characterized by a membership greater than zero in the fuzzy set $\tilde{A}(\mu_{\tilde{A}}(x) > 0)$. The border has a non-zero but not complete membership for the fuzzy set $\tilde{A}$, the range contains the elements whose membership is between 0 and 1 so $0 < \mu_{\tilde{A}}(x) < 1$.

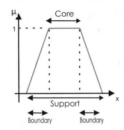

Fig. 1. Characteristics of the membership function

Defuzzification

Is the process of converting to a quantifiable value in classical logic [17, 19], given fuzzy sets and their corresponding degrees of membership, in other words, it transforms from a fuzzy set to a classical set. The function reduces the collection of values of membership functions into a single accumulated amount called a centroid as shown in Fig. 3. Hilera-Gonzales [18] mentions that the output in the fuzzy union, as shown in Fig. 2, can be expressed mathematically with the Eq. (2):

$$C_n = \sum_{i=1}^{R} C_i = \tilde{C}. \tag{2}$$

Regarding the calculation of defuzzification, there are seven methods that can be used with fuzzy output functions: the maximum membership principle, the centroid method, weighted average, maximum mean membership, the center of sums, the center of the largest area and the first of maxima or last of maxima; of these, the method applied in this research project is the centroid method.

Centroid Method

Hilera-Gonzales [18] argues that the use of the centroid method as shown in Eq. (3) is the most used which is defined in algebraic form by

$$x^* = \int \frac{\mu_{\tilde{C}}(x).x.dx}{\mu_{\tilde{C}}(x)}. \tag{3}$$

where the integral symbol is used for algebraic integration.

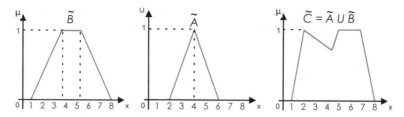

Fig. 2. Typical fuzzy output

Fig. 3. Centroid x* of the map of fuzzy sets

2.3 Mamdani

Ebrahim Mamdani's method [20] states that the membership functions in the output are fuzzy sets in which after the aggregation process, there is a fuzzy set for each output variable that needs defuzzification.

Zadeh [21] mentions that a fuzzy conditional statement provides a basis for using fuzzy algorithms more systematically and, therefore, more effectively than was possible in the past with traditional methods. Therefore, fuzzy algorithms become an important tool for an analysis of systems and decision processes that are too complex for the application of conventional mathematical techniques. The formal characterization of a fuzzy algorithm is expressed in terms of the notion of a fuzzy turing machine or a Markoff Fuzzy algorithm. Thus the diffuse algorithm instructions that are included in the following three classes were used: Assignment statements, diffuse conditional statements and conditional action statements.

The rules under these fuzzy algorithms have the following logical structure:

$$\mathfrak{R}_i : If\ x_1\ is\ A_{i1}\ and\ x_n\ is\ A_{in}\ then\ u\ is\ B_i\,,\ \ i = 1, 2, 3, \cdots, 125$$

3 Methodology

3.1 Applicant Selection Method According to Enrollment Profile

The method of Fig. 4 is developed under the criteria established by Peruvian university law 30220 in article 98, where it states that [3, 21] "... the university admission selection process consists of a knowledge examination as a compulsory and main process as well as an assessment of aptitudes and attitudes in an optional complementary way ..." and the SINEACE requirements described in standard 18.

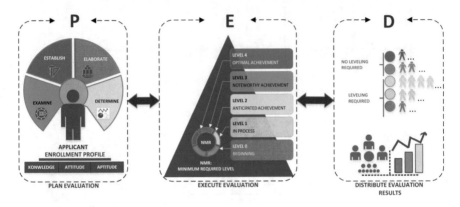

Fig. 4. Selection method for applicants according to the enrollment profile

The method establishes mechanisms for the selection of applicants with the objective of evaluating the degree of compliance between the enrollment profile and the standard established by the rules of the institution, and this consists of three phases: the first, brings together the processes of planning evaluation, when the study program evaluates the knowledge of the applicants through percentage weights assigned to subjects that are more linked to the profile and the characteristics of the study program, where a greater percentage means a greater link; then, the study program establishes a minimum percentage that the applicant must reach to be admitted to the program to which he/she is applying, this percentage is in accordance with the infrastructure capacity, available vacancies and other internal aspects that the program considers pertinent; subsequently scales are formulated in 5 intervals, where the baseline is a minimum percentage required from which it identifies whether an applicant entered or not; finally, the levels reached are located by means of rubrics of evaluation in relation to the intervals where the minimum level is zero and the maximum is 4, with level 3 proving that the student reached the mark.

In the second phase, the evaluation is carried out considering the criteria established in the planning phase. The results obtained by an applicant in the exam are used first to identify whether or not he/she can enter compared with the minimum level required and secondly to determine the level reached. Finally, the last phase consists of preparing statistical reports by classifying them by level to be distributed and disseminated to whom they concern. The results can be used by the study program to establish leveling strategies for the entrants.

3.2   Formulation Sequence of the Fuzzy Model

First of all, the method of selection for applicants was analyzed, in terms of the process, its instruments and procedures. Based on this analysis, the input linguistic variables were identified with their membership functions. Then, the ranges in the domain of the membership functions were identified for each input linguistic variable that corresponds to the skills of the knowledge dimension. Then, the rules for the inference engine were formulated with the Mamdani method in a combinatorial way.

Regarding the application specifically of the Mamdani method, it involved carrying out the following procedures. In the first place, the antecedents of the identified rules of the input variables were evaluated according to their minimum range and maximum range. Then, from the consequent of each rule and the value of each antecedent obtained in step 1, a fuzzy operator of implication was applied, obtaining for each rule a new fuzzy set linked with the AND operator since some combinations satisfy the profile of the entrant, excluding those rules that do not comply with their established ranges with their respective membership functions. Third, the conclusions for each rule in step 2 were added in such a way that they are combined in a single fuzzy set using an aggregation operator called MAXIMUM to get its highest value. Finally, a solution was obtained by acquiring an output with the centroid method and the defusing technique whose result is a number and not a fuzzy set; for this the following logical criterion was applied for each rule used in the investigation: If (Competency Level 1) and (Competency Level 2) and (Competency Level 3) then (Entrant Profile Level)

Finally, the model was validated as follows: (1) Admission experts were selected and the structure of the model was delivered with the data from a pilot sample of applicants for an admission period, (2) the experts entered data into the model through the matlab interface and compared the values of the PED method and the fuzzy logic method according to the levels reached by each of the applicants and (3) the experts expressed their value judgment through an instrument to then use the Aiken V formula and establish the respective validation.

4 Results

4.1 Fuzzy Model Design for the Selection of the Enrollment Profile

The fuzzy logic model is adaptable to any study program considering its own evaluation parameters and weights. Below are the modeled specifications for a study program.

This model has three input variables: competency 1 (CAPAS), demonstrates ability to analyze, abstraction and synthesis, competency 2, has basic knowledge in science (COBACI) and competency 3 has basic knowledge in letters (COBALET) that correspond to the competencies of the enrollment profile, and one exit variable: profile of the entrant (PERIngresante).

The fuzzy inference system [22] processes the inputs through the Mamdani method, as shown in Fig. 5 and its reasoning is shown in Table 1. In this method, the input values are introduced with respect to the linguistic variables named in the previous paragraph, processed by means of If-Then style rules and a single value in the output is found.

Each one of the input variables and the value of the parameter output variable, as can be seen in Table 1, have a linguistic value associated with their respective

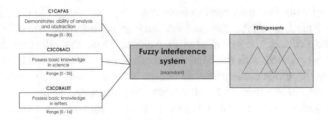

Fig. 5. Fuzzy logic model to evaluate the enrollment profile

parameters and membership function. Thus competency 1 has the following linguistic values: C1N0INI (Competency 1 at level 0 → Beginning), C1N1PRO (Competency 1 at level 1 → In process), C1N2PRE (Competency 1 at level 2 → Anticipated achievement), C1N3LDEST (Competency 1 at level 3 → Noteworthy achievement) and C1N4LOPT (Competency 1 at level 4 → Optimal achievement); as well as the parameters for the domain of the membership functions by rank in its associated membership function. Figure 6 shows the structure of linguistic variables CAPAS y PERIngresante, each with its levels respective to the domain of the membership functions X.

Mathworks in a publication about the use of their tools [22] states that human knowledge is expressed in terms of fuzzy rules, in the style *If-Then* <Fuzzy Proposition> in which the types of fuzzy propositions are atomic, such that x is A, where x is a linguistic variable and A is a linguistic value, and compound propositions are fuzzy atomic propositions linked with logical connectors.

Table 2 shows the core of the fuzzy logic model through 125 Mamdani rules, product of the combination of the 5 levels of the 3 competences (CAPAS, COBACI and COBALET) of the income profile, which gives rise to 125 possible outputs of the entrant profile (PERIngresante), of which not all comply with the profile required by a study program, therefore the AND operator was used to join the values of the input variables, which together act as criteria to exclude rules that are not met. This method evaluates background rules with logical connectors *AND* or *OR* to find a single value in the consequent output variable. This method is applied instead of others [22] because the output variables are not linear or constant, and the *singleton* transfer function is not used. The output variables are functions of triangular and trapezoidal right and left transfers, justified by the method that is being used (Mamdani).

4.2 Results of the Fuzzy Logic Model

Before using the elaborated fuzzy model, the structure was submitted to expert assessment to determine its consistency with the theoretical aspects and the procedures of the admission process with the PED method. The experts qualitatively mentioned that the model is consistent according to the mentioned criteria and quantitatively that it is valid through Aiken's V with a value of 0.87 which is considered acceptable.

Below are the values of some input and output results. It is worth noting that to take into account that the outputs of the fuzzy rules are centered on the linguistic variable x, resulting in a value x which must belong to a degree of membership [19] $\mu(x) \in [0\,;1]$.

Table 1. Input/output of the fuzzy logic model

Input/output	Domain	Linguistic value	Parameters	Membership
Competency 1 Demonstrates ability of Analysis, abstraction	[0–30]	C1N0INI	[0 0 15 19]	Trapezoidal
		C1N1PRO	[15 19 22.5]	Triangular
		C1N2LPRE	[19 22.5 25.5]	Triangular
		C1N3LDEST	[22.5 25.5 30]	Triangular
		C1N4LOPT	[25.5 30 0 0]	Trapezoidal
Competency 2 Possess basic knowledge in science	[0–35]	C2N0INI	[0 0 17.5 20.5]	Trapezoidal
		C2N1PRO	[17.5 20.5 24]	Triangular
		C2N2LPRE	[20.5 24 28.5]	Triangular
		C2N3LDEST	[24 28.5 35]	Triangular
		C2N4LOPT	[28.5 35 0 0]	Trapezoidal
Competency 3 Possess basic knowledge in letters	[0–16]	C3N0INI	[0 0 2.5 5.5]	Trapezoidal
		C3N1PRO	[2.5 5.5 8.5]	Triangular
		C3N2LPRE	[5.5 8.5 11.5]	Triangular
		C3N3LDEST	[8.5 11.5 16]	Triangular
		C3N4LOPT	[11.5 16 0 0]	Trapezoidal
Output Entrant profil00e	[0–81]	PFLevel0	[0 0 35 45]	Trapezoidal
		PFLevel1	[35 45 55]	Triangular
		PFLevel2	[45 55 65.5]	Triangular
		PFLevel3	[55 65.5 81]	Triangular
		PFLevel4	[65.5 81 0 0]	Trapezoidal

In the following lines we can see an example of the introduction of values in the fuzzy inference system as inputs for the linguistic variables (competencies) and their result in the output variable. The results of those values served to validate the model and estimate the effectiveness respect the method PED

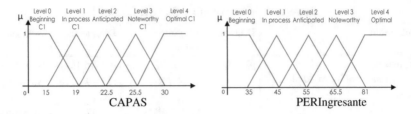

Fig. 6. Graphic representation of the linguistic variable CAPAS and PERIngresante

Table 2. Rules of the fuzzy inference system

Competency 01	Competency 02	Competency 03: C3COBALET				
C1CAPAS	C2COBACI	C3N0INI	C3N1PRO	C3N2PRE	C3N3DEST	C3N4OPT
C1N0INI	C2N0INI	PFLevel0	PFLevel0	PFLevel0	PFLevel0	PFLevel1
C1N0INI	C2N1PRO	PFLevel0	PFLevel1	PFLevel1	PFLevel1	PFLevel1
C1N0INI	C2N2PRE	PFLevel1	PFLevel1	PFLevel1	PFLevel1	PFLevel1
C1N0INI	C2N3DEST	PFLevel1	PFLevel1	PFLevel1	PFLevel1	PFLevel2
C1N0INI	C2N4OPT	PFLevel1	PFLevel1	PFLevel1	PFLevel2	PFLevel2
C1N1PRO	C2N0INI	PFLevel0	PFLevel0	PFLevel0	PFLevel0	PFLevel1
C1N1PRO	C2N1PRO	PFLevel1	PFLevel1	PFLevel1	PFLevel1	PFLevel1
C1CAPAS	C2COBACI	C3N0INI	C1CAPAS	C2COBACI	C3N0INI	C1CAPAS
C1N1PRO	C2N2PRE	PFLevel1	PFLevel1	PFLevel2	PFLevel2	PFLevel2
C1N1PRO	C2N3DEST	PFLevel1	PFLevel2	PFLevel2	PFLevel2	PFLevel2
C1N1PRO	C2N4OPT	PFLevel2	PFLevel2	PFLevel2	PFLevel2	PFLevel2
C1N2PRE	C2N0INI0	PFLevel0	PFLevel0	PFLevel1	PFLevel1	PFLevel1
C1N2PRE	C2N1PRO	PFLevel1	PFLevel2	PFLevel2	PFLevel2	PFLevel2
C1N2PRE	C2N2PRE	PFLevel2	PFLevel2	PFLevel2	PFLevel2	PFLevel2
C1N2PRE	C2N3DEST	PFLevel2	PFLevel2	PFLevel3	PFLevel3	PFLevel3
C1N2PRE	C2N4OPT	PFLevel2	PFLevel3	PFLevel3	PFLevel3	PFLevel3
C1N3DEST	C2N0INI	PFLevel0	PFLevel1	PFLevel1	PFLevel1	PFLevel1
C1N3DEST	C2N1PRO	PFLevel1	PFLevel1	PFLevel2	PFLevel2	PFLevel2
C1N3DEST	C2N2PRE	PFLevel2	PFLevel2	PFLevel2	PFLevel3	PFLevel3
C1N3DEST	C2N3DEST	PFLevel3	PFLevel3	PFLevel3	PFLevel3	PFLevel3
C1N3DEST	C2N4OPT	PFLevel3	PFLevel3	PFLevel3	PFLevel3	PFLevel4
C1N4OPT	C2N0INI	PFLevel1	PFLevel1	PFLevel1	PFLevel1	PFLevel1
C1N4OPT	C2N1PRO	PFLevel1	PFLevel2	PFLevel2	PFLevel2	PFLevel2
C1N4OPT	C2N2PRE	PFLevel2	PFLevel2	PFLevel3	PFLevel3	PFLevel3
C1N4OPT	C2N3DEST	PFLevel3	PFLevel3	PFLevel3	PFLevel4	PFLevel4
C1N4OPT	C2N4OPT	PFLevel3	PFLevel3	PFLevel4	PFLevel4	PFLevel4

Rule no.66. If (C1CAPAS is C1N2LPRE) and (C2COBACI is C2N3LDEST) and (C3COBALET is C3N0INI) then (PERIngresante is PFNivel2) (1)
[21;24 > and [26;31 > and [0;4 > then [50;60 > any degree of membership from 0.40 to 0.6. Input matlab: [22;27;3], output matlab [53.2] which means that it is at knowledge level 2, in other words anticipated achievement.
Rule no.74. If (C1CAPAS is C1N2LPRE) and (C2COBACI is C2N4LOPT) and (C3COBALET is C3N3LDEST) then (PERIngresante is PFNivel3) (1)
[21;24 > and [31;35] and [10;13 > then [60;71 > any degree of membership from 0.60 to 0.8. Input matlab: [21;32;10], output matlab [62.7] which means that it is knowledge level 3, in other words noteworthy achievement.
Rule no. 84. If (C1CAPAS is C1N3LDEST) and (C2COBACI is C2N1PRO) and (C3COBALET is C3N3LDEST) then (PERIngresante is PFNivel2) (1)
[24; 27 > and [19; 22 > and [10; 13 > then [50; 60 > any degree of membership from 0.40 to 0.60. Input matlab: [25; 20; 11], output matlab [53.1], which means that it is knowledge level 2, in other words anticipated achievement.

The graphical interface of MatLab v17a allows for a graphic display of the model. The user can enter a set of values corresponding to the characteristics of the membership function used [22] for the input variables in the ranges of the domain of the membership functions (3 first columns) and in the same way for the output variable (Fig. 7).

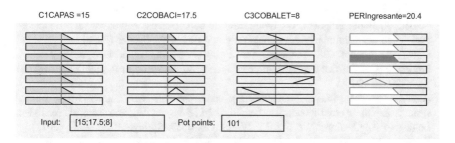

Fig. 7. Defuzzified outputs of the fuzzy inference system

Finally, to evaluate the effectiveness of the fuzzy model, we considered the data of 128 candidates to a study program as a random sample in an entrance examination, which were entered into the fuzzy model to determine its effectiveness from the nonparametric test of Wilcoxon, made in SPSS 24.0. The value $Z = -2.111$ shows that the model was effective with respect to the traditional PED model with p value = $0.035 < 0.05$, demonstrating in this way that the model gives better results with respect to the selection of the candidates of a university program according to their admission profile.

5   Conclusions

This research project has developed a fuzzy model for the selection of applicants to university study programs, to admit them or not. The model has incorporated the evaluation of knowledge competencies to obtain the outputs. It handles all possible and necessary evaluation rules, constituting a support tool for evaluators, providing more accurate information when evaluating the profile of the applicant.

This fuzzy model is effective because it provides better results with respect to the selection of candidates for a university program based on their entrance profile, in terms of the accuracy of the level that corresponds to the applicant.

A recommendation for future work is to implement a distributed and responsive web application that allows for the entry and consumption (outputs) of data in real time.

References

1. Audretsch, D.B.: From the entrepreneurial university to the university for the entrepreneurial society. J. Technol. Transf. **39**(3), 313–321 (2014)
2. OECD: Making development happen. OECD Dev. Cent. **3**, 36 (2016)
3. SINEACE: Modelo de Acreditación para Programas de Estudios de Institutos y Escuelas de Educación Superior, p. 36, 2016
4. Davidovitch, N., Soen, D.: Predicting academic success using admission profiles. J. Int. Educ. Res. **11**(3), 125–142 (2015)
5. Montero Morales, J.A., Gómez Urgellès, J., Alías Pujol, F., Garriga Berga, C., Vicent Safont, L., Badía Folguera, D.: Evaluación de competencias subjetivas. experiencia en la evaluación del rendimiento del trabajo en grupo de los estudiantes (2008)
6. Houe, R., Grabot, B., Tchuente, G.: Fuzzy logic in competence management. In: Proceedings of 7th Conference European Society on Fuzzy Logic Technology, LFA 2011, pp. 651–656 (2011)
7. Alonso, J.M., Magdalena, L.: Generating understandable and accurate fuzzy rule-based systems in a Java environment. In: Fanelli, A.M., Pedrycz, W., Petrosino, A. (eds.) Fuzzy Logic and Applications. WILF 2011. LNCS, vol. 6857. Springer, Heidelberg (2011)
8. Kozae, A.M., Elshenawy, A., Omran, M.: Intuitionistic fuzzy set and its application in selecting specialization: a case study for engineering students. Int. J. Math. Anal. Appl. **2**(6), 74–78 (2015)
9. Lisi, F.A., Straccia, U.: Towards learning fuzzy DL inclusion axioms. In: Fanelli, A.M., Pedrycz, W., Petrosino, A. (eds.) Fuzzy Logic and Applications, vol. 6857, pp. 58–66. Springer, Heidelberg (2011)
10. Voskoglou, M.: Fuzzy logic as a tool for assessing students' knowledge and skills. Educ. Sci. **3**, 208–221 (2013)
11. Jevšček, M.: Competencies assessment using fuzzy logic. J. Univers. Excell. **5**(2), 187–202 (2016)
12. Ma, J., Zhou, D.: Fuzzy set approach to the assessment of student-centered learning. IEEE Trans. Educ. **43**(2), 237–241 (2000)
13. Ruvalcaba Coyaso, F.J., Vermonden, A.: Lógica difusa para la toma de decisiones y la selección de personal. Univ. Empres. **17**(29), 239–256 (2016)
14. Zadeh, L.A.: Fuzzy sets. Inf. Control **8**(3), 338–353 (1965)

15. Bellman, R.E., Zadeh, L.A.: Decision-making in a fuzzy environment. Manag. Sci. **17**(4), 141–164 (1970)
16. Zadeh, L.A., Fu, K.S., Tanaka, K., Shimura, M., Negoita, C.V.: Fuzzy sets and their applications to cognition and decision processes. Syst. Man Cybern. IEEE Trans. **7** (February), 122–123 (1977)
17. Sivanandam, S.N., Sumathi, S., Deepa, S.N.: Introduction to fuzzy logic using MATLAB (2007)
18. Liu, J.: Intelligent control design and MatLab simulation (2018)
19. Escolano Ruiz, F., Carzola Quevedo, M.Á., Alfono Galipienso, M.I., Colomina Pardo, O., Miguel Ángel, L.O.: Inteligencia-Artificial Modelos, Tecnicas y Areas de Aplicacion. Primera, España (2003)
20. Hilera González, J.R., Martínez Hernando, V.J.: Redes Neuronales Artificiales Fundamentos, modelos y aplicaciones. Primera, Minneapolis (1995)
21. Ley Universitaria N° 30220, p. 68 (2014)
22. Mathworks, C.: Fuzzy Logic Toolbox™. USA (2018)

The Digitization of the Russian Higher Education

Irina Mavrina[1]([envelope]) and Anna Mingaleva[2]([envelope])

[1] Ural Federal University named after the first President of Russia B. N. Yeltsin, Ekaterinburg, Russia
iraika@bk.ru
[2] VShEM, Ural Federal University named after the first President of Russia B. N. Yeltsin, Ekaterinburg, Russia
mingaleva.ann@yandex.ru

Abstract. Today the digital economy is a system of economic, social and cultural relations based on the use of digital information and communication technologies. Russia has created the Digital economy program, which is aimed at introducing digital technologies in all spheres of life: in the economy, in business, in public administration, in the social sphere and in the municipal economy. The paper investigates the features of digitization in the Russian education sector and education system of foreign countries. Such changes in the country are quite real, but the rapid development of the digital economy in the Russian education today is hampered by new challenges and threats. In order to overcome the obstacles to the development of digital in the in the Russian education sector, universities should join forces with the state, research centers and other organizations, whose activities are aimed at accelerating the commercial development of innovative technologies in the education sphere. The systematization of all aspects of the economy is an important step towards the conscious growth and innovative development of the country. The digitization of the Russian education is the most actual direction of development today.

Keywords: Higher education · Digitization · Digital tools of education

1 Introduction

Mobile devices acquire high importance for the modern man, who strives for convenience and mobility in all spheres of society. The Russian Federation ranks 41st on the readiness for the digital economy with a significant gap from the dozens of leading countries such as Singapore, Finland, Sweden, Norway, the United States, the Netherlands, Switzerland, the United Kingdom, Luxembourg and Japan [1].

Such a significant lag in the development of the digital economy from world leaders is due to the gaps in the regulatory framework for the digital economy and the insufficiently favorable environment for doing business and innovation and, as a consequence, the low level of digital technology use by business structures.

However, the number of training and compliance of educational programs with the needs of the digital economy is insufficient in Russia. There is a serious shortage of

T. Antipova and A. Rocha (Eds.): DSIC 2018, AISC 850, pp. 134–142, 2019.
https://doi.org/10.1007/978-3-030-02351-5_17

personnel in the educational process at all levels of education. Digital tools of educational activities are not sufficiently applied in the final attestation procedures of students. The learning process is not included integrally into the digital information environment. All this requires the active implementation of digitization approaches to the training of higher education specialists and the development of digitization of the universities themselves. This situation can't be corrected without training specialists in all areas of skills in the digital economy. Possession of skills of computer and digital technologies, and not the simplest ones (like working with digital devices as ordinary users), but more advanced, is necessary for future entrepreneurs. This is provided, including the Program "Digital Economy of the Russian Federation", approved in 2017 [1].

2 Theoretical and Institutional Background

The interrelationship between globalization and the education system is analyzed in many scientific papers and strategic documents [2–4]. At the suggestion of the World Economic Forum, the latest version of the international network readiness index, presented in the report "Global Information Technologies" for 2016, is used to assess countries' readiness for the digital economy. The improved index measures how well countries use digital technologies to improve competitiveness and well-being, and also assesses the factors affecting the development of the digital economy [5]. Digital literacy in education issues are also widely discussed [6, 7].

The problems of motivating students to study are currently the most relevant and are being actively studied in foreign and domestic science [8–10]. Brown, Molesworth and others believe that the students' motivation for learning will increase if we consider students as consumers of the product of higher education [11, 12].

An analysis of the world scientific literature on the issues of digitization of training has shown that there is a lot of theoretical criticism around the use of digital technologies in education, including the higher education [13–17], but the body of critical appraisal of national digital learning and teaching strategies is more limited [18]. On the contrary, many foreign researchers note that the "national strategies play a crucial role in framing how digital technologies are enacted in higher education" [19]. So the policy-makers should extend their consideration of universities entrepreneurial activity to include the development of human capital [20]. For example, Munro describes in his article the some findings of an analysis of thirteen digital teaching and learning strategies issued by government departments and non-departmental public bodies in the UK between 2003 and 2013. "It demonstrates that, across the strategies, digital technologies are depicted as tools for advancing the marketisation of UK higher education" [19].

In order to manage the development of the digital economy in Russia, the Program "Digital Economy of the Russian Federation" was adopted, which defines the five basic directions for the development of the digital economy in the Russian Federation for the period up to 2024. The basic directions include normative regulation, personnel and education, the formation of research competences and technical facilities, information infrastructure and information security. The main goals and objectives of the direction concerning personnel and education are:

- creation of key conditions for the training of digital economy personnel;
- the improvement of the education system, which should provide the digital economy with competent personnel;
- the creation of a motivation system to develop the necessary competencies and the participation of personnel in the development of the digital economy of Russia.

Solving these goals will improve the labor market, which must be based on the demands of the digital economy. The "The Strategy of scientific and technological development of the Russian Federation until 2035" is another document that determines the prospects for the development of the digital economy in Russia. The increase of the competitiveness of the national economy and the effectiveness of the national security strategy based on leadership in research and development, the creation of innovative products and the high rate of creation and mastering of new knowledge are mentioned as key tasks of The strategy of scientific and technological development of the Russian Federation until 2035 (paragraph 8 of the Strategy) [21]. The main subjects intended to ensure the implementation of the provisions of the Strategy are organizations that directly carry out scientific, technical and innovative activities and use the results of such activities (Clause 6 of the Strategy). Among them, a special place is occupied by scientific and educational organizations. They provide the society with human and information resources for creating and using the latest technologies and innovative solutions that most effectively meet the "big challenges" of modern development. Regarding international documents in the field of digitalizing education, among them are the Recommendations on Mobile Learning Policies, published in 2015 by the UNESCO Institute for Information Technologies in Education. These recommendations state that "mobile technologies allow to significantly expand and improve learning opportunities in a variety of conditions …" [22].

3 Introduction of Digital Education System in Russia

3.1 A Current Level of Digitization in Russian Education

At present, methods of strengthening computer and digital profiling in the preparation of students of IT directions are already being developed and tested quite extensively. The course of informatics and information and communication technologies in general education programs is standardized, technologically and substantively, and personnel for the digital economy are being trained. Educational organizations have access to the "Internet" network and are represented there on their websites in accordance with state requirements. The dynamics of the main indicators of digitization of higher education institutions is presented in Table 1 [23].

Along with traditional technologies of teaching, mobile technologies are used that open new opportunities for students and teachers: application of communication; access to large educational resources and making quick feedback between students and teachers. Also, the use of digital devices allows the discussion of information and

Table 1. The main indicators of digitization of higher education

	2010	2011	2012	2013	2014	2015	2016
The number of students of higher education per 10000 pop., people	493	454	424	394	356	325	300
The number of students of higher education in the direction "Computer science and computer technology" per 10000 pop., people	3	9	9	9	9	10	11
Number of personal computers used for educational purposes for 100 students of state education institutions, thing	7	8	11	12	13	13	14
Number of personal computers used for educational purposes, which are part of local area networks, per 100 students of high education, thing	15	17	19	21	20	14	22
Number of personal computers used for educational purposes with access to the Internet, per 100 students of higher education, thing	14	16	18	21	20	14	22
The share of educational institutions that have a website on the Internet, in the total number of institutions to higher education, %	96.3	98.2	99.8	99.8	97.5	100	95.1
The share of educational institutions implementing educational programs using distance educational technologies in the total number of institutions to higher education, %	49.8	52.8	59.0	60.2	57.4	78.2	42.8

familiarization with all participants of the educational process on-line. And more and more important is the fact that the mobile devices (which include mobile phones, tablet computers, electronic readers, portable audio players) become an indispensable tool for solving a large number of tasks in business: data exchange, quick access to up-to-date information, mobile payment services, customer support, etc.

However, for successful digitization (transfer to digital technologies), business leaders and entrepreneurs must themselves have the skills to work in the digital economy. One way to achieve this goal is the training of personnel in the field of entrepreneurial activity in accordance with the requirements of digital economy [24, 25]. And educational institutions that form the scientific and educational potential of the national and regional economy should ensure this first of all [26]. The modern process of learning is a process that takes place in an uncertain and continuously changing environment. Today, it is impossible to imagine an education system without the use of information and communication technologies (ICT) [27]. Teaching students using mobile technologies improves their practical skills, facilitates access to educational materials and information exchange between all participants of the educational process, provides conditions for the introduction of methods of inclusive education, allows new conditions, in comparison with traditional education, for creating an accessible educational environment.

There are various mobile applications for educational use that are related to real business. For example, a restaurant business. Recently, the enterprises of this sector have faced the problem of strategic and tactical management of the development of the ability to survive in the growing competition and the introduction of new digital technologies has become a good way to succeed. Most residents of our country (especially large cities) are already familiar with such digital tools as their own applications of large restaurant chains and individual restaurants, cafes and bars, with which they talk about their services, innovations, promotions and support feedback from customers. Moreover, such implemented mobile applications with loyalty programs are increasingly replacing loyalty cards. Such digital technologies allow restaurants and cafes to create their own chips that increase the competitiveness of the institution, attract visitors, thereby enabling entrepreneurs to earn money correctly, and consumers to maximize utility [25].

The effect of networks flexibility and reliability is achieved by "digitization" of educational infrastructure. The problem of digitization of the economy as a whole and infrastructure sectors in particular has become even more important in connection with the taken course for the implementation of the program "Digital economy of the Russian Federation", approved by the order of the Russian Federation Government on July 28, 2017 [1]. On the basis of the "road map", an action plan will be developed that will describe the activities necessary to achieve the specific "milestones" of this Program, indicating who is responsible for implementing the activities, sources and amounts of funding. The action plan will be approved for three years, which implies its annual renewal. Technology of education, including the digital technology, "needs to enhance student choice and meet or exceed learners' expectations [28].

3.2   Prospects for Digitalizing the Russian Higher Education

The "road map" for the development of the educational sector in Russia is presented in the Program "Digital economy of the Russian Federation". In the "road map" three main stages of the development of the directions of the digital economy are singled out, according to the results of which it is planned to achieve the target status for each of the directions: 2018 – the stage 1; 2020 – the stage 2 and 2024 – the stage 3. The planned values of the proportion of Russian citizens who have improved literacy in the field of information security, media consumption and the use of Internet services are also listed in this "road map" (see Fig. 1 [1]).

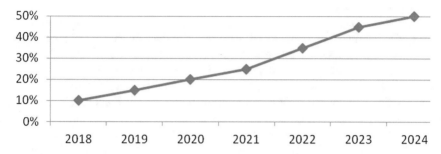

Fig. 1. The proportion of citizens who have improved literacy in the field of information security, media consumption and the use of Internet services, %

The indicators of achievement of the planned characteristics of the digital economy of the Russian Federation by 2024 present in the Fig. 2 [1].

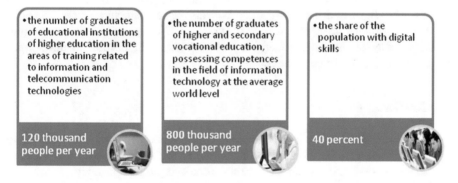

Fig. 2. Indicators of achievement of the planned characteristics of the digital economy of the Russian Federation by 2024

This puts forward new requirements for the training of future student and all Russian citizens for the formation of their knowledge and competencies that meet the requirements of the functioning of the digital economy.

Thus, the actual need for the training of modern personnel in the field of higher education is the development of programs and teaching methods that change the approach to educational activity and are conducive to the formation of professional information and telecommunication competencies for future graduates of training and retraining courses for entrepreneurs and all Russian citizens [23].

These planned values can be achieved on the basis of the widespread introduction of new special subjects into the education system. These are the following training courses: "Management of electronic commerce (digital entrepreneurship)"; "Management in the digital business"; "Marketing and advertising in the digital business" (as an option - "Marketing and advertising in e-commerce"); "Management of digital corporations"; "Information Systems Management and Cybersecurity"; "Financial Cybersecurity"; "Digital technologies in the high-tech sector"; "Digital service as a new segment of the economy"; "Digitization of traditional industries and spheres of the economy"; "Digitization of financial services of universal banks"; "Strategic management of digital holding and cross-border corporate structures"; "Digital and electronic technologies of universal banking"; "Integrated electronic banking and digital financial instruments (crypto-currencies, software mining, etc.)"; "Investment electronic banking and financial analytics for online trading" and others [24]. This subjects have been taught extensively for a long time in foreign universities, but its which are almost impossible to meet in Russian education programs.

Some of these training courses can also be developed into independent areas of master's or professional training. For example, "Management in the digital business" with its more in-depth study in functional areas: financial management, personnel management, production management, innovative management, marketing, etc., as well as in business areas and areas can become an independent training course or retraining of personnel within the framework of special programs for training entrepreneurship, forming the necessary competencies.

This is all the more important if a person intends to organize business or to work in IT areas, which take an increasing share both in the real structure of the economy. However, traditional areas of business are no less subject to the influence of digitization, and targeted training in methods and methods of digitization should become mandatory in the all programs of education.

4 Conclusions

Digital technologies are becoming increasingly important, as the mobile applications (special additions) are developing, which are a tool that allows to optimize the communication processes, create an information space and promote learning.

So, the modern digital learning technologies are becoming more and more in demand in the educational process of various target groups, especially among entrepreneurs and professionals, and in all higher education institutions. The programs of digital education will always be relevant in many businesses will constantly improve and develop. In order to overcome the obstacles to the development of digital in the in the Russian education sector, universities should join forces with the state, research centers and other organizations, whose activities are aimed at accelerating the commercial development of innovative technologies in the education sphere. Also, Russian universities should more widely implement educational courses and subjects related to digitization of education.

References

1. The program "Digital Economy of the Russian Federation", approved by the order of the Government of the Russian Federation on 28 July, 2017, No. 1632-p (2017)
2. Butcher, N.: Technologies in higher education: mapping the terrain. UNESCO IITE Policy Brief (2014)
3. Lane, A.: Global trends in the development and use of open educational resources to reform educational practices. UNESCO IITE Policy Brief (2010)
4. Mingaleva, Zh., Mirskikh, I.: Globalization in education in Russia. Procedia Soc. Behav. Sci. **47**, 1702–1706 (2012)
5. Baller, S., Dutta, S., Lanvin, B. (ed.): The Global Information Technology Report 2016. Geneva Copyright by the World Economic Forum and INSEAD (2016)
6. Fairclough, N.: Language and Globalization. Routledge, Oxon (2006)
7. Karpati, A.: Digital literacy in education. UNESCO IITE Policy Brief (2011)
8. Brown, R. (ed.): Higher Education and the Market. Routledge, Oxon (2011)
9. Mingaleva, Zh., Bykova, E., Lobova, G.: Motivation processes of educational activities in Russia. Procedia Soc. Behav. Sci. **83**, 985–989 (2013)
10. McGettigan, A.: The Great University Gamble: Money, Markets and the Future of Higher Education. Pluto Press, London (2013)
11. Brown, R., Carasso, H.: Everything for Sale?: The Marketisation of UK Higher Education. Routledge and the Society for Research into Higher Education (SRHE), London (2013)
12. Molesworth, M., Nixon, E., Scullion, R. (eds.): The Marketisation of Higher Education and the Student as Consumer. Routledge, Oxon (2011)
13. Bayne, S.: What's the matter with 'technology-enhanced learning'? Learn. Media Technol. **40**(1), 5–20 (2015)
14. Bulfin, S., Johnson, N.F., Bigum, C.: Critical Perspectives on Technology and Education. Palgrave Macmillan, Basingstoke (2015)
15. Kirkwood, A., Price, L.: Technology-enhanced learning and teaching in higher education: What is 'enhanced'and how do we know? A critical literature review. Learn. Media Technol. **39**(1), 6–36 (2014)
16. Selwyn, N.: Technology and education—why it's crucial to be critical. In: Bulfin, S., Johnson, N.F., Bigum, C. (eds.) Critical Perspectives on Technology and Education, pp. 245–255. Palgrave Macmillan, Basingstoke (2015)
17. Guri-Rosenblit, S.: Digital Technologies in Higher Education: Sweeping Expectations and Actual Effects. Nova Science Publishers, Inc., Hauppauge, New York (2009)
18. Selwyn, N., Gorard, S., Furlong, J.: Adult Learning in the Digital Age: Information Technology and the Learning Society. Routledge, Oxon (2006)
19. Munro, M.: The complicity of digital technologies in the marketisation of UK higher education: exploring the implications of a critical discourse analysis of thirteen national digital teaching and learning strategies. Int. J. Educ. Technol. High. Educ. **15**(1), 1 (2018)
20. Ratzinger, D., Amess, K., Greenman, A., Mosey, S.: The impact of digital start-up founders' higher education on reaching equity investment milestones. J. Technol. Transf. **43**(3), 760–778 (2018)
21. The strategy of scientific and technological development of the Russian Federation until 2035. http://sntr-rf.ru/materials. Accessed 16 Mar 2018
22. UNESCO Policy Guidelines for Mobile Learning. https://iite.unesco.org/pics/publications/ru/files/3214738.pdf. Accessed 17 Mar 2018
23. Rosstat. https://www.gks.ru/free_doc/new_site/business/it/monitor_rf.xls. Accessed 19 Apr 2018

24. Mavrina, I.N., Mingaleva, A.D.: Entrepreneurship education as a factor of regional economic development. In: Insights and Potential Sources of New Entrepreneurial Growth, pp. 370–381. Filodiritton Publisher, Bologna (2017)
25. Mingaleva, A.D.: Development of Information Competencies in Training Entrepreneurship in the Digital Economy. Economics and Management in the Conditions of Digitization: State, Problems, Foresight, pp. 381–387. Publishing Polytechnic University, St. Petersburg (2017)
26. Mavrina, I.N., Mingaleva, A.D.: Formation of entrepreneurial competencies in the different programs of entrepreneurial education and training. In: Education for Entrepreneurial Business and Employment, pp. 45–59. Compass Publishing, Newton Abbot (2017)
27. Selwyn, N.: The use of computer technology in university teaching and learning: a critical perspective. J. Comput. Assist. Learn. 23(2), 83–94 (2007)
28. HEFCE: Report to HEFCE by the online learning task force collaborate to compete. Seizing the opportunity of online learning for UK higher education. Higher Education Funding Council for England (HEFCE), Bristol. http://www.hefce.ac.uk/pubs/year/2011/201101 (2011). Accessed 13 Mar 2018

M-Learning Didactic Strategy for Children Diagnosed with Dyslexia

Daniela Benalcázar Chicaiza$^{(\boxtimes)}$, Mayra I. Barrera G.,
María Cristina Páez Quinde, and Marcelo E. Pilamunga P.

Universidad Técnica de Ambato, Huachi Chico 180206, Ecuador
{da.benalcazar, mayraibarrerag, mc.paez,
em.pilamunga}@uta.edu.ec

Abstract. This research of M-learning as a teaching strategy for children diagnosed with dyslexia, obtained results through the survey of therapists who do not implement technological resources to improve cognitive and motor skills of children in the teaching process. By a DST-J test applied to children it was established that there are three levels of risk on specific tasks of literacy: high, moderate and slight. With mobile learning as a foundation that provides an interactive environment that facilitates access to different content and supports the socialization of judgments, opinions, knowledge, contributes to dealing with the challenges of teaching and learning for students with special abilities and creates educational tasks, this research aims to demonstrate the relationship between M-learning as a didactic strategy and children with dyslexia, in order to use innovative strategies that help in their development.

Keywords: Mobile applications · Dyslexia · Mobile devices
Didactic strategy · Writing reading process · M-learning

1 Introduction

At the national level, current education requires in a relevant way the use of technological tools so that children with learning disorders better understand the content taught by teachers, with the purpose of implementing an innovative model of education in education through mobile applications [1].

In such a virtue, it is necessary that educational institutions and those that treat children with learning difficulties, as well as their teachers or professionals, adapt curricular contents and learning environments with technological resources. The M-learning will provide a better and active teaching-learning process according to the needs of children with learning difficulties.

New techniques should be used with dyslexic children that allow them to adapt multi sensory learning with the use of mobile technology to improve the pedagogical process between the student and the therapist, with the sole objective of efficiently contributing to self-esteem, knowledge and development of skills of children with dyslexia [2].

[3] In the scientific article entitled " With you in the distance: the tutorial practice in virtual training environments", it is indicated that the management of health institution managers should define strategies and technological pedagogical models to apply in the

care for children with dyslexia problems, which should likewise be taught to their facilitators for its correct application.

1.1   Information and Communication Technologies in the Educational Field

Information and communication technologies are the set of techniques, advances and modern devices that make up storage, processing and transfer of information through the Internet to be used in various areas such as education, health, finance, tourism, and social interaction, among others [4]. Information and communication technologies in the educational field play a very important role since they have allowed for innovation in the teaching-learning process [5]; in turn they have directly involved the student in his/her self-learning through dynamic and interactive virtual environments. On the use of didactic material and information and communication technologies (TICs), it is noted that to improve the academic scope "In the field of education, ICTs have wide applicability as educational portals or websites, virtual classrooms for teaching - learning, videoconferencing, educational software and material with multimedia support that can be distributed through the Internet" [6].

Virtual environments and virtual media have revolutionized education since they have adapted educational contents and materials to allow access from any place and at any time through technology with the purpose of improving student learning [7]; M-learning represents an innovative component in the teaching and learning process, therefore, it is necessary to know the main characteristics of its usefulness [8]; mobile learning offers different benefits to learning, where teachers and students can interact without limitations, and there is ease of access to educational resources [9]. M-learning encourages interaction through virtual learning environments, where teachers and students can engage in extracurricular activities in order to obtain feedback on some specific content so that they reinforce their knowledge through interactive and innovative learning spaces supported by the technologies of information and communication.

The educational modality facilitates the construction of knowledge, the resolution of learning problems and the development of skills or abilities [10]. Mobile devices within the educational environment allow students to build their knowledge interactively thanks to technology. The exchange of information, ideas and contributions is facilitated [11]. Mobile technology provides a variety of tools applicable to educational processes, as well as M-learning that can be applied in various ways in the teaching and learning process; in this way interactions and continuous activities are influenced; With the help of the diversity of information that exists on the web, students keep up to date, achieve cooperative learning, develop skills that allow them to improve their learning processes and their academic performance.

2   Methodology

Within the study of M-learning as a didactic strategy for working with children diagnosed with dyslexia, the most appropriate methodology in the development of the application is ADDIE which consists of predominant phases such as data analysis, the design of the App, software development, its implementation and evaluation in children diagnosed with dyslexia between the ages of 9 and 11 years of age. This process complies with sequential phases that provide adequate use in both grammar and significant learning to deal with the disorder.

In the development of the present study, a total population of 93 participants was considered; data that are detailed in Table 1.

Table 1. Population

Object of study	Frequency	Percentage
Children diagnosed with Dyslexia	90	97%
Therapists	3	3%
Total	93	100%

Although information was gathered from three therapists working with children diagnosed with dyslexia, the data analysis was based on the results of the DST-J Test [12], in order to generate confidence in the estimation of the evaluated characteristics from the perspective of educational psychology; it allows evaluation of learning processes for the detection of dyslexia. The test in its second version mentions that it was originally called the Dyslexia Screening Test (DSJ) and that it was published in the United Kingdom in 1996. The main objective of the test was to discover those children who were at risk of failure in reading so that they could receive extra support at school.

When proposing this test in the present study, it is intended to detect the difficulties children present in reading and writing. Also, its application was considered because it is a public test and does not need the author's permission; it involves three tests of direct evaluation: reading, copying and dictation, as well as nine tests of indirect evaluation: names, coordination, postural stability, phonetic segmentation, inverse digits, meaningless reading, verbal and semantic fluency and vocabulary. The performance of this test includes an assessment, which indicates the existence of the risk of dyslexia in the child, as well as the extent of this risk, which may be mild, moderate or high. Likewise, the DST-J provides information on the strengths and weaknesses in the child's performance in the different tasks assigned.

2.1 Results

Of the twelve tests, the direct assessment tests (reading, copying and dictation) were considered as they relate directly to the reading and writing process. Table 2 shows the scores obtained:

Table 2. Dyslexia screening test

Items	Alternatives				
	Strong point	Absence of risk	Slight risk	Moderate risk	High risk
(3) Reading (R)	1	10	5	10	64
(5) Dictation (D)	2	8	5	0	75
(8) Copying (C)	2	18	12	0	58

After processing these data, they are next observed graphically (Figs. 1, 2 and 3):

Table 3. Reading, dictation and copying results

Alternatives	Kolmogorov-Smirnov[b]			Shapiro-Wilk		
	Stadistical	gl	Sig.	Stadistical	gl	Sig.
Reading						
Absence of risk	0.096	10	.200*	0.97	10	0.892
Slight risk	0.136	5	.200*	0.987	5	0.967
Moderate risk	0.096	10	.200*	0.97	10	0.892
High risk	0.063	64	.200*	0.955	64	0.021
Dictation						
Strong point	0.26	2				
Absence of risk	0.105	8	.200*	0.975	8	0.933
Slight risk	0.136	5	.200*	0.987	5	0.967
High risk	0.062	75	.200*	0.955	75	0.009
Copying						
Strong point	0.26	2				
Absence of risk	0.078	18	.200*	0.961	18	0.629
Slight risk	0.089	12	.200*	0.967	12	0.876
High risk	0.064	58	.200*	0.955	58	0.032

[b]Shows the normality and the selection of the statistician, in this case Kolmogorov-Smirnov for the representation of the level of significance.
*Correction of significance of alternatives less than 0.05.

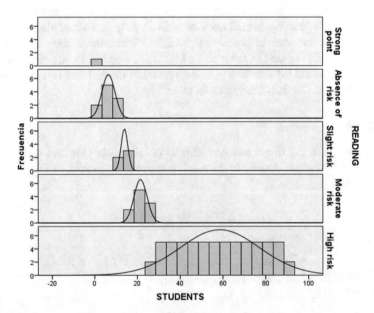

Fig. 1. Reading test results with the alternatives: strong point, absence of risk, slight risk, moderate risk and high risk. Population of 90 children.

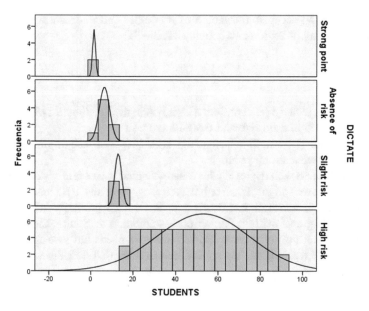

Fig. 2. Dictation test results with the alternatives: strong point, absence of risk, slight risk and high risk. Population of 90 children.

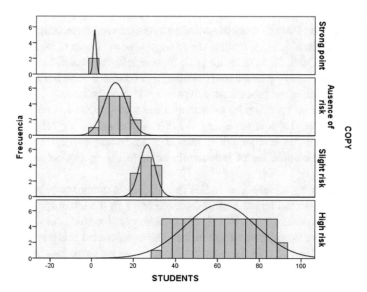

Fig. 3. Copying test results with the alternatives: strong point, absence of risk, slight risk and high risk. Population of 90 children.

Upon developing the distribution analysis of the three items, in which the behavior of the data can be measured, it is concluded that the data is distributed symmetrically taking into consideration the scales within the reading, dictation and copying tests. That is, the data meet an acceptable standard of normality and its frequency is notable at high risk in reading, dictation and copying (Table 3).

3 Conclusions

58.5% of children with dyslexia present a high risk in the Reading test (R), 21% present moderate risk, 14% in mild risk and 6.50% have no risk.

In the Dictation (D) test, 53% have a high risk, 26% have a slight risk, 19% have no risk, while 2% have a strong point.

61.5% of children with dyslexia who were evaluated have a high risk in the Copying test (C), 26.5% have a slight risk, 2% have a strong point and 10% have no risk.

These results demonstrate that dyslexia presents itself primarily in activities associated with reading and writing. Furthermore, traditional teaching systems where students with dyslexia do not receive support for their needs but instead are forced to adapt to inflexible teaching methods, may lead to increased difficulties in the reading and writing process.

4 Discussion

M-learning facilitates the construction of knowledge, the resolution of learning problems and the development of skills [8]; mobile devices within the educational field stand out when used in the teaching-learning process, where students construct their knowledge interactively [11]; mobile technology promotes accessibility to education and motivates individualized learning [13]; mobile learning is a didactic strategy that contributes to the teaching of students, works in accordance with the requirements of students, and enriches the process in a dynamic and innovative way [14].

[15] Mobile learning provides autonomy thanks to portability and improves academic performance; [16] it is necessary to create a plan describing school activities to be undertaken and the technological resources to be employed; [17] when using technology in the completion of homework, students are motivated and spend more time studying.

[18] The planning of activities builds meaningful learning in students, being that the use of a didactic and technological resource helps in the learning process.

Educational psychology is a branch of psychology responsible for studying human learning specifically in educational institutions, through detailed analysis of how to learn and teach in order to find effectiveness in the processes performed by students [19].

References

1. Esbrí, J.S.: Herramientas informáticas para la orientación y el asesoramiento vocacional. Scielo, pp. 201–232, 5 de 11 de 2013
2. Colomer, C.: Aplicación de m-learning para alumnado con capacidad intelectual límite en el área de matemáticas en primaria. Educar, 15–222015 (2015)

3. Martínez, M.T., Briones, S.M.: Contigo en la distancia: la práctica tutorial en entornos formativos virtuales. Pixel-Bit. Revista de Medios y Educación, pp. 81–86, 26 de Marzo de 2014
4. Mastromatteo, E.: Tecnologías de la Información y la Comunicación. Glossarium-BITri (2014)
5. Morales, E., Morales, X., Ocaña, J.: Desarrollo de un espacio virtual en Kichwa para el aprendizaje de ofimática con software libre. Universidad Técnica de Ambato 4(11), 8 (2017)
6. Bautista Sánchez, M.G., Martínez Moreno, A.R., Hiracheta Torres, R.: El uso de material didáctico y las tecnologías de información y comunicación (TIC's) para mejorar el alcance académico. Ciencia y Tecnologia - Facultad de Ingeniería Mecánica y Eléctrica de la Universidad Autónoma de Nuevo León, pp. 183–194 (2014)
7. Galvis, Y., Barona, J., Cardona, J.: Prevalencia de dislipidemias en una institución prestadora de servicios de salud de Medellín. Universidad de Antioquia. Universidad de Antioquia, p. 11, 9 de Marzo de 2013
8. Moreira, C., Delgadillo, B.: La virtualidad en los procesos educativos: reflexiones teóricas sobre su implementación. Dialnet. Tecnología en Marcha, 28(1), 121–129 (2015)
9. Meneces, N.: Ventajas y desventajas de la educación virtual y presencial. Universidad del Valle, p. 6, 6 de Abril de 2017
10. Hernández, S., Moreno, J.: El M-Learning como recurso fundamental en la asignatura de Fundamentos de Metodología de la Investigación. Universidad Autónoma del Estado de Hidalgo, Instituto de Ciencias Básicas e Ingeniería, Área Académica de Ingeniería, Pachuca, Hidalgo. México, p. 18, 15 de Septiembre de 2015
11. Vidal, M., Gavilondo, X., Rodríguez, A., Cuéllar, A.: Aprendizaje móvil. Moving Learning. Escuela Nacional de Salud Pública 29(3), 11 (2015)
12. Fawcett, R., Fernández, S., Santamaría, P.: DST-J. Test para la Detección de la Dislexia en Niños. Tea ediciones. Editoral Médica Panamericana, p. 29 (2016)
13. Rivero Cárdenas, I.G.: Tecnologías educativas y estrategias didácticas: criterios de selección. Revista Educación y Tecnología (3), 190–206 (2013)
14. Guevara, O.: Servicio de Innovación y Apoyo Téc. a la docencia y a la investigación. Obtenido de M-Learning: características y ventajas. UNIVERSITAS, 26 de Abril de 2016
15. Pediguer, E.: ¿Qué es el M-learning? Obtenido de Sumando Historias, 11 de Noviembre de 2015. http://www.sumandohistorias.com/reportajes/que-es-el-m-learning/
16. Avello Martínez, R., Duart, J.: Nuevas tendencias de aprendizaje colaborativo en e-learning. Claves para su implementación efectiva. Estudios Pedagógicos, XLII 1, 271–282 (2016)
17. Fernández, I.: Las TICS en el ámbito educativo. Obtenido de EDUCREA (2014). https://educrea.cl/las-tics-en-el-ambito-educativo/
18. González, P.: Guioteca. Obtenido de Beneficios de usar TIC en la educación, 15 de Marzo de 2013. https://www.guioteca.com/educacion-para-ninos/beneficios-de-usar-tic%E2%80%80 99s-en-la-educacion/
19. Regader, B.: Psicología educativa: definición, conceptos y teorías. Obtenido de Psicología y Mente (2014). https://psicologiaymente.net/desarrollo/psicologia-educativa#!

A Comparative Study on Maturity Models for Information Systems in Higher Education Institutions

João Vidal Carvalho[1]([⊠]), Rui Humberto Pereira[1], and Álvaro Rocha[2]

[1] Polytechnic of Porto/CEOS.PP - R. Dr. Jaime Lopes de Amorim,
4465-004 São Mamede de Infesta, Portugal
{cajvidal, rhp}@iscap.ipp.pt
[2] Department of Informatics Engineering, University of Coimbra,
Pinhal de Marrocos, Coimbra, Portugal
amrocha@dei.uc.pt

Abstract. In the last decades the Higher Education Institutions (HEI) have been faced with new challenges, requiring profound changes in their internal and external processes. The HEI have handled this new reality by means of the dematerialization of those processes. Thus, capable Information Systems (IS) are required to support such complex processes. Another aspect is related with the lack of standardization in the HEI' academic management processes, each HEI works according to its own internal regulations, putting a strong barrier for the adoption of standard packages of software, as happens in enterprises with the ERP. These two factors combined, turns the work of IT managers very difficult in terms of management and knowing how really the IS is capable of support those HEI's processes.

In the education sector, the Maturity Models (MM) have been used to evaluate HEI in several dimensions, such as ICT, management, process management, course curricula, course/HEI accreditation, e/m-learning, online courses and pedagogical strategies. Based on the guidelines of a methodology for a systematic literature review, the MM of different subareas of education were identified and categorized in a previous paper. In the present paper, we perform a comparative study of those MM previously identified in the scope of our research, the HEI.

Keywords: Stages of growth · Maturity models · Higher Education Institutions
Education information systems · Management

1 Introduction

In the last decades, we have seen many changes in the HEI. In the past, higher education was much restricted in terms of capacity of students. The massification of this kind of education, puts HEI under enormous pressure for providing the required and capable means. In addition, the new teaching paradigms (Bologna Treaty), the new perspective of the student as a "costumer", the opening of universities to enterprises, in terms of knowledge transference, the HEI rankings and HEI competition, forced the

© Springer Nature Switzerland AG 2019
T. Antipova and A. Rocha (Eds.): DSIC 2018, AISC 850, pp. 150–158, 2019.
https://doi.org/10.1007/978-3-030-02351-5_19

HEI to reinvent higher education and adopt agile management methodologies, in order to be capable of adapting to the constant environment changes.

The Information Systems (IS) are the primarily weapon that HEI have used, performing the desmaterialization of processes that involve entire internal academic community (students, teachers, staff and managers), as well as other external institutions in the educational and others sectors. Todays HEI have Information Ecosystems (IE) composed by a large spectrum of platforms, such as: Academic Management ERP, Financial ERP, Student Relationship Management (SRM), Learning Management System (LMS), Content Management Systems (CMS), Survey tools, Business Intelligence (BI), Current Research Information System (CRIS) and repository of publications, among many others. Due to this scenario, HEI must have an integrated vision of all these individual platforms as a unique information system capable of supporting their transversal organizational processes.

Another aspect is related with the lack of standardization in the HEI' academic management processes, each HEI works according to its own internal regulations, putting a strong barrier for the adoption of standard packages of software, as happens in the enterprises with the ERP. Fortunately, nowadays in some areas some commercial and open source products are a *de facto standard*, the Moodle (https://moodle.org) is a good example. Some initiatives for HEI interoperability [1] and online services such as ORCID make visible the will to change this scenario, by the need of unified processes among HEI.

Managing such complex ecosystems of platforms and processes, requires powerful tools to evaluate and guide HEI in terms of capability to support these organizational processes and high level of IS integration. Thus, we started our research work in the scope of MM of those Information Ecosystems.

In a previous work [2], a literature review was done, in order to find any eventual gap on the existing models of maturity. Aiming to conduct a comprehensive and wide literature review, it was necessary to define a strategy [3] in order to identify and analyse systematically the available literature on MM of education IST. This literature review carried out [2], was based on the strategies of Webster and Watson [4] and Tranfield et al. [5]. Due to space restrictions, here we cannot discuss the methodology that we applied. In [2] we detailed present that literature review methodology.

After following that systematic methodology of literature review, we consider a few set of research works that we consider as related with ours. Giving continuity to this previous work, in this paper we perform a comparative study of those MM that were identified.

In the next section, is presented a brief overview of the MM in IS area (second section). Then, in the third section we discuss each one of the identified MM, performing a comparative approach. We finish this paper presenting a summary and the closing remarks.

2   Maturity Models in IST Management

The MM are available to respond to many different challenges. These models provide information for organizations to address the problems and challenges in a structured way, providing both a reference point to assess the capabilities as a roadmap for improving [6]. In other words, the MM offer an orientation through an evolutionary process, incorporating the procedures for improving activities [7].

Various MM have been proposed over time, both for the development of individuals and for the general evolution of organizations or the particular evolution of the IS management function. These models mainly differ in terms of a number of stages, variables of evolution and focus areas [7–9]. Each of these models identifies certain characteristics that specifically define the objectives of the next stage of growth. These types of models can be applied situationally within education in order to strategically planning for IST maturation, based on the degree of alignment between the educational organization (e.g. HEI) strategy and the selected growth path, as well as associated investments and improvement activities.

3   Results and Comparison

As a result of the previous work of literature review, it was found that the MM for education IST are developed involving different types of entities, including national and international education companies, research organizations as well as academic experts in this domain.

It was also found that there are two approaches: in one hand, the highly specialized models that have focused in one education subsystem and on the other hand, the more comprehensive models, i.e. models representing the educational institution IS as a whole. Also, it was found that most of the analysed MM does not disclose the design process nor the research options for development and validation, thus compromising the researcher work.

Within the HEI IST domain, which is the main focus of our research, several MM have been proposed, although these models are still at an early stage of development. These models have an important focus on the management of IS of an educational institution, either in a global perspective or by defining one of its dimensions. Thus, in our literature review we considered five MM which are described below. Regarding the number of maturity stages, there are models from 3 stages as the case of eQETIC [10] up to 8 stages (ICTMMEI-DV [11]). It is precisely these two that are not based on the CMM [12], while the remaining three have this model as reference.

They all suggest attributes that the organization should possess to be positioned at each stage. However, most IST MM do not explicitly identify any assessment tool. Only the ICTE-MM provide a tool to assess the fulfilment of requirements, to effectively place an organization in a certain level.

Next, a brief description will be made about each maturity models, whereas in Table 1, synthesizes strengths and weaknesses of all these maturity models, and identifies some gaps in implementation at an HEI.

Maturity Model for ICT in School Education (ICTE-MM) [13]: The ICTE-MM has three elements supporting educational processes: information criteria, ICT resources, and leverage domains. Changing the traditional and exclusive focus on ICT, five leverage domains are defined: Infrastructure, Educational Management, Administrators, Teachers and Students.

Despite its large spectrum of coverage, this MM does not explicitly considers issues such as business process definition/documentation and IS capability for supporting such processes. This is a MM based on international standards for assessing the school's development regarding to the use of ICT and not a MM for accessing the IS capability for supporting the school's management and teaching/learning processes. Issues like software for academic management, financial management and teaching/learning process management are superficially approached, applying only three variables, as well as in School Management, by means of six Critical Variables, none of them covering the business process definition. We consider that there are other missing Critical Variables which are fundamental to achieve a more comprehensive MM for accessing the use of ICT in schools. Additionally, this is a generic MM for school educational processes not focused in HEI.

Capability Maturity Model for Quality Education (CMM-QE) [14]: CMM-QE is a framework for quality education assessment and process improvement with five maturity levels. The CMM-QE evaluates the Education system engineering process from the multi perspectives of academic, infrastructure, administration, facilities etc. CMM-QE use critical factors (Key Indicators) to be quantified to assess the maturity level of the Educational institutions.

Despite those authors' goals, the proposed MM is not clearly presented. Although a number of variables grouped in four measurement models covering several areas of the educational institution is referred, none of them have a concise and systematic description. Only an apparently unordered and unrelated list of characteristics is presented. In our point of view, this missing systematization of the assessed attributes compromises the reader's full understanding regarding the proposed authors' framework, as well as its applicability in the real world practice. This lack in the description of the CMM-QE, is not compensated with any previous work presenting the authors' framework. As far as we know, this model was not adopted in subsequent studies and the academic community has not significantly referenced it.

Online Course Quality Maturity Model Based on Evening University and Correspondence Education (OCQMM) [15]: This model, proposes to assess the quality of online courses in Evening University and Correspondence Education. OCQMM can guide the institutions that engaged in adult education to meliorate the implementary process, so that the implementation quality of online course will be improved. OCQMM divided online courses quality maturity in evening university and correspondence education into four maturity evolving ladder levels, each low-level is a basis that achieve a higher level.

We consider the proposed MM sufficiently comprehensive in terms of key areas, addressing relevant quality issues of online courses. However, there are important missing issues such as teacher motivation and pedagogical practices that are not considered. More important, we also consider insufficient the level of detail in which the six key process areas are described, making very difficult to replicate the authors' experience of testing the proposed MM in other institutions. This limitation is not mitigated by means of any other previous publication where the authors sufficiently present their model. Regarding the model systematization, the authors do not provide any methodology or analytic methods for determining the school's maturity level in

Table 1. Strengths and weaknesses of maturity models for HEIs' ISTs

MM	Strengths	Weaknesses	Gaps in HEIs
ICTE-MM	• Involves the entire academic community (in terms of vision, digital literacy and efficiency of use of ICT) • Based on International standards • Adopts a web support tool to perform automatic maturity evaluation • Based on a known and established reference model such as CMMI • Suggests critical variables (and weights) for a key domain to be positioned at a specific maturity level	• Considers neither the business process definition nor documentation in any of the dimensions or variables • Does not adopt a known methodology for the development of the maturity model	• Focused on the use of ICT, not access to IST capability in supporting the institution's business process • Generic maturity model for school educational processes, thus not focused on HEIs
CMM-QE	• Presents a methodology for modelling the relationships between latent variables • The model uses critical factors (key indicators) to be quantified in order to assess the maturity level of educational institutions • Based on a known and established reference model such as CMM	• The dimensions and variables are not cleared presented • Does not provide an automatic maturity evaluation tool • Does not adopt a known methodology for the development of the maturity model • Difficult to replicate in other institutions	• Generic maturity model for school educational processes, thus not focused on HEIs
OCQMM	• In particular, the context of online courses and correspondence education is comprehensive in terms of key areas • Based on a known and established reference model such as CMM	• Insufficient level of detail in terms of how the six key process areas are discussed • Does not provide any methodology or analytical approaches for determining the maturity of courses or the school	• Despite being a model for course quality, it does not consider facilities for students, administrative support or other specific IS aspects of HEIs in their successful pursuit of their mission and duties

(continued)

Table 1. (*continued*)

MM	Strengths	Weaknesses	Gaps in HEIs
	• Structured across multiple dimensions, which aim to cover all processes in the academic arena	• Does not provide an automatic maturity evaluation tool • Does not adopt a known methodology for the development of the maturity model • The model has not yet been fully validated	
ICTMMEI-DV	• Based on International standards • The model is intended to be prescriptive, advocating best practice in ICT infrastructure development in an education institution • Improvement is obtained in a staged and progressive way • Suggests attributes for an entity to be positioned at a specific maturity level	• Not adapted for developed countries • Does not present process areas and maturity practices that encompass the various entities • Does not provide an automatic maturity evaluation tool • Does not adopt a known methodology for the development of the maturity model • The model has not yet been fully validated	• Strictly focused on ICT • Issues related to management process definition and other relevant aspects of ISs are not considered • Despite HEIs being the subject of this model, it is aligned with primary educational levels and not well suited for HEIs
eQETIC	• Based on international standards, as well as other research works and frameworks issued by associations and governments • In particular, the context of online education is comprehensive in terms of key areas • The model structure is defined, as well as its mode of application • Considers the principles of continuous process improvement • Suggests implementation rules for an entity to be positioned in a specific maturity level	• Does not provide an automatic maturity evaluation tool • Does not adopt a known methodology for the development of the maturity model • The model has not yet been fully validated, as it has only been developed in the course of exploratory research	• Despite being a model for digital educational solutions, it does not consider blended learning and traditional face-to-face teaching, nor facilities for students, administrative support or other specific IS aspects of HEIs in successful pursuit of their mission and duties

each key process area. Either they provide a way for determining the maturity level of an online course or the school as a whole.

Maturity Model for ICT in Educational Institutions in Developing Countries (ICTMMEI-DV) [11]: This proposal aims to provide guidance for ICT infrastructure planning and to create a reference model to the necessary development phases for the efficient use of these resources. The model defines the ICT infrastructure resource levels required to achieve primary organizational objectives expressed in the form of student learning outcomes. The levels in this model show management, teaching and technical staff, as well as donors how to make most efficient use of ICT resources by maximizing opportunities for student learning.

Despite the lack of discussion regarding those three important levels, we can conclude that this MM is strictly focused on ICT. Issues related to management process definition and other relevant aspects of IS are not considered in this model. This MM was specially designed for education institutions of developing countries, in which the resources are very limited. Such context, is very different from the ones that exist in developed countries, making this MM not well suited for institutions in these countries. Additionally, this model intends to cover a broad type of educational levels, which have distinct educational goals. In our opinion, the author's proposal is aligned with primary educational levels, and not well suited for HEI.

eQETIC: A Maturity Model for Online Education [10]: This is a model capable of supporting steps that guide the planning, development, and maintenance of digital educational solutions. eQETIC model follows a continuous process improvement approach, whereas the implementation of processes in a developer organization of these types of solutions favours the development lifecycle and the quality of these solutions. The model allows the organization to implement the processes belonging to each level at a given time, and these levels and processes are organized in six common entities.

This model is focused on the quality of the product development process, including the learning process, the environment and aspects that condition the success of the education institution in terms of quality of the specific scope of solutions (distance education, e-learning and learning objects). Despite being a comprehensive model on such type of solutions, it does not consider other types of teaching such as blended learning and traditional face-to-face teaching, as well as facilities for students, administrative support, or other specific IS aspects of HEI in their full achievement of mission and duties.

4   Summary and Closing Remarks

In this paper we continue the exploratory phase of our work in order to identify and classify the existing MM focused on the evaluation of the IS capability to support the HEI's processes. As starting point, we consider the MM identified in a previous work based on a systematic literature review. You should notice that these MM are a sub set of those MM, which are focused on the ICT and IS dimensions of HEI.

Most of these studied models are still in an early development stage and in a premature phase of affirmation and consolidation, being proposed by their authors through exploratory studies. In fact, few of the identified models are adopted in a large scale, nor are significantly referenced by the academic community. Additionally, most of these models are not sufficiently explicit in the way they were developed and validated, and especially because they are poorly detailed, they do not provide tools to determine the maturity stage nor structure the characteristics of maturity stages. In the case of the adoption of a tool for assessing the system's maturity, it was found that most of the models, besides focusing on the assessment of the system's maturity, pay attention to an improvement path of such maturity. However, not all have a properly systematized process to move to a higher maturity level. Also, the authors did not apply weights to each of the influencing factors (or dimensions), that is, in the assessing process of the overall maturity of education IST, all influencing factors have the same importance. Based on the analysis, it was possible to verify that no model was developed based on the guidelines of the development methodologies of MM [16–18].

Additionally, these studied MM are very focused in technology itself and not in its capability of supporting todays HEI (as well as other types of Educational Institutions), in their challenges: dynamic and agile management, new teaching strategies, flexible formative portfolio and knowledge management. We are convicted that all the remain dimensions in HEI can benefit if their Information Ecosystem stays optimized.

As a result of this study, as far as we know, none of the identified models has a sufficiently focused on the capability of the IS support complex, diversified, interoperable and dynamic organizational processes of HEI. In this perspective, a new model to fill the gap should be designed. This new model, should include the main influence factors with different weights depending on their relative importance and its development should be supported by rigorous scientific methods of conceptualization and validation. This model should also identify the IST key strategical areas of HEI and apply international standards for management of IST and HEI as well. Such MM will enable the evaluation of the HEI, in terms of their practices and strategies of IST, for supporting their institutional processes at all levels: organisational strategies, management, operative management, teaching and research. Thus, it empowers the capabilities of the HEI and its human capital (administrators, staff, teachers and students).

References

1. Ribeiro, L.M., et al.: Interoperability between information systems of Portuguese higher education institutions. In: EUNIS 22nd Annual Congress Book of Proceedings (2016)
2. Carvalho, J.V., Pereira, R.H., Rocha, A.: Maturity models of education information systems and technologies: a systematic literature review. In: 2018 13th Iberian Conference on Information Systems and Technologies (CISTI). IEEE (2018)
3. Carvalho, J.V., Rocha, Á., Abreu, A.: Maturity models of healthcare information systems and technologies: a literature review. J. Med. Syst. **40**(6), 1–10 (2016)
4. Webster, J., Watson, R.T.: Analyzing the past to prepare for the future: writing a literature review. MIS Q. **26**(2), 13–23 (2002)
5. Tranfield, D., Denyer, D., Smart, P.: Towards a methodology for de-veloping evidence-informed management knowledge by means of systematic review. Br. J. Manag. **14**, 207–222 (2003)

6. Caralli, R., Knight, M.: Maturity Models 101: A Primer for Applying Maturity Models to Smart Grid Security, Resilience, and Interoperability. Software Engineering Institute, Carnegie Mellon University, Pittsburgh (2012)
7. Mettler, T., Rohner, P.: Situational Maturity Models as Instrumental Artifacts for Organizational Design. In: DESRIST 2009, Malvern, PA, USA (2009)
8. Rocha, Á.: Evolution of information systems and technologies maturity in healthcare. Int. J. Healthc. Inf. Syst. Inform. 6(2), 28–36 (2011)
9. Carvalho, J.V., Rocha, Á., Abreu, A.: Main influence factors for maturity of hospital information systems. In: Information Systems and Technologies (CISTI), Gran Canaria, España, vol. 1, pp 1059–1064 (2016)
10. Rossi, R., Mustaro, P.N.: eQETIC: a maturity model for online education. Interdiscip. J. e-Skills Lifelong Learn. 11, 11–24 (2015)
11. Bass, J.M.: An early-stage ICT maturity model derived from Ethiopian education institutions. Int. J. Educ. Dev. Inf. Commun. Technol. 7(1), 5–25 (2011)
12. Paulk, M., et al.: Capability Maturity Model for Software Version 1.1. Software Engineering Institute, Carnegie Mellon University, CMU/SEI-93-TR-024 (1993)
13. Solar, M., Sabattin, J., Parada, V.: A maturity model for assessing the use of ICT in school education. J. Educ. Technol. Soc. 16(1), 206–218 (2013)
14. Manjula, R., Vaideeswaran, J.: A new CMM-Quality Education (CMM-QE) framework using SEI-CMM approach and calibrating for its process quality and maturity using structural equation modeling-PLS approach. Int. J. Softw. Eng. Appl. 6(4), 117–130 (2012)
15. Gu, D., Chen, J., Pu, W.: Online course quality maturity model based on evening university and correspondence education (OCQMM). In: 2011 IEEE 3rd International Conference on Communication Software and Networks (ICCSN), pp. 5–9. IEEE (2011)
16. de Bruin, T., et al.: Understanding the main phases of developing a maturity assessment model. In: 16th Australasian Conference on Information Systems (ACIS 2005) (2005)
17. Becker, J., Knackstedt, R., Pöppelbuß, J.: Developing maturity models for IT management—a procedure model and its application. Bus. Inf. Syst. Eng. 1(3), 213–222 (2009)
18. Mettler, T.: Supply Management im Krankenhaus: Konstruktion und Evaluation eines konfigurierbaren Reifegradmodells zur zielgerichteten Gestaltung. Ph.D. thesis. Institute of Information Management. St. Gallen, University of St. Gallen (2010)

Automatic Test Generation on the Basis of a Semantic Network

Elena Dolgova, E. V. Eriskina[✉], Rustam Faizrakhmanov,
E. A. Kasyanova, D. S. Kurushin, N. M. Nesterova, and O. V. Soboleva

State National Research Politechnical University,
Komsomolsky Av 29, 614000 Perm, Russia
eriskina.katena@mail.ru

Abstract. The paper is devoted to the problem of automatic generating of tests based on the natural language texts. The methods of generating natural language questions are considered. The domain model presented in the form of a denotatum graph and the experiments verifying the model are described. The program algorithm for generating tests based on the texts in natural language and pseudocode are given. The analysis of the results of this program and the main shortcomings are demonstrated. The algorithm of the program "finalization" using n-grams and the search of associative connections and its pseudocode are shown. The test generated as a result of the algorithm usage is presented.

Keywords: Test generation · Automatic test generation · Natural-language text
Domain model · Denotative graph · Generation of test templates
Trees AND/OR · n-grams · Associative links · Python · Pymystem
Rutermextract

1 Introduction

Automatic test generation is one of the areas of applied computational linguistics, which is developing quite rapidly. In addition to the tasks related to the simplification of the educational process, the applications developed for automatic generating of tests based on natural language texts are oriented to the problems of processing of such texts and carrying out its complete analysis. The difficulties that arise in the development of such analysis are caused by the fact that in practice not all the formulated requirements are realized. The complexity of parsing and the complexity of creating an expert knowledge base, reflecting a full scale model of the real world are the main problems in automatic text processing [1].

The idea of creating applications for automatic generation of tests based on natural language arose in the course of the work on the project aimed to the development of the model of the text denotative analysis. The result of this analysis must be so-called denotatum graph reflecting the structure of the text content. The process of constructing such a graph includes extraction of language units from the text, referring to denotata and their relations. At the stage of this model verification it was necessary to check the correctness of information in the standard denotatum graph. The problem of adequate representation of a domain model in the form of denotatum graph is that the main process of the graph formation is the understanding of the basic content [2].

© Springer Nature Switzerland AG 2019
T. Antipova and A. Rocha (Eds.): DSIC 2018, AISC 850, pp. 159–165, 2019.
https://doi.org/10.1007/978-3-030-02351-5_20

1.1   Verification of Domain Model

The problem of adequacy can be resolved if we can prove the unambiguity and the cohesion of the domain model, presented in the form of a denotatum graph. In order to verify the model against these criteria, on the basis of denotatum graph, you can use testing methods similar to the approach used in educational programs. It was suggested to the selected working group to answer the test and the answers to the questions of the text had to be based only on the data presented in the graph, not based on our own experience and knowledge. Then, the same test was solved automatically on the basis of denotative model. In result it was found out that if a test question is formulated correctly and the description of the domain model corresponds to the test, an intelligent system builds a chain of reasoning that is quite similar to the logical reasoning of a man. Moreover, the system answers correlate with the answers of the control group.

The final stage of verification of a domain model is to create a program which will generate a test based on the analysis of a natural language text. This method of checking the model adequacy will help to define how semantically correctly built the relationships between denotata and whether it is possible to generate the correct person-oriented statements on its basis. The result can also be active dialogue system that can not only answer questions but ask them as well.

1.2   Methods of Automatic Test Generation

The attempt to solve the problem of test generation based on natural language texts is not new. There are many methods for generating test questions. For example, Tar-asenko [3] gives a comparative analysis of the main methods of generating test questions. These methods are the following: (1) on the basis of trees AND/OR templates, (2) by rebuilding sentences and (3) on the basis of a text corpus. As a result of simple comparison of methods according to a number of criteria, it was concluded that for the compilation of complex questions it is advisable to use the method of generating tree-based AND/OR. This method is good because it does not depend on the form of presentation and the volume of generated text and gives you the opportunity to generate questions for almost any part of the sentence. In order to diversify the forms of questions a method based on trees AND/OR needs to be combined with the method of using templates, which will increase the final number of questions. Several methods of generating test questions are given in the article by Balashova [4]. One of those ways is the generation based on formal grammars, the essence of which is to select a random output in the grammar and use it to generate mathematical tests and tests of general intelligence. Another way is based on the generation of semantic networks, which have become the basis for the program described above, in the experiment on verification of a domain model and for creation of a test generation program.

2   The Denotatum Graph as a Domain Model

The denotatum graph reflecting the hierarchical model of denotata and their relations, which corresponds with the model of a fragment of the real situation, was used as the domain model. The method of constructing such a graph was developed by Novikov [5]. The domain model based on denotatum graph is presented by the "denotatum pairs" – the "denotatum-relation-denotatum" chains. And the attribute "weight of connection" characterizing the "importance" of connections is used in this model [6]. The authors have built a test graph of domain model "Artificial intelligence", based on the domain model thesaurus and literature on the chosen topic, a fragment of which is shown on Fig. 1.

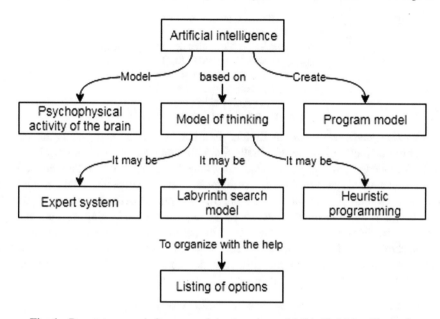

Fig. 1. Denotatum graph fragment of the domain model "Artificial intelligence".

The model includes 40 denotatum pairs. Of course, this number of terms is not enough to describe the chosen domain model, but it is enough for testing a software model. "Understanding" of the domain model includes not only revealing of its basic terms and searching for their equivalents in thesaurus, but also revealing and fixing the connections between these concepts, and applying these relationships to build the "denotatum-relation-denotatum" chains [7].

2.1 The Algorithm for Generating Test Programs

The program of natural-language text processing and generating the text on the basis of test questions will help to test the adequacy of the constructed graph. The algorithm of this program is described by a set of functions. These functions describe how to load the program, which is written in advance on the basis of the reference/standard graph of the knowledge base, to generate test questions, to generate answers and to output a structured test. The basic functions of the program are the generation of a question and an answer. Function generate_question searches in the knowledge model for a match with Eq. (1):

$$T = \left(W_1^q, \ldots, W_n^q, D^1 \ldots, D^m\right) \tag{1}$$

where W_i^q—question words, D^j—denotata that can participate in the question, T—a tuple of values, i.e. the question must begin with the word W_1^q.

The function extracts the first denotatum, then moves the graph up to the level above and extracts the "root" denotatum for an already extracted one and then puts them in the right place in a pre-defined structure. The pseudocode of question generation function is presented in Program Listing 1.

```
def generate_question(den, relations, template):
        denotats = [den]
            for den1, rel, den2 in kb.keys():
                if  den2 == den and
                rel in relations and
                kb[(den1, rel, den2)]:
                    denotats.add(den1
            return tokenize(denotats, template)
```

Function generate_answers is divided into two parts. The first part is the correct answer to the question. To do this, firstly the set of correct answers, that must be found in the knowledge base, is given. Then searching for a match in the knowledge base with the generated question in the first denotatum and relationship is taking place. If more mathes than one are found, only the first one remains and the rest are removed. The second part of the function describes the algorithm of searching for the "wrong" answers. The difference between the second part of the function and the first part is that the matching in the knowledge base occurs at the match of relations between denotata and the mismatch between the initial (first) denotatum. The pseudo code of generation function is shown in Program Listing 2.

```
def generate_answers(den1, rel, den2, n_incorrect,
n_correct):
            var = 'a', res = {}, keys_to_delete = []
            for i in range(n_correct):
              correct = []
              for d1, r, d2 in kb.keys():
                if d1 == den1 and r == rel and kb[(d1, r,
d2)]:
                    correct += [d2], keys_to_delete += [(d1,
r, d2)]
            res.update({var:{"is_correct":1, "text":
correct}})
            var = chr(ord(var)+1)

            incorrect = [], need_to_generate =
            for d1, r, d2 in kb.keys():
              if d1 != den1 and r == rel and kb[(d1, r,
d2)]:
                    incorrect += [d2]
                    if randint(2,4):
                      res.update({var:{"is_correct":0, "text":
incorrect}})
                        var = chr(ord(var)+1)
                        incorrect = []
                        need_to_generate -= 1

            del(keys_to_delete)
            return res
```

Loading the knowledge base into the program and output of structured text are performed by standard means of the language Python [8] and Json [9] respectively

The output filtering based on n-grams is implemented to eliminate lexical and grammatical inaccuracies (the names of denotata, not grammatically corrected, are displayed?). The denotata that make up the question (interrogative words are also considered as denotata) of the generated question are searched for in the model. The comparison allows to find all the duplicates and those that meet the requirements of answer generating are included in the output. The search algorithm for n-grams is presented in pseudocode in Program Listing 3.

```
def ngramm_sentence(tokens, model):
    t0, t1 = sentence_beak, used = [], phrase =
'', i = 0
    while True:
        t0, t1 = t1, find_pair(model, t0, t1,
tokens[i], used)
        i += 1
        if t1 == end_phrase or not t1: break
        if t1 in delimiters or t0 == end_phrase:
          phrase += t1
        else:
          phrase += space + t1
        used += [t1]
    return phrase.
```

For correct operation a well-trained model of n-grams is required. This will not only increase the number of generated questions, but also give a resource for generating linguistically correct sentences that, in fact, is the purpose of the created program. For this we have developed a module, which according to a request based on a specific concept, traverses recursively encyclopedia (for example wiki) and generates the corpus, in which it puts all the texts found and associated with the entered word. The module operation is organized with the help of libraries Pymystem3 [10] and Rutrmextract [11].

3   The Results of Testing of the Source Code

Having realized all the described above functions, we launch control testing program [12]. The output is automatically generated, structured test consisting of questions drawn up based on the model of the domain model "Artificial intelligence". Due to the using of search according to trained n-grams, the questions in this text represent a linguistically correct sentences (see Table 1). An example of the test generated by software model, see in Applications.

As you can see, this model successfully performed a semantically correct transformation "['chto', 'takoe', 'prolog']" → "chto predstavljaet soboj prolog?"

Table 1. The tuple of denotata names and the texts that correspond with them.

№	The tuple of denotations, see (1), Russian	Question, Russian	Translation
1	['chto', 'vkljuchat'', 'dvoichnyj', 'kod']	chto vkljuchaet v sebja dvoichnyj kod?	what includes binary code in itself?
2	['chto', 'vkljuchat'', 'blok-shema']	chto vkljuchaet v sebja blok-shema?	what is included in the block diagram?
3	['chto', 'takoe', 'prolog']	chto predstavljaet soboj prolog?	what the prolog is?
4	['chto', 'vkljuchat'', 'nepreryvnyj', 'fizicheskij', 'velichina']	chto vkljuchaet v sebja nepreryvnaja fizicheskaja velichina?	what includes the continuous physical quantity?

Questions, generated by model (russian), correct answer marked with 1:
(1) chto vkljuchaet v sebja nepreryvnaja fizicheskaja velichina?
(a) signal, 1
(b) svjazannyj blok, 0
(v) dannye, 0
(g) chelovekoorientirovannyj jazyk, 0
(2) chto vkljuchaet v sebja blok-shema?
(a) svjazannye bloki, 1
(b) dannye, 0
(v) chelovekoorientirovannyj jazyk, 0
(g) estestvennyj jazyk, 0
(d) dannye, 0
(3) chto vkljuchaet v sebja signal, kak jelement dvoichnogo koda?
(a) dannye, 1
(b) chelovekoorientirovannyj jazyk, 0
(v) estestvennyj jazyk, 0
(4) chto vkljuchaet v sebja psevdokod?
(a) chelovekoorientirovannyj jazyk, 1
(b) estestvennyj jazyk, 0
(v) dannye, 0
(g) signal, 0
Definitely, the creation of a domain model is a fundamental task for application creation. We are at the start of the denotative analysis study as a method of constructing a software model of the domain model. We have discovered that correct formulation of the problem makes possible to 'force' an intellectual system to construct linguistically correct chains (sentences). Further research in this area could lead to the creation (appearing) of a new way of constructing models of domains based on denotative analysis of a domain model.

Acknowledgments. This paper is prepared within work of the grant RFFI № 17-47-590128.

References

1. Bolshakova, E.I., Klysinski, E.S., Lande, J.V., Noskov, A.A., Peskova, O.V., Yagunova, E. V.: In: D.T.B.V.A. Galaktionov, M.: The Automatic Processing of Natural Language Texts and Computer Linguistics. MIEM (2011)
2. Nesterova, N.M.: Abstract translation: the problem of semantic closure and semantic adequacy. Bull. Chelyabinsk State Univ. **25** (2011)
3. Tarasenko, S.V., Ryazanova, N.Y.: Analysis of methods of automatic generation of questions in natural language. Electron. Sci. Tech. J.Eng. J. **12** (2015)
4. Balashova, I.Y., Volyn, K.I., Makarychev, P.P.: Methods and means of generation of test tasks from natural language texts. Models Syst. Netw. Econ. Tech. Nat. Soc. **1** (2017)
5. Novikov, A.I., Nesterova, N.M.M.: Abstract Translation of Scientific and Technical Texts. USSR Academy of Sciences, Institute of linguistics (1991)
6. Hertha, N., Kurushin, D.S., Nesterova, N.M., Soboleva, O.V.: Experimental studies denotative models of understanding in applications of automatic text summarization. Eng. J. Don **4** (2015)
7. Luger, G.F.: The Problems of Semantic Networks Natural Language Understanding and Semantic Modeling. Publishing house "Williams" (2003)
8. Programmers in Python. https://www.ibm.com/developerworks/ru/library/l-python_part_1/index.html. Accessed 05 Dec 2017
9. Javaskript.ru. https://learn.javascript.ru/json. Accessed 05 Dec 2017
10. Processing and marking up the collection of texts. Grammatical parser MYSTEM (library "pymystem3" for the Python programming language). https://vuzlit.ru/399009/obrabotka_razmetka_poluchennoy_kollektsii_tekstov_grammaticheskiy_parser_mystem_biblioteka_pymystem3_yazyka. Accessed 23 Nov 2017
11. Rutermextract 0.3. Python. https://pypi.python.org/pypi/rutermextract. Accessed 23 Nov 2017
12. Github. https://github.com/eriskina/test_generator. Accessed 05 Dec 2017

Data Literacy as a Compound Competence

Alex Young Pedersen[✉] and Francesco Caviglia

Centre for Teaching Development and Digital Media, Aarhus University,
Jens Chr. Skous Vej 4, 8000 Aarhus C, Denmark
alex@tdm.au.dk

Abstract. Data literacy can be defined as a compound competence consisting of some level of competence in statistics, data visualization and more generic competencies in problem-solving using different data. Data literacy is closely related to data science but differs in the level of competence. While data science is a specific domain for trained specialists, data literacy is suggested as a central element in education preparing all young people to become citizens in an information society. In presenting two exemplars of resources and practices that both rely on and foster the attainment of data literacy it is proposed that data literacy is best defined as a compound competence that first and foremost can be ascribed to a community of practice rather than the single individual. The definition, therefore, calls for new and further interdisciplinary collaboration that integrates different competencies and levels of skill.

Keywords: Data literacy · Data science · Education

1 Defining Data Science

The current development towards a more 'data-driven society' [17] and the proliferation of 'datafication' [16] whereby phenomena in nature and society are increasingly quantified with the goal of gaining new insights through data analysis pose a societal challenge [2]. Datafication and the growing volume of publicly available data on the internet is in part the result of our extensive use of digital technologies which is accompanied with both positive and negative consequences [19]. On the one hand datafication have the potential to help in our transition to a more sustainable future [20] but there is also the risk that data can be used against weak citizens [11] or as means of unwanted electronic evaluations and surveillance [15]. These developments open up the question of which competences are necessary for active citizenship in a society where data is central in both individual and collective decisions making. Figure 1 shows the composition of competences involved in practices of data science according to Conway [6].

The figure that is composed like a Venn-diagram also shows the compound character of data science that emerges in the overlapping between *Hacking Skills*, understood as e.g. being able to manipulate text files at the command-line, understand vectorized operations, thinking in algorithms etc., *Math & Statistics Knowledge* which includes the ability to apply appropriate math and statistics methods e.g. being familiar with descriptive statistics and regression analysis and being able to interpret the results,

T. Antipova and A. Rocha (Eds.): DSIC 2018, AISC 850, pp. 166–173, 2019.
https://doi.org/10.1007/978-3-030-02351-5_21

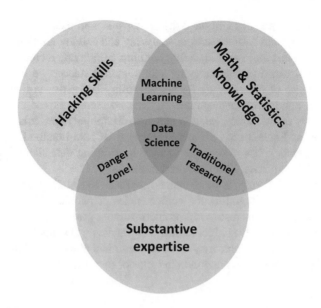

Fig. 1. Competencies for data science according to Conway [6].

and *Substantive expertise* which covers the traditional researchers ability to motivate questions and generate hypotheses. The combination of hacking skills with knowledge of math and statistics gives machine learning, but it is not data science. Data science also involves inquiry and discovery through means of posing questions about the world in such a way as to enable testing them empirically using math and statistics. This is where much traditional research takes place unfortunately without giving much attention or credence to understanding technology. The overlap between skills and expertise is the most problematic area. It is populated with people that are capable of extracting data, run a linear regression and report the coefficients without understanding what they mean and how to rightfully interpret them. They can produce seemingly legitimate reports that can be used to manipulate or willfully misled. They know enough to be dangerous. One way of counterbalancing this danger and at the same time promoting productive knowledge building is to develop and foster data literacy [6].

2 From Data Science to Data Literacy

It is important to distinguish between data science and data literacy. Data science is the domain of specialists that require extensive training and education. Data literacy on the other hand is a compound competence that involves coming to know about data science and being able to utilize that knowledge in action. Data literacy has been succinctly defined as "the ability of non-specialists to make use of data" [9]. The practices and tools of data literacy are changing at a fast pace, thereby making data literacy into one of the 'literacies of the digital" that require lifelong competence development [13]. More specific definitions of data literacy that have been used as basis for educational intervention range from focusing on skills in statistics and data visualization to more general dispositions and competences in analyzing and solving problems with the help of data. We adopt this more encompassing definition, involving a combination of skills in relation to real world problem-solving and an ethical stance (see Fig. 2) [26].

The importance of data literacy for employability and active citizenship has been suggested by a number of meta-studies [1, 18] and courses aimed at acquiring data literacy is already being implemented in new interdisciplinary curricula, for example as core subject in the Bachelor of Arts and Sciences at UCL in London [23]. Educational reform and innovation has a key role to play in bringing data literacy to the broad educated population across disciplines and curricula.

Furthermore, data literacy is more than an approach to fostering individual learning. As we discuss below, integrating the competencies of people with different areas of expertise lies at the core of real world problem-solving practices. To this purpose, we propose two cases of practice in 'community informatics' [11] aimed at expanding data literacy within a community that may serve as exemplars of real world problem-solving contexts that highlights the compound character of data literacy.

3   Cases

In the following we present two *exemplars*–that is excellent examples–of real world contexts that serves as an empirical basis for the definition of data literacy as a compound competence of a community of practice [25]. The cases also serve as possible contexts for raising the level of data literacy in the general population and for constituting spaces for knowledge building and community building.

3.1   Case 1 – How Much Money Do Italians Spent on Slot Machines?

Slot machines account for over 50% of the money spent on legal gambling in Italy [22], where regulation of gambling is a recurring theme of public debate. Communities, through local administrators or grassroots organizations, routinely try to limit the diffusion of slot-machines through local regulations or boycotting campaigns. The debate on the issue is typically based on principles of freedom and economic interest vs. public health and less frequently on concrete data about the phenomenon. Italian national statistics institutions provide aggregated data on legal gambling [22] but local details about how much money was spent on slot-machines in a given town were not available, until the national tax agency was forced to release the data to a consortium of data-journalists under the 'Freedom of Information Act' issued in 2016 [4]. After overcoming considerable technical obstacles–the data were delivered as 10,000 pages in PDF format–the consortium eventually published an interactive web page [10] allowing users to find the number of slot machines and the average amount of money gambled for each Italian 'comune'[1]. Thanks to the web page it is possible to see local

---

[1] The 'comune' is the basic administrative unit in Italy ranging from cities to municipalities. In 2017 there was 7,978 units in Italy.

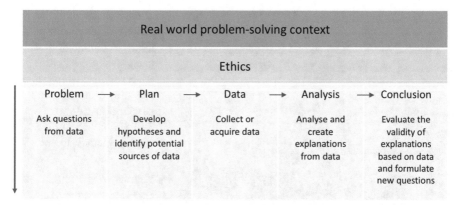

Fig. 2. The space of data literacy skills. The downward pointing arrow reflect more specialized data handling skills and a deepening of understanding in relation to each competence [26].

data for money spent per capita in slot machines and compare the data with other comunes or nationwide.

For example, in 2015 Verbania, a town in Piedmont with 30,000 inhabitants, spent €1,606 on slot machines per capita, equivalent to 8% of its taxable income per capita (see graph in Fig. 3). The municipal administrators of Verbania had previously in vain tried to limit the activity hours of slot machines [21]. The graph shows the relationship between the average income per capita (horizontal axis) and the average amount used on slot machines per capita (vertical axis) for all Italian municipalities (represented by red crosses) in 2016. On average each Italian gamble 6% of his annual income in slot machines. The graphs is divided the municipalities into four quadrants (gray lines): the municipalities with the lowest per capita income with the lowest per capita play ("they earn little and play a little"), those with lower income, but played higher ("they earn little, but gamble a lot"), those with higher income and higher play ("they earn a lot and gamble a lot") and finally those with higher income, but gambled lower ("they earn a lot, but gamble little") [10].

The web page compares the figures for each town to national indicators and provides additional data about the destination of the money, of which approximately 70% goes back to winners, 17% to the state, 6% to the shop-owner and 5% to the company which owns the slot machine. Both the data on the individual comune as well as the distribution of the money flow were little known to the larger public.

In the wake of this investigation, many local newspapers produced articles to help make sense of the data. For example, high values may not be meaningful for places which are popular tourist destinations, while in other contexts much lower values do reflect serious social problems. At a more general level, the investigation has raised the level of the public awareness and instigated public discussion to include facts on which all actors are now forced to agree (although these same facts do not by any means dictate a solution). The investigation was made possible because of the convergent effort of people with different backgrounds and competencies, from legal matters to data acquisition, data analysis, data visualization, web publishing and, ultimately,

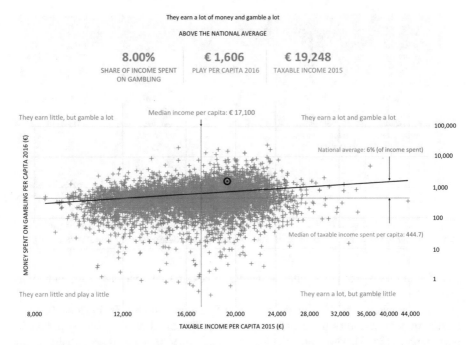

Fig. 3. Ratio between income and per capita spending on slot-machines for the town of Verbania, Italy, our translation [10].

traditional writing. These competences can flourish in contexts where data journalism meets 'community informatics' [11], often referred to as 'civic tech' [7].

3.2   Case 2 – Civic Data Hackathons for Non-profit Organizations

'Hackathons' or 'hacking-marathons' are events in which a group of computer programmers engage in a coordinated effort at producing code for a project, which can be related to a technology or to a cause (e.g., developing apps to help people with a specific disability). For example, hackathons have been a way of mobilizing the competences of developers in the early days of data journalism [24].

'Civic data hackathons' (e.g. DataKind[2] and Open Data Day[3]) are hackathon-type events that bring together non-profit organizations (henceforth, NPOs) and 'data volunteers' with the goal of performing collaborative analyses of data that support data-driven decisions or better understanding of civic issues [5]. Members of NPOs typically possess a deep understanding of their domain-specific context and content but may lack expertise in data analysis and the resources necessary for hiring external data experts. Civic data hackathons constitute promising contexts for integrating the expertise of NPOs with data experts, but at least in the practices observed by Chou, Li & Sridharan

---

[2] http://www.datakind.org/.

[3] http://opendataday.org/.

[5] collaborative data analysis did not proceed beyond an initial, exploratory stage because the experts in the domain could not sufficiently grasp the methods applied by data experts.

Hou & Wang [12] have reported on two civic data hackathons organized by graduate students at Michigan University and involving 9 NPOs and 40 data volunteers. The problem of connecting different communities of practice was addressed by organizing groups of 'knowledge brokers' [17, 25] with the role of serving as bridges between NPOs and data expert by facilitating translation, coordination, and alignment between perspectives of the two groups and matching the various competencies of data volunteers with the needs of the NPOs. Knowledge brokers strived to keep the focus on actionable collaborative analysis but had to mediate between the dual goals of civic data hackathons, that is to develop actionable knowledge for the NPOs and to improve data literacy in the community [12, 14].

4 Discussion

Both the investigation on slot machines in Italy and the civic data hackathons describe practices in which heterogeneous groups that to a varying degree connect with core competencies in data science have addressed complex real-world issues through collaborative inquiry. In both cases, the notion of data literacy emerges "as a property of a community as opposed to an individual, with members of the community making different contributions" [26]. None of the two cases involves formal learning, although civic data hackathons are indeed organized by universities, while investigative data journalism can be an ideal partner for 'connected curricula' at universities that see themselves as 'facilitators of community-based learning' [8]. Both cases also highlight the initial interdependence of the different participants in solving complex issues and the need for someone to act as 'knowledge brokers' in order to attain the dual goal of knowledge building and community building.

5 Conclusion and Future Work

Major universities primarily in the US and UK are offering courses in various branches of data science on the Coursera platform for MOOCs[4]. Google has made freely available advanced tools and computational power for experimenting with data science. With universities and corporate research institutions investing heavily in developing study degrees and online learning resources for educating data scientists the need and demand for specialists in data science is being addressed. But the expanding role of data and data science in society also highlights a need for educating citizens who are not meant to become specialists, but should become able to connect to methods, tools and results of data science. For example, professionals such as journalists, business

---

[4] https://www.coursera.org/browse/data-science.

employees or members of NPO's who need to understand data and analyses, or teachers who wish to integrate data-based reasoning in their disciplines.

. To encompass data literacy in education will in many cases result in the teachers working at the boundaries of their competences and sometimes beyond. Therefore, it is important for the members in the communities of practice to support each other through e.g. supervision and peer learning [3]. Also, in an educational context it can be productive and practical to view data literacy as a compound competence not just in relation to the individual domains of skill but also in relation to the heterogenous social composition of the group that as a whole may possess data literacy. In the context of learning it is therefore important focus on the compound character and to choose interdisciplinary issues that in a substantial way rely on multiple competences.

References

1. Bhargava, R., Deahl, E., Letouzé, E., Noonan, A., Sangokoya, D., Shoup, N.: Beyond data literacy: reinventing community engagement and empowerment in the age of data (2015)
2. Bell, G., Hey, T., Szalay, A.: Beyond the data deluge. Science **323**, 1297–1298 (2009)
3. Boud, D., Cohen, R., Sampson, J.: Peer learning and assessment. Assess. Eval. Higher Educ. **24**, 413–426 (1999)
4. Bruschi, M.: L'Italia delle slot. Ovvero: di Foia, pdf, database e inchieste collettive (Slot-Machines in Italy: About Freedom-of-Information-Act, pdf, Database and Collective Inquiry) (2017). https://medium.com/visuallab/litalia-delle-slot-ovvero-di-foia-pdf-database-e-inchieste-collettive-562686eb2f02
5. Chou, S., Li, W., Sridharan, R.: Democratizing data science. In: Proceedings of the KDD 2014 20th ACM SIGKDD International Conference on Knowledge Discovery and Data Mining, New York, NY, USA. pp. 24–27. Citeseer (2014)
6. Conway, D.: The Data Science Venn Diagram. http://drewconway.com/zia/2013/3/26/the-data-science-venn-diagram
7. Donohue, S.: What Can Civic Tech Learn from Social Movements? | Omidyar Network. https://www.omidyar.com/spotlight/what-can-civic-tech-learn-social-movements
8. Fischer, G., Rohde, M., Wulf, V.: Spiders in the net: universities as facilitators of community-based learning. In: Carroll, J.M. (ed.) Learning in Communities: Interdisciplinary Perspectives on Human Centered Information Technology, pp. 17–20. Springer, London (2009)
9. Frank, M., Walker, J., Attard, J., Tygel, A.: Data Literacy: what is it and how can we make it happen? Editorial. JoCI. **12**, 4–8 (2016)
10. GEDI: L'Italia delle slot—Scopri quanto si gioca nel tuo Comune. https://lab.gedidigital.it/finegil/2017/italia-delle-slot/
11. Gurstein, M.B.: Open data: empowering the empowered or effective data use for everyone? First Monday **16** (2011)
12. Hou, Y., Wang, D.: Hacking with NPOs: collaborative analytics and broker roles in civic data hackathons. Proc. ACM Hum. Comput. Interact. **1**, 1–16 (2017)
13. Littlejohn, A., Beetham, H., McGill, L.: Learning at the digital frontier: a review of digital literacies in theory and practice. J. Comput. Assist. Learn. **28**, 547–556 (2012)
14. Markauskaite, L., Goodyear, P.: Epistemic Fluency and Professional Education: Innovation, Knowledgeable Action and Actionable Knowledge. Springer, Amsterdam (2017)

15. Mau, S.: Das metrische Wir: über die Quantifizierung des Sozialen (The Metric We: About the Quantification of the Social). Suhrkamp Verlag, Frankfurt am Main (2017)
16. Mayer-Schönberger, V., Cukier, K.: Big Data: A Revolution That Will Transform How We Live, Work, and Think. Houghton Mifflin Harcourt, New York (2013)
17. Pentland, A.: The data-driven society. Sci. Am. **309**, 78–83 (2013)
18. Ridsdale, C., Rothwell, J., Smit, M., Ali-Hassan, H., Bliemel, M., Irvine, D., Kelley, D., Matwin, S., Wuetherick, B.: Strategies and best practices for data literacy education: knowledge synthesis report (2015)
19. Samuels, R.: Auto-modernity after postmodernism: autonomy and automation in culture, technology, and education. In: McPherson, T. (ed.) Digital Youth, Innovation, and the Unexpected, pp. 219–240. The MIT Press, Cambridge (2008)
20. Sharif, R., Van Schalkwyk, F.: Introduction. JoCI. **12**, 4–8 (2016)
21. Stella, G.A.: Il Tar multa il sindaco anti slot-machine (The Regional Administrative Court fines the Mayor against slot machines). http://www.corriere.it/cronache/12_marzo_23/stella-il-sindaco-anti-slot-machine_c5b92a22–74af-11e1-9cbf-6c08e5424a86.shtml
22. Talamo, G., Manuguerra, G.: The gambling sector: a socio-economic analysis of the case of Italy. East. Eur. Bus. Econ. J. **2**, 315–330 (2016)
23. UCL: Arts and Sciences (BASc) Programmes. http://www.ucl.ac.uk
24. Vermanen, J.: Harnessing external expertise through hackthons. In: Gray, J., Bounegru, L., Chambers, L. (eds.) The Data Journalism Handbook: How Journalists Can Use Data to Improve the News, pp. 44–47. O'Reilly Media Inc, Sebastopol (2012)
25. Wenger, E.: Communities of Practice: Learning, Meaning, and Identity. Cambridge University Press, Cambridge (1998)
26. Wolff, A., Gooch, D., Montaner, J.C., Rashid, U., Kortuem, G.: Creating an understanding of data literacy for a data-driven society. JoCI. **12**, 9–26 (2016)

Particularities of Language Classes in a Multi-cultural Context

Belyasova Julia[1]($\boxtimes$) and Teleshova Raisa[2]($\boxtimes$)

[1] Public Social Welfare Centre, Brussels, Belgium
coppeebelyasova@hotmail.com
[2] Novosibirsk State Pedagogical University, Novosibirsk, Russia
raissa.telechova@gmail.com

Abstract. The paper is devoted to the disclosure of the particularities of teaching languages in a multicultural context, as well as the implementation of a culture dialogue on the example of foreign language classes in Russian and Belgian educational institutions. We present two youth novels and three tasks in the reading activity. Based on these tasks, an educational experiment was realised in four classes with Belgian, Slovak and Russian students and teachers Relationship between the overall understanding of each novels and teaching practice of three teachers (Russian, Slovak and Belgian) is presented in this article. Through the analysis of reflexive pause booklets and questionnaires completed by the learners, we were able to see that the instruments developed and applied by the teacher at each stage of the process of knowledge mediation (epistemic, heuristic, pragmatic) helped students readers to acquire autonomy in reading, provided they become aware of the usefulness of the proposed device. The terms "multicultural education", "dialogism", "cross-cultural competence", "multiculturalism" are explained.

Keywords: Dialogue of cultures · Multicultural education
Cross-cultural competence · Universal human values · Multiculturalism

1 Introduction

1.1 Teacher and Student: Mutual Influence

In the modern educational space, communication between students and teachers issued from different cultures becomes quite common in Russia and in Belgium. In this regard, there is an increased interest of researchers in the problems of multicultural education which realizes the mechanism of the dialogue of cultures. This problem has already been considered in our articles [8]; nevertheless, we would like to note similarities in the views of Russian and Belgian teachers-researchers.

The methodological basis of our research is the theory of multicultural education (E.R. Khakimov, A.V. Kolesnikov), the concept of dialogue culture (M.M. Bakhtin), as well as the theory of multiculturalism (M. Barre, P. Hadermann, L. Robert). In the context of the growth of cultural diversity and interaction, the Russian and Belgian educational systems need innovative methodological developments which determine the relevance of our article.

© Springer Nature Switzerland AG 2019
T. Antipova and A. Rocha (Eds.): DSIC 2018, AISC 850, pp. 174–187, 2019.
https://doi.org/10.1007/978-3-030-02351-5_22

We recognize that the main goal of students is to learn to communicate in a foreign language, but it is impossible to deny that culture has its place in the foreign language classes. Being a teacher, we pinpoint one of the main tasks - to participate in the personal cultural enrichment of our students. This axiom is correct and independent of the teaching subject, but it is especially confirmed when a person studies in a foreign language class where he is directly connected to an other culture, different from his own.

Indeed, the foreign language studying is also associated with the discovery of a new culture, different ways of life and ways of thinking. Thus, the integration of the cultural dimension with the language dimension allows learners to open up to other cultures and thereby contribute to more objective, more tolerant and more respectful perception of the realities. Consequently, students can challenge stereotypes and common features of the studied culture, guided by learned cultural elements (for example, lifestyle, habits, behavior).

In our daily teaching practice, we put emphasis on cultural links transmitted by the media language, advertising, as well as the most used phraseological units and everyday expressions that convey the brightness and value of the studying culture, so that students can interpret and use them for a better understanding of the foreign country and its language.

Agreeing with many academic researchers in this field [22], we affirm that culture is the support of the language and its important component, as it allows us to open new horizons and expand our thinking.

1.2   Students: Intercultural Mutual Enrichment

Teaching a foreign language for several years in a group consisting of students, speakers of different cultures (African, Asian, European, Arabic, Russian, etc.), we realize that in this process, the student's personal culture interacts and interweaves with the culture of other students and with the culture of the studied language through our mediation. This fact also requires additional efforts from the teacher in organizing a meaningful aspect of the lesson, in holding a discussion/dialogue between representatives of different countries.

Thus, we agree with E.R. Khakimov who argues that the essence of multicultural education "is the combination of several cultural traditions in content, methods and organizational forms of education which leads to the recognition of cultural diversity as a social norm and personal value, as well as the appropriation of cultural and human images as a result of creative intercultural mutual enrichment" [20, p. 8].

We stressed in our previous articles [11] that multicultural education assumes the strengthening of the role of the educational moment in the learning process in terms of the formation of a value attitude to the cultural identity of one's own and another's country. The success of the implementation of the multicultural educational model depends largely on the readiness to recognize the multiculturalism of the modern world and to share cultural values; that requires an improvement in the quality of education in everyday pedagogical practice. Multicultural education involves the implementation of a dialogue of cultures in the learning process.

1.3   Training of Future School Teachers in Multi-cultural Context

The program-concept of communicative foreign language education "Development of individuality in the dialogue of cultures" [19] considers national and universal values as components of the educational aspect of foreign culture, along with the cognitive aspect.

Formation of universal values with preservation of national identity is considered in our article as a necessary condition for entering in the world educational space.

Moreover, we find that students should be as familiar as possible with the culture of the studying language because "communication in a foreign language consists first in mastering of the language, and then in the ability to adapt the received language skills in situations of communication and in the contexts; these abilities are acquired through culture "[22]. Thus, we are really confronted with the fact that the linguistic and cultural elements are interrelated. So it is impossible to ignore the cultural aspect in the current process of teaching/learning a foreign language.

The success of a multicultural educational process is determined by the degree of formation of general cultural skills, which, along with cross-cultural competence, make up the pedagogical culture [5] of foreign language teachers. Sharing T.A. Kolosovskaya's point of view, we consider the teacher's cross-cultural competence as an integral quality of the individual, including knowledge of the comparison and adaptation of cultural differences and universals, the ability to retranslate an onather different culture " [5].

2   Cross-Cultural Skill: Theoretical Framework

The formed cross-cultural skill means "to be communicative, contact in various social groups, to realize the need to explore not only the cultural origins of representatives of different peoples, but also the specifics of the psychology of their personality, patterns of behavior manifested in conflict situations, as well as demographic characteristics and life experience" [5, p. 8].

The importance of developing a cross-cultural skill is also highlighted by foreign specialists of the learning/teaching of foreign languages. We equate the concept of "multiculturalism" [5], developed by M. Barré and his colleagues, to the concept of "cross-cultural competence". They argue that the motivation for multiculturalism is important for learners studying a foreign language, both in their own country and in the language environment where they often experience the traditions and customs of a foreign country; this is also an inevitable aspect for any teacher who must be ready to "adapt to the sociocultural heterogeneity of students" [5, p. 19].

To help a teacher in this approach, the Belgian universities plan, in the process of preparing teachers, activities to develop a multicultural aspect. M. Barré and his colleagues present an interesting methodological suggestion, developed for future French foreign language teachers. This method consists of three steps that can be applied to the teaching of any language:

1. "to know oneself analyzing one's perceptions";
2. "to understand the value system and the world view of the Other";

3. "learn to hold discussions in order to create a common intercultural platform" [5, p. 19].

Belgian Experience. In Belgian universities, for the mastery of this method, future teachers are encouraged to create a "journal of surprises" during the internship [5, p. 20], which reflects their positive and negative impressions of communication/teaching in a multicultural environment. These effects must be accompanied by self-reflective criticism, self-esteem in relation to another person.

The purpose of this methodological approach is to learn the teacher to digestion, to question his own pedagogical activity and, especially, "to analyze the elements, the impressions gathered by the reflexive approach" [5, p. 20] to develop and progress as a competent teacher in a multicultural interactive context.

We set up an educational experience to observe, in a communication process, the impact of teaching methods of reading on the reception and perception of novels by Belgian students, on the one hand, and Russian, Romanian and Slovak learners, on the other hand. We experimented with mediation instruments in five classes, with 62 students in total. We were able to validate our interfaces from the point of view of their use, to see that the observables collected were representative of students' knowledge and that they were sufficiently rich and reliable to allow a diagnosis. We found correlations between the methods used by the teachers interviewed and the level of comprehension and the interpretation of the novel read by the students.

We considered the teacher of the Jumet Technical School (in Belgium) as an avid reader who makes every effort to pass on her passion to the students. They are not experts in analytical reading but they read, understand and interpret intuitively guided by their first impressions. By working with a weak class in reading, this teacher puts aside the learning of literary analysis techniques. She focuses the teaching of reading on the development of literary taste and the emotional reception of the novel.

Following our analysis of the methodological practices declared by the teachers, we think that the teacher of the Decroly school (Brussels, Belgium) and her colleague from the technical school of La Louvière (Belgium) are teachers who put a lot of will to teach the techniques of textual analysis. Nevertheless, the results of their students are not the same. In the first case (Decroly school), the criticism of the text is more accentuated. It often focuses on classical works and is accompanied by written work. In the second case (Technical School of La Louvière), the textual, stylistic and referential analysis of youth novels is usually done orally and students are not used to expressing their understanding of the text by writing.

Thus, comparing these two classes, we have seen that the comprehension-interpretation of the novel by the students of the technical school of La Louvière seems more vague, less explicit. As for the pupils of the Decroly school, they did a good job of textual interpretation, although they did not really appreciate the proposed novel. In this case, we notice that this teacher devotes a lot of time to learning literary analysis which is very effective in this context. It seems to us, moreover, that the shift towards the rational side of reading is a little detrimental to the emotional reception of the novel.

Russian Point of View. A joint article of the teachers of the State Pedagogical University in Novossibirsk is devoted to a comparative analysis of universal values in the cross-cultural aspect. [16] They argue that in order to increase the motivation for language education, to improve mutual understanding between representatives of different countries and cultures, additional efforts are required from teachers in organizing the content aspect of their lesson, in organizing and conducting discussion/dialogue between representatives of different countries.

And in order to train teachers as best as possible for the real conditions of teaching in Russia, within the framework of the disciplines in State Pedagogical University of Novosibirsk (Country Studies, History and Geography of Foreign Countries), the formation of universal values along with the formation of skills to understand others' views on the discussed problems takes place. The future teachers are also able to achieve a certain mutual understanding in the face of differences in points of view and beliefs. The universal constants characteristic of all nations (nature, geography, family, religion, traditions, and others) are realized by different forms in the cultures of different peoples; that can lead to certain problems in communication [11].

In our opinion, the Russian, Romanian and Slovak teachers focus their teaching of literary reading on the psychological evolution of the characters. Their goal is to teach students to express their critical attitude to the novel, to explain the behavior of the characters, to justify their opinions.

Through the analysis of the work of their students, we noticed that lexico-grammatical obstacles are felt even by the learners qualified as good readers. For this reason, teachers aspire to teach their students techniques that would allow them to overcome their language difficulties. On the other hand, Russian, Romanian and Slovak teachers tend to neglect the teaching of analytical reading of works. Therefore, the students' interpretations usually focus on the explanation of the behaviors of the characters and their feelings. The copies of these readers contain almost no elements of purely literary analysis.

The interest of our research is all the higher as it straddles two didactic fields: that of French as a first language and that of French as a foreign language. Our analysis of the data shows that low-level first-language readers and foreign-language readers have similar comprehension-interpretation difficulties. Therefore, the use of the same methodological instrument (the reflective pause booklet) could perhaps be effective in the two different contexts.

3   Table of Intervention Devices

Inspired by Séverine De Croix's [11, p. 286] overall plan of didactic devices, we adapt it to our own research to highlight the steps, the goals and the time of each tool used in this experiment. Following De Croix [11], these devices are useful for collecting data concerning five axes of our reading model, namely the reading attitude, the value judgment, the reading systems, the identification or distancing from the novel and textual comprehension-interpretation.

We cross this table of devices with Martin's meta-communication model [14] to clarify the place of each device (mediating instrument) in the teaching-learning process described in our work. More specifically, we suggest that teachers use the exchange of class notes, the reflective pause booklet that accompanies students 'individual reading, and the questionnaire at the end of the novel in order to facilitate the students' comprehension-interpretation of the novel and to enable them to acquire a reading autonomy with the literary text. We asked for a preliminary interview with each teacher to explain our experience.

We gave them the liberty to organize the reading activity of the novel according to their methods, the time available and the motivation of the students. We nevertheless specified the importance of respecting the three tasks necessary for the correct progress of this experience (Table 1).

4 Relationship Between the Overall Understanding of the Novel and Teaching Practice

The Case of the Pupils of the Technical School of Jumet (Belgium) by Analyzing of a Youth Open-Ended Novel "Julie" by B. Coppée. Using of these intervention devices by teachers in different classes helps to establish the relationship between the overall understanding of the novel and teaching practice of teacher.

By analyzing the reflexive pause booklets and the French native speaker student's questionnaire, we notice that she has some difficulties in the overall understanding of the novel. This may be due to the lack of a practice of analysis of the literary text, as well as to gaps in the capture of stereotypes. On the other hand, it succeeds in interpreting in a rather coherent way some chosen sentences which can contain a general idea of the novel. For example, the sentence "Life is touching sometimes. She screams "("Julie" by B. Coppée) is interpreted as: Life, it's beautiful peaceful, we keep what we have in ourselves but sometimes it has to happen (the French native speaker student). We think she's talking about Julie (main caracter of Coppée's novel), who keeps her husband's disdain and disgust, for fear of revealing them until the moment when his whole world falls over...

The detailed analysis of Justine's reflexive pause booklets, as well as the careful study of the work of other Belgian pupils, allows us to confirm certain prejudices concerning the status of the family in Western Europe. Paillet's [15] statements about the current situation of households, the role of women in the family and in society can be verified by reading the responses of young readers. The perception of the family by the children of divorced parents, evoked by Smits [17], is confirmed by the critical examination of questionnaires and reflexive notebooks of Belgian pupils.

We note the same difficulties of interpretation among the other students in this group. Thus, only two readers out of fifteen manage to interpret certain passages of the novel read. It seems, however, that their reading is rather intuitive: they are not really aware of the way in which they have been able to interpret the text.

Seven other students in this group reach the first understanding, very general and very unstable. In other words, they include a certain part of the text to get a rough idea about the development of the story. Their linguistic and cultural background allows them to identify textual references in an intuitive way. They take as banal words ("this beauty"), a whole phrase ("a man who finds you beautiful") or they explain only the situation described ("the husband works andcome back late and his wife takes care of the

Table 1. Three tasks for pedagogical experience.

Task 1: take knowledge of the tasks to be done, discuss the first impressions before and during the reading, propose assumptions before and after. It is an epistemic mediation which, through the mediator instrument (oral exchanges), makes the connection between the questions asked (as a goal to be attained) and the resources (the novels to be read)

Steps	Goals	Support
1. "Formulation of an opinion on the text" [6] 2. "Oral explanation of the answers given in the questionnaire" and comment on his comments in the reflexive pause booklet	* "Developing of the capacity to appreciate a text (expressing one's appreciation)" [6] * "Describing one's procedures (to understand and process a questionnaire (and a reflexive pause booklet)) and to self-evaluate one's understanding [6]	- A youth novel to open end "Julie" by B. Coppée - A youth novel-mirror "Since your death" by F. Andriat

Task 2: read individually a youth novel, verbalize one's impressions, understanding, interpretation in the reflective pause booklet. It is a heuristic mediation which is presented as a fusion between the process of reading and the reflections fed by the mental representations specific to each reader. The reflective pause booklet as a mediator helps students to learn the skills

Steps	Goals	Support
1. Reading a rather lengthy youth novel and practice the reflexive pause booklet 2. Expression of first impressions 3. Creation of the first reading hypotheses and return on the advanced declarations and their correction 4. "Pointing of reading difficulties" (Ibid.) 5. "Prediction of the evolution of the narrative" [6] 6. "Elaboration of the possible end of the novel" [6] 7. "Individual written reformulation of the end of the novel" [6] 8. Exchanges in class (information, impressions)	* "Mobilize knowledge, collect clues, connect them to formulate hypotheses and create a waiting horizon" [6] * Exercise certain reading processes: mental representation, memorization of information, textual induction * Integrate collective exchanges in the reformulation of one's hypotheses	- A youth novel to open end "Julie" by B. Coppée - A youth novel-mirror "Since your death" by F. Andriat - Two reflexive pause booklets corresponding to the two mentioned novels

Task 3: read a youth novel and share one's understanding of the specific elements in a questionnaire. It is a pragmatic mediation that develops and acquires the skills of readers. The mediator instrument (the questionnaire) focuses on students' difficulties with texts and makes it possible to explain their skills acquired or not

Steps	Goals	Support
1. Description of one's relation to reading, in general, and to the proposed novel, in particular, (pleasure, interest); Description of one's reading mode (place, time, average); Description of one's own reading activity (degree of understanding, difficulties encountered) 2. Individual reading of the novel 3. "Treatment of a reading questionnaire" [6] aimed at a global understanding of the text, the characters, their relationships and their changes during the novel	* "Understand the text globally" [6]. * Clarify the role and evaluation of the characters * "Describe one's relationship to reading and share one's representations" [6]	- A youth open-ended novel "Julie" by B. Coppée - A youth novel-mirror "Since your death" by F. Andriat - Two questionnaires of global understanding that correspond to the two mentioned novels

children, the house, the household, the kitchen"). In our opinion, their intuitive grasping is not sufficiently guided so that they can reach a higher level of understanding.

To find explanations, we turn to the written statements of their teacher. We find that, according to her, she does not really adopt a fixed strategy to study novels with her students. Moreover, she avoids offering them long classic texts. She favors a study of extracts, a comparison of several texts, a reading of images. She does not teach the notion of stereotype for fear of her students not understanding the stereotype and not applying it correctly. This teacher, like her other colleagues who work with rather weak classes, relies on the emotional reception of the book to develop the taste of reading rather than to learn literary analysis. Nevertheless, she tries at each study of a new text to solve the problems of comprehension and to answer the questions of her pupils about their reading.

Eight students in this class have difficulty understanding the novel. Through their responses, we note that these readers arrive at a first understanding only thanks to constant help from the teacher. The individual discovery of the text is often done at home. Students are then overwhelmed by their positive or negative emotions and are not able to create the necessary ground for understanding. On their own, these students find it difficult to establish "topics" on the basis of common knowledge, to explain those already given in the questionnaire, to understand why they like or do not appreciate the passage of the text, the behavior of the character. They need frequent exchanges organized in class by the teacher. They can then express orally their impressions and reactions, ask questions. However, the professor tells us that these readers do not make much progress in their reflections and often answer the questionnaire in the same way. They also tend to judge the characters, and the teacher must lead them to explain, to justify their point of view every time they speak.

The Case of Slovak and Russian Pupils by Analyzing of a Youth Novel to Open end "Julie" by B. Coppée. Alica, a Slovak pupil whose work we have chosen as an example to compare with that of the Belgian pupil, seems to be the most experienced reader: her explanations are more accurate, her understanding deeper. Even in her interpretation, she goes further than the Belgian reader. For example, in two lines she takes up a story of the main characters, drawing a conclusion that life represents a complex whole filled with happiness, joy, but also sadness, hardship and difficult choices: "life takes place well sometimes, but from time to time, everything is destroyed, it seems that nothing remains, everything becomes difficult but then, normal life resumes and we continue to live as we were used.

The stripping of the reflexive notebooks of other Slovak and Russina pupils gives us the same impression: they are readers very attentive to the development of the plot between the characters. They create a lot of hypotheses about the continuation of history, ask themselves pertinent questions. Nevertheless, for most of them, the general understanding of the novel escapes them. This phenomenon is explained by the fact that these readers read in a foreign language and the capture of the innuendoes is not always done correctly.

Moreover, it is surprising to see that their answers reveal that they are wisely mobilizing familiar, stereotyped understandings. For example, they talk about the woman who keeps the unity of the family or the selfish man who thinks only of the satisfaction of his desires: "I believe that this is the way we see men in Tim's flaws, men who are very masculine; it is often said that they think in a simplified way, "explains a student.

Concretely, the group of Slovak and Russsian learners includes, first of all, six pupils who do not show a good understanding. By asking ourselves the question, "why? We realize that the construction of stereotypy is compromised. In fact, the "virtual referents" [7] needed for the creation of the latter are not identified by these students. They have lexical difficulties specific to people reading texts in a foreign language. These obstacles prevent the first understanding, the basis of the textual interpretation. Despite multiple attempts to create hypotheses about the continuation of history, their "topics" [7] collapse. These readers do not manage to locate "places of certainty" [7], necessary for a general understanding, stable and correct.

Then we distinguish seven students who testify to their more or less good understanding of the novel. The path of reflection of these readers enters the schema that we adopted to estimate the level of the understanding of the text. Thus, they build a stereotype to base their narrative assumptions. The location of "topics" [7] allows them to confirm or change them. In any case, these readers reach the level of the second understanding. However, we sometimes find some whites in their explanations or one or the other wrong interpretation.

The testimonies of the Slovak teacher give us some insights. She has a different vision of reading skills to teach her students. She says she has focused on developing language and argumentative skills. We noticed that his students have a fairly good level of language and that they feel very comfortable to express their opinions, to argue their remarks. This professor also insists on the knowledge of different types of texts and their identification signs, as well as on the possibility of comparing and observing a psychological evolution of the characters. These taught techniques are reflected through the responses of his students. This explains the fact that their copies contain almost no elements of purely literary analysis. They read with their heart and their emotions and show a rather participative reading.

Another element that characterizes the Slovak teacher is the fact that she favors the emotional reception of the book. This is an essential step in her work with her students on the literary text. For the rest, the discovery of the text is generally done in class by the explanation of the vocabulary, the revelation of the theme and the plot. She often organizes debates and exchanges to explain the characteristics of the characters, and she also discusses with her students about the style and choice of the lexicon by the author. She says: "I make sure to teach young people to" read correctly ", in other words, to take pleasure in reading, to appreciate the literary text, to understand it. She doesn't work in depth the textual interpretation and teaches only a few elements of literary analysis.

The Reading Attitude of Belgian and Romanian Pupils by Analyzing of a Youth Novel-Mirror "Since Your Death" by F. Andriat. To make the comparison, we take three target classes: eleven Belgian pupils of the technical school of La Louvière, eleven Belgian pupils of the Decroly general education school and ten Romanian high school students. We believe that the confrontation between two Belgian classes, from backgrounds that are often opposed, may reveal a different appreciation of the same novel and provide some interesting explanations. The parallel between the Belgian public and the Romanian public shows their similarities and their differences which stem from the diversity of the first languages and the cultures of origin.

We would like to present reader's thoughts on the definition of death. They give us the opportunity to grasp the general reading trends of each class and also to focus on the evocative dissimilarities of culturally significant perceptions.

The question: "What does death represent for Ghislain? reveals that Belgian readers of the technical school place more emphasis on abstract values. Their definitions of death are limited to short words without explanation: "woe", "hell", "fatality" and others. They do not seek to justify their statements by the ideas of the text. The same type of answer is given by 27% of students at Decroly School. This percentage is less, but nevertheless important. Readers of the technical school read evaluatively, putting forward the ideal and abstract values, those of the Decroly school in a global way with the pessimistic perception of the reality described in the novel. This generalizing perception of the text proposed by Decroly's readers (27%) is reflected in the response that evokes "the disappearance of a loved people, the transition to another world and something terrible that scares everyone. as well as the end of the life of loved people. These global ideas are not unique to the character and can be attributed to anyone. In this way, these pupils implement general categories by which they imagine death. Romanian readers, in 30% of cases, mention death as "something incomprehensible and appalling that causes a lot of sadness and pain". Their testimonies show that they consider the approach of mourning as an individual and punctual action. This finding is explained by a personal approach to the reading mode that is specific to these respondents. We do not notice the same trend in the responses of Belgian students from two groups.

On the other hand, they raise the feeling of injustice that appears in the words of the young hero. Romanian high school students do not adopt this response strategy. 18.5% of Belgian pupils in technical education as well as 18.5% of Belgian general education students emphasize the feeling of inequity expressed by the adolescent's angry and aggressive behavior. According to them, the hero conceives death as the injustice that gives reason to his irascible behavior. This ability to correlate the facts described and their causes constitutes the characteristic of synthetic or sociological reading. According to our observations, it is more specific to our Romanian public than to the Belgian public. But the answer received does not reveal this correlation. We find the justification for this if we look at this from another angle. We believe that the expla-nation is due to the fact that it is, in this case, injustice. This notion is related to the society to which each individual belongs. The feeling of injustice is more developed among Belgian pupils in general education, in our opinion. They are ready to defend their opinion in the name of justice and to judge what seems unjust to them.

Thus we find again that the dissimilarity is explained by "an area of indetermi-nacy". It also shows that there is no really limits between different types of reading. We can only say that this or that part of the public takes hold of one or the other form of reading.

In the conclusion, we say that the testimonies of the Romanian readers have allowed us to note that their reading can be described as synthetic or sociological. They can make a judgment both positive and negative about the hero or his actions, but they do not condemn him, and seek, on the other hand, the link between the facts and their causes. We also notice the personal involvement of these readers and their identification with the character. They follow the progress of the adolescent in his maturity of mind and judgment. We emphasize that this personalizing approach to the conception of grief is lived by everyone individually in their own way. In addition, some Romanian readers make comments that make them think that they adopt the phenomenal reading of Belgian pupilsof Decroly.

As for the Belgian pupils of the Decroly general education school, their reading seems to us to be characterized as phenomenal. Its peculiarity is to treat the reality described in the text as normal, that is to say that the rebellious, aggressive and sometimes shocking reaction of the hero as well as his rebellion against death are judged acceptable by the society and justified by its norms. This reading is also characterized by the global look and pessimistic that it poses on the described reality, as well as by its generalizing perception of the text. Indeed, the readers concerned accurately capture the author's thoughts on four stages of acceptance of death. They use them as global ideas that they do not only attach to the concrete character but attribute to anyone. In this case, we also notice the shift of a minority of these respondents towards the synthetic reading proper to Romanian students.

As for the Belgian pupils of the technical school of La Louvière, we think that the evaluative reading characterizes them. This type of reading highlights the ideal and abstract values of the readers who stand out in their definition of death as well as in their critical judgment on the aggressive behavior of the character. However, a small percentage of these readers report a phenomenal reading, like the students at Decroly School.

We think that the synthetic reading is specific to the majority of Romanian readers, but that the testimonies of the Belgian pupils also present the characteristics of this type of reading. Moreover, they express emotional attachment to the character, which is rather the peculiarity of the Romanian reading. We see that these two groups of readers, with their differences, are getting closer and sometimes have similar points of view on how to mourn. Explanations can come from the fact that these students belong to the "elite" of the society to which they belong.

Decroly School is renowned as an institution that only accepts children from affluent families with university parents. The Romanian College "Petru Rares", meanwhile, also offers high quality education accessible to students of high social level. Both groups of readers received a good education, an open mind and an open outlook on the world.

We also want to highlight inconsistencies between the two Belgian groups. Readers of the technical school are, in our opinion, subject to evaluative reading, but at the same time we notice that they see the character in relation to the society around him. We also observe that they judge the inappropriate behavior of the teenager as normal. These characteristics (linked to the standards set by society and perception of normality in the text) are peculiar to the phenomenal reading that dominates readers of Decroly School. The rapprochement of these two groups is obvious: they belong to the same European society and can, in principle, have the same points of view on the question of mourning. However, the two publics are differentiated by their social level, by their access to education and by their perception of studies, which is more evident in ideological differences.

In this way, we have shown the peculiarities of each of our target groups, their dissimilarities and their common points, which express themselves through their own values. In the following chapter, we will examine the issue of identifying readers with characters during reading and how they distance themselves from the events described in the novel.

5 Conclusion

Studying a language in a multicultural context, learners often operate with categories of "one's own" and "another's", representing a binary opposition of world culture as a whole. Our personal experience in teaching foreign languages in Russian and Belgian educational institutions shows that students are happy to analyze and compare the cultural and historical characteristics of their and other countries (France, Belgium, Germany, Great Britain, USA, Syria, Afghanistan, Ukraine, Latvia and others).

In this context, the question of the qualitative selection of didactic material for the adequate realization in situations of intercultural communication arises. The universal and cultural-national component of a foreign language reflects the national-cultural specifics through the selection of the most used phraseological units.

In the case of foreign language lessons, the dialogue of cultures is interpreted as an approach that reveals the national specificity of people's thinking by comparing linguistic units, various arts belonging to different national cultures and stereotypes of speech behavior of different languages speakers [8].

The last decades have been marked by significant social changes related to globalization and increased mobility of people. Undoubtedly, these changes have influenced the world of education because teachers face increasingly diverse sociocultural classes [22].

Based on this, it became necessary to adapt the training skills and improve the teachers' professionalism in order to offer to the future teachers appropriate initial training and to already confirmed teachers, targeted training aimed at understanding the multicultural educational process as an active dialogue of cultures.

The didactic analysis of the pupils' readings and the choices made by their teachers seemed precious to us. Researching the methods used by teachers, establishing why and how they use them, making teaching practices explicit and coherent has enabled us, by mobilizing theoretical axes, to contribute to the effectiveness of pedagogical devices to make concrete didactic proposals.

We have crossed the points of view of Béguin, Rabardel [22] and Martin [14] and applied their model of the teaching-learning process to our field of teaching, namely the reading of the literary work in a multicultural school context. We therefore conceive the reading exercise in our experiment as an activity mediated by instruments - oral exchanges in class, questionnaire after reading, reflexive pause book. Concretely, this last instrument appeared to us as an active principle of the pragmatic mediation which unfolds in two stages during the activity of reading. First, the reflexive pause book reflects students' ideas fueled by mental representations and textual comprehension externalized through classroom oral exchanges. Then, readers deliver their responses that interpret the novel via the quiz after reading.

Through the analysis of reflexive pause booklets and questionnaires completed by the learners, we were able to see that the instruments developed and applied by the teacher at each stage of the process of knowledge mediation (epistemic, heuristic, pragmatic) helped students readers to acquire autonomy in reading, provided they become aware of the usefulness of the proposed device.

As a conclusion to the results of this research, we will say that teachers' attitudes or practices, accompanied by a knowledge-mediating instrument, make it easier for students to express their value judgments and their ways of perceiving and learning. interpret the literary text.

References

1. Abdelaziz, N.F.: Pourquoi sensibiliser les apprenants à la culture de la langue cible en classe de FLE. https://arlap.hypotheses.org/8184
2. Arkhipova, I.V.: Deutsch plus Landeskunde: educational-methodical manual for higher educational institutions in the field of "Pedagogical Education": ext. UMO high schools of the Russian Federation. Publishing House of the National Pedagogical University, Novosibirsk (2014)
3. Arkhipova, I.V.: Landeskunde: Educational-Methodical Manual. Publishing House of the National Pedagogical University, Novosibirsk (2015)
4. Berré, M., Hadermann, P., Robert, L.: La formation des enseignants de FLE/S en Belgique: un état des lieux. In: Dialogues et cultures 61, pp. 160–173 (2015). http://fipf.org/sites/fipf.org/files/d_c_ndeg61_-_2015.pdf
5. De Croix, S.: I read, I write: the practice of the journal of reading. Enjeux **50**, 121–129 (2001)
6. Federal State Educational Standard of Higher Education. Direction of training 44.03.05 "Pedagogical education (with two training profiles)". Qualifications: Academic Bachelor. Application bachelor, http://www.rsu.edu.ru/wordpress/wp-content/uploads/2014/02/44.03.05_Pedobrazovanie_2_profilya.pdf
7. Groux, F.B.D., Porcher, L.: French as a Foreign Language. L'harmattan, Paris (2011)
8. Khakimov, E.R.: Designing the practice of multicultural education on the basis of the polyparadigmatic approach: author's abstract. dis. ... of the doctor ped. Sciences: (13.00.01). Izhevsk (2012). http://elibrary.udsu.ru/xmlui/bitstream/handle/123456789/8820/hakimov%202012.pdf?sequence=1
9. Kolesnikov, A.S.: The analysis of methodological directions of research of continuous pedagogical education. Cross-cultural approach in science and education, pp. 4–13 (2013)
10. Kolosovskaya T.A.: Formation of cross-cultural competence of future teachers: author's abstract. dis. ... cand. ped. sciences. Chelyabinsk (2006)
11. Lepage, F.: The Youth Literature in 1970–2000. Fides, Montreal (2003)
12. Martin, M.: What mediations and mediatizations for all teaching-learning situations with or without T.I.C. In: 16th Congress of the Society. French Institute of Information and Communication Sciences: Affirmation and Plurality, Compiegne, France (2008)
13. Paillet, P.: The status of the evolving family. Soc. Inf. **139**, 45–46 (2007)

14. Passov, E.I.: Program-concept of communicative foreign language education "Development of individuality in the dialogue of cultures". Enlightenment, Moscow (2000). http://www. prosv.ru/Attachment.aspx?Id=8719
15. Smits, J.: Statistics of divorce in the European Union. E-education (2010). http://www.veille-education.org/2010/07/24statistics-du-divorce-in-lunion.html
16. Teleshova, R.I.: Country Studies: A Textbook. Publishing House of the National Pedagogical University, Novosibirsk (2014)
17. Teleshova, R.I.: Francophone World and Francophones: A Tutorial. Publishing House of the National Pedagogical University, Novosibirsk (2015)
18. Teleshova, R.I.: Dialogue of Cultures in the Classroom on a Foreign Language. Publishing House of the National Pedagogical University, Novosibirsk (2015)
19. Kostina, E.A., Kretova, L.N., Teleshova, R.I., Tsepkova, A.V., Vezirov, T.G.: Universal human values: cross-cultural comparative analysis. In: Procedia—Social and Behavioral Sciences, pp. 1019—1028. Publishing House of the National Pedagogical University, Novosibirsk (2015)

Digital Engineering

Intuition and Quantification of Mental Variables of Cognition Subjects in the Processes of Comprehension of the Surrounding World

Valery Kharitonov, Alexander Alekseev, Darya Krivogina,
Varvara Spirina$^{(\boxtimes)}$, Roman Shaydullin, and Nikita Safonov

Perm National Research Polytechnic University,
Komsomolsky av. 29, 614990 Perm, Russia
spirina@cems.pstu.ru

Abstract. Methodological fundamentals for the quantification of preferences of economic entities include the classification of the steps of the corresponding procedure, allowing the formation of context sequences that correspond to the proposed technologies of the tasks being solved. When building a technology, the known approaches to the interpretation of the concept *"intuition"*, differing in orientation to the left and right hemispheres of the human brain, are used jointly. Model examples of the technology of comprehension of research results by subjects of control objects that perform ranking and selection on a set of objects as a preliminary and final task, are given. At this stage, the possibilities of the targeted use of the software that has passed the official registration, are illustrated.

Keywords: Quantification · Onomasiology · Semasiology · Semantics
Economic entities · Mental variables · Preferences · Intuition
Subject-oriented control · Mathematical modeling · Ranking
Artificial intelligence · Non-manipulated choice

1 Introduction

Mankind has always paid enough attention to the procedures for measuring the parameters of the surrounding world, establishing standards and units of measurement, the necessary tools and techniques for using them to ensure the necessary accuracy of the results [1, 2]. These quantization procedures of physical, or phase variables are designed to compare real-world objects to each other. They are the primary step of the broader concept of "quantification", making the transition to the language of numbers, that is, making variables measurable taken from the area of semantics [3–5], responsible for the semantic side of the language - words, phrases, sentences. The problems of constructing the quantification processes of preferences and the use of intuition [6–9] of economic entities therein are of great importance in problems of subject-oriented control (SOC) [10–13]. However, the questions of semantics (comprehension) of the results of comparing objects on panoplies of alternatives by methods of modeling preferences have not been fully solved.

The article discusses the most important problems of quantification, including methodological fundamentals for quantifying the preferences of economic entities, examples of technology for comprehending the results of research of control objects by subjects, solving ranking and selection problems on a panoply of objects, and also the possibilities of using the original software complex are illustrated - the platform of preferential models built on a compromise of two "opposite" approaches to the interpretation of the concept *"subjectivity"*.

2   Methodological Fundamentals for the Quantification of Preferences of Economic Entities

According to Ozhegov *"meaning"* is the inner content, the meaning of something, the results to be comprehended [14]. To think is to know, understand, represent a goal, a rational basis, an internal logical representation of something, an understandable rational basis.

In the procedures of quantification, its steps (Fig. 1) can alternate in any order, preserving its individual destiny.

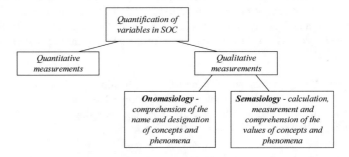

Fig. 1. Classification of the stages of quantification procedure.

Onomasiology [15] is a division of semantics, a science that studies the ways of naming and designating concepts and phenomena of the surrounding reality.

Semasiology [16] is a division of semantics, a science that studies the ways of measuring values and changing the meaning of concepts and phenomena.

Quantification technologies rely on the emerging opportunities for the most complete use of natural mental abilities of a person, relating to the innate and continuously developing human ability in the field of "intuition". There are two fundamentally different approaches to this concept. In accordance with the first approach, when *"intentional thinking* resembles reasoning, it has a critical and analytical nature", since it involves the connection of the left hemisphere of the human brain. So, in the dictionary of Krysin [17], based on the literal translation of *"intuition"* (French) is to look intently, closely, intuition means "flair, subtle understanding, penetration into the very essence of something", so that possible ways of developing the original agenda are "called and analyzed" from the depths of the brain. At the same time, the development of intuition itself as a mental tool is especially effective.

This approach is opposed to another that is based on the inclusion of the right hemisphere of the human brain, **the direct inclusion of the natural mechanism of intuition** in the thought process associated with the manifestation of motivation for decision-making is used (D. Maers' western school) [6]. In Webster's dictionary we can find a statement of this point of view: "Intuition is our ability to direct knowledge, immediate insight, without preliminary observations or reasoning." Psychologist Daniel Kahneman argues: "Intuitive thinking is like perception, it passes quickly and effortlessly" [18].

Both approaches are legitimate and compatible, which will be illustrated in Sect. 3 by their synthesis in a single software tool used as an intellectual support in decision-making. This result, connected with the balancing of both approaches, is quite consistent with the well-known remark attributed to Albert Einstein: "Not everything that can be counted can be relied on, and not everything that can be relied on lends itself to the calculation".

3 Model Examples of the Technology of Comprehension of Research Results by Subjects of Control Objects

3.1 Solving the Problem of Ranking a Set of Matching Objects

The two-part formalization of the task of ranking a set of matching objects using intuition in the interpretation of the first approach is expedient to implement, first by constructing a necessary and sufficient sequence of quantification steps on a full set of alternatives representation, ensuring the non-manipulability of the resultant semantics.

(a) Naming and designation of targets of research (class of objects), for example, shopping facilities A, subclass A', for example, shopping facilities of the city of Perm, $A' \in A$.

(b) Naming and designation of a set of phase characteristics of representation for $\bar{x}$ objects in the named subclass A', for example, area, sq.m., x_1, transport accessibility, min., x_2, assortment of goods, number of units, x_3.

(c) Naming and designation of a set of qualimetric characteristics of representation for $\bar{X}$, for example, trade significance, X_1, transport accessibility level, X_2, variety of goods, X_3, in the qualimetric scale X_1, X_2, $X_3 \in [1, 4]$.

(d) Naming and designation of a set of the reduction function of phase representation characteristics to qualimetric $\bar{X} = \bar{\sigma}(\bar{x})$ in the field of their determination: $\left[x_1^{\min}, x_1^{\max}\right]$; $\left[x_2^{\min}, x_2^{\max}\right]$; $\left[x_3^{\min}, x_3^{\max}\right]$.

(e) Quantitative intuitive measurements of integer values of the reduction functions $\bar{\sigma}_i$, $i \in \overline{1, 3}$ in the field of the determination of phase characteristics $\bar{x}$ (cerificates): $(x_1(1), x_1(2), x_1(3), x_1(4))$, $(x_2(1), x_2(2), x_2(3), x_2(4))$, $(x_3(1), x_3(2), x_3(3), x_3(4))$.

(f) Qualitative intuitive measurements of continuous values of non-linear reduction functions $\bar{\sigma} \in [1, 4]$ in the entire field of the determination of phase characteristics $\bar{x}$ (interpolation, smoothing): $X_1 = \sigma_1(x_1)$, $X_2 = \sigma_2(x_2)$, $X_3 = \sigma_3(x_3)$.

The second part of the problem assumes the specifics - the use of the represented set of alternatives A', which is ranked using the previously constructed model of the behavior of the economic entity when ranking objects in class A, which plays the role of non-manipulated "artificial intelligence" for the subclass A'.

(g) Quantitative measurements $\{\overline{x*}\}_{A'}$ of phase characteristics $\bar{x}$ in the subclass A': size matrix $|A'| \times 3$ with the number $|A'|$ of units (x_1^*, x_2^*, x_3^*).

(h) Naming and designation of the type R for the convolution of qualimetric variables $\bar{X}$, $\hat{X} = R(\bar{X})$, for example, linear $\hat{X} = \sum (\bar{X} \times \bar{k})$, $\bar{k}$ – weighted coefficients of qualimetric variables.

(i) Intuitive construction of the convolution R, also naming and intuitive measurement of its parameters, for example, weighted coefficients k_1^*, k_2^*, k_3^*.

(k) Qualitative calculation of the convolution $\{\hat{X}^* = R(\overline{X^*})\}_{A'}$ in the subclass A' and placement of the results in the qualimetric scale $[1, 4]$.

(l) Designation of elements in the ranked series $\overline{1, |A'|}$ subclass A'.

(m) Deduction of a subset A'' subclass A' by excluding from subsets of the subclass A' with a non-strict order relation of objects with a larger (lower) number.

General set of semantics: $\{A, \ A', \ \bar{x}, \ \bar{X}, \ \bar{X} = \bar{\sigma}(\bar{x}), \ \{\overline{x*}\}_{A'}, \ \hat{X} = R(\bar{X}),$ $\{\hat{X}^* = R(\overline{X^*})\}_{A'}, \ \overline{1, |A'|}, \ \bar{\sigma}_i, \ i \in \overline{1, 3}, \ \bar{\sigma} \in [1,4]\}$.

Summary semantics: an ordered subset of the class of shopping facilities A'' with a strict order relation (Fig. 2).

3.2   Solving the Selection Problem on a Ranked Set of Matching Objects A'' with a Strict-Order Relation

The selection problem is solved in accordance with the variational semantics specified by naming and designating the mathematical construction in the form of predicates, using intuition in the interpretation of the first approach:

(1) identifying *a more preferred* (by R) object $\exists! \, a \in A'' -$

$$\forall(b)(b \in A'')(\exists! a \in A'')P(R_a \geq R_b); \tag{1}$$

(2) identifying a subset $A''' \subset A''$ with the cardinal number $|A'''|_{3a\partial}$ *the most preferred* object from $A'' -$

$$\forall(a, \ b)(a \in A''')(b \in A'')P(R_a \geq R_b), \ |A'''| = |A'''|_{3a\partial}; \tag{2}$$

(3) identifying *the least preferred* (by R) object $\exists! \, a \in A'' -$

$$\forall(b)(b \in A'')(\exists! a \in A'')P(R_a \leq R_b); \tag{3}$$

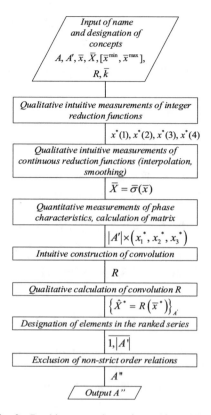

Fig. 2. Ranking procedure of matching objects.

(4) identifying a subset $A''' \subset A''$ with the cardinal number $|A'''|_{3a\partial}$. **the least preferred** objects from A'' –

$$\forall(a,\, b)(a \in A''')(b \in A'')P(R_a \leq R_b),\ |A'''| = |A'''|_{3a\partial}; \qquad (4)$$

(5) identifying **the most preferred** (by R, $R \geq R_{\min}$) object $\exists!\, a \in A''$, –

$$\forall(b)(b \in A'')(R_b \geq R_{\min})(\exists!a \in A'')P(R_a \geq R_b); \qquad (5)$$

(6) identifying a subset $A''' \subset A''$ with the cardinal number $|A'''|_{3a\partial}$. **the most preferred** (by R, $R \geq R_{\min}$) objects from A'' –

$$\forall(a,\, b)(a \in A''')(b \in A'')(R_b \geq R_{\min})P(R_a \geq R_b),\ |A'''| = |A'''|_{3a\partial}; \qquad (6)$$

(7) identifying **the least preferred** (by R, $R \leq R_{\max}$) object $\exists!\, a \in A''$ –

$$\forall(b)(b \in A'')(\exists!a \in A'')(R_b \leq R_{\max})P(R_a \leq R_b); \qquad (7)$$

(8)   identifying a subset $A''' \subset A''$ with the cardinal number $|A'''|_{3a\partial.}$ **the least preferred** objects by R, $R \leq R_{\max}$ from A'' –

$$\forall(a,\ b)(a \in A''')(b \in A'')(R_b \leq R_{\max})P(R_a \leq R_b),\ |A'''| = |A'''|_{3a\partial.};\quad (8)$$

(9)   identifying **the most preferred** (by R) object $\exists!\ a \in A''$ with limitations for the characteristic x_b –

$$\forall(b)(b \in A'')(x_b \geq x_b^{\min})(\exists!a \in A'')P(R_a \geq R_b);\quad (9)$$

(10) identifying a subset $A''' \subset A''$ with the cardinal number $|A'''|_{3a\partial.}$ **the most preferred** objects from A'' with limitations for the characteristic x_b –

$$\forall(a,\ b)(a \in A''')(b \in A'')(x_b \geq x_b^{\min})P(R_a \geq R_b),\ |A'''| = |A'''|_{3a\partial.};\quad (10)$$

(11) identifying **the most preferred** (by R) object $\exists!\ a \in A''$ with limitations for a subset of characteristics $\rho^{\bar{x}} \in B(\bar{x})$, $|B| = 2^{|\bar{x}|}$, where B – boolean operator from the subset $\bar{x}$, $\rho^{\bar{x}} \succ \rho^{\bar{x}}_{\partial on.}$ –

$$\forall(b)(b \in A'')(x_b \geq x_b^{\min})(\rho^{\bar{x}} \succ \rho^{\bar{x}}_{\partial on.})(\exists!a \in A'')P(R_a \geq R_b);\quad (11)$$

(12) identifying a subset $A''' \subset A''$ with the cardinal number $|A'''|_{3a\partial.}$ **the most preferred** objects from A'' with limitations for a subset of characteristics $\rho^{\bar{x}} \in B(\bar{x})$, $|B| = 2^{|\bar{x}|}$, where B – boolean operator from the subset $\bar{x}$, $\rho^{\bar{x}} \succ \rho^{\bar{x}}_{\partial on.}$ –

$$\forall(a,\ b)(a \in A''')(b \in A'')(x_b \geq x_b^{\min})(\rho^{\bar{x}} \succ \rho^{\bar{x}}_{\partial on.})P(R_a \geq R_b),\ |A'''| = |A'''|_{3a\partial.}.$$
$$(12)$$

And so on.

4 Solving the Problems of Ranking and Selection with the Support of a Specialized Software Package

The specialized software package, which has been officially registered as a computer program [19], is a platform for modeling a wide range of applied problems of ranking and selection. It technologically combines the previously described approaches (Sect. 2) to the interpretation of the concept "intuition". Below (Figs. 3, 4, 5 and 6), for example (Sect. 3), the fragments of how the preference model works, are illustrated when solving the problem of ranking, though the fragments (Figs. 3, 4 and 5) suggest the possibility of using intuition in the interpretation of the second approach.

Fig. 3. Setting a characteristic of the object domain in the software package «Jobs-Decon».

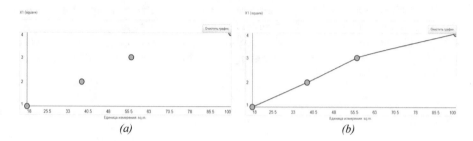

Fig. 4. Setting discrete (a) and continuous (b) reduction functions for the characteristic of object x1 in the software package «Jobs-Decon».

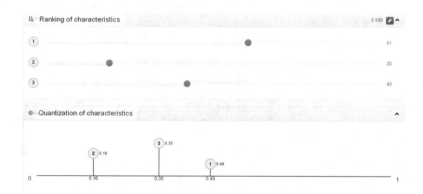

Fig. 5. Procedure of characteristic ranking in the software package «Jobs-Decon».

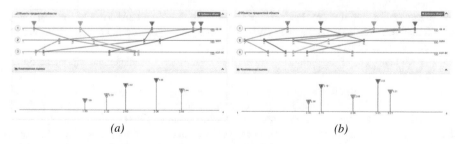

Fig. 6. Procedure of the integrated assessment of characteristics in the software package «Jobs-Decon».

5   Conclusion

The results presented in the article a significant novelty in the field of quantification of mental variables, the type of preferences, and the construction of new platforms of preference models, built on a compromise between two "opposite" approaches to the interpretation of the concept "*subjectivity*". These results, including [16], have no analogues in modern literature [1, 2, 6, 9, 10] and for the first time make it possible to solve complex multi-alternative and multi-factor ranking problems and selecting with a special emphasis on establishing a strict-order relation on the set of matching objects and the non-manipulability of the resultant semantics. In this case, a sufficiently high level of justification of management decisions by economic entities is established, excluding the forced episodes of "guessing" their best alternatives and/or manifestations of subjectivity.

References

1. Kothari, C.R.: Research methodology: methods and techniques. New Age International (2004)
2. Sargent, R.G.: Verification and validation of simulation models. J. Simul. **7**(1), 12–24 (2013)
3. Vol'f, E.M.: Funktsional'naya semantika otsenki (Functional semantics of evaluation). URSS, Moscow (2009)
4. Carnap, R.: Introduction to Semantics. Harvard University Printing Office, Cambridge (1948)
5. Jackendoff, R.: Semantics and Cognition, vol. 8. MIT Press, Cambridge (1983)
6. Myers, D.: Intuition. Series Masters of psychology, Saint-Petersburg (2018)
7. Duenes, D.: Let intuition be your guide, pp. 113–115, 134, 141. New Woman (1998)
8. Miller, G.A.: Psychology: the Science of Mental Life. Harper & Row, New York (1962)
9. Gilovich, T.: How we know what isn't so: the fallibility of human reason in everyday life. Free Press, New York (1991)
10. Kharitonov, V.A. et al.: Intellektual'nye tehnologii obosnovanija innovacionnyh reshenij: monografija (Intellectual technology studies innovative solutions: monograph). In: Kharitonov, V.A. (ed.) Perm, Izd-vo Perm.gos. tehn. un-ta (2010)
11. Kharitonov, V.A., Belyh, A.A.: Tehnologii sovremennogo menedzhmenta (Modern management technologies). In: Kharitonov, V.A. (ed.) Perm, Izd-vo Perm.gos. tehn. un-ta (2007)
12. Fleischmann, A., Kannengiesser, U., Schmidt, W., Stary, C.: Subject-oriented modeling and execution of multi-agent business processes. In: IEEE/WIC/ACM International Joint Conferences on Web Intelligence (WI) and Intelligent Agent Technologies (IAT), vol. 2, pp. 138–145. IEEE (2013)
13. Fleischmann, A., et al.: Subject-Oriented Business Process Management. Springer, Berlin (2012)
14. Ozhegov, S.I., SHvedova, N.Y.: Tolkovyj slovar' russkogo yazyka: 72500 slov i 7500 frazeologicheskikh vyrazhenij (Explanatory dictionary of the Russian language: 72500 words and 7500 phraseological expressions). Az, Moscow (1992)
15. Grzega, J.: Onomasiological approach to word-formation. A comment on Pavol Štekauer: Onomasiological approach to word-formation. SKASE J. Theor. Linguist. **2**, 76–80 (2005)

16. Koch, P.: Lexical Typology. Language Typology and Language Universals: An International Handbook, pp. 1142–1178 (2001)
17. Krysin, L.P. (ed.): Tolkovyj slovar' russkoj razgovornoj rechi (Explanatory dictionary of the living Russian language), Moscow, Yazyki slavyanskoj kul'tury, vol. 1 (2014)
18. Kahneman, D., Hertwig, R.: Do frequency representations eliminate conjuction effects? An exercise in adversarial collaboration. Psychol. Sci. **12**, 269–275 (2001)
19. Alekseev, A.O., Vychegzhanin A.V., Dmitryukov M.S., Krivogina D.N., Melekhin M.I., Kharitonov V.A., Shaydullin R.F.: Avtomatizirovannaya sistema sub"ektno-orientirovannogo resheniya linejnykh zadach ranzhirovaniya/vybora na osnove soedineniya kreativnosti i tekhnologichnosti («Dzhobs-Dekon») (Automated system of subject-oriented solution of linear ranking/selection tasks based on the connection of creativity and technology («Jobs-Decon»)). Certificate of state registration of computer programs Rossiiskaia Federatsiia no. 2018614405 (2018)

Two-Diode Model Parameter Evaluation from Dark Characteristics of Back-Contact Back-Junction Solar Cells

Luis Chuquimarca[1](✉) ⓘ, Ximena Acaro[2](✉) ⓘ,
Alfonso Gunsha[3](✉) ⓘ, Luis Villamagua[4](✉) ⓘ,
and David Sánchez[1](✉) ⓘ

[1] Universidad Estatal Península de Santa Elena, 240204 La Libertad, Ecuador
{lchuquimarca,dsanchez}@upse.edu.ec
[2] Universidad de Guayaquil, 090514 Guayaquil, Ecuador
ximena.acaroc@ug.edu.ec
[3] Universidad Nacional de Loja, 110103 Loja, Ecuador
alfonso.gunsha@unl.edu.ec
[4] Grupo de Fisicoquímica de Materiales, Sección de Fisicoquímica y
Matemáticas, Departamento de Química y Ciencias Exactas,
Universidad Técnica Particular de Loja, Apartado, 1101608 Loja, Ecuador
lmvillamagua@utpl.edu.ec

Abstract. The back-contact back-junction (BC-BJ) solar cells is a novel structure that increases the optoelectronic performance of the device. The two-diode model provides information regarding the different process involved in the BC-BJ solar cell operation, through the behavior of the dark current-voltage (I-V) curve, at different distances (from 200 μm to 900 μm) and widths of the emitter-Al (from 50 μm and 100 μm). Simulated results indicate that smaller width increase the performance and efficiency. In the analyzed model we get into the performance, qualities and dimensions through of electrical simulations, using the Technology Computer-Aided Design (TCAD). The saturation current densities and parasitic resistance are determined through the adjustment and optimization of the dark (I-V) curve, resulting that the longer distance the emitter increases the dark series resistance; also, the dark saturation current density is greater by decreasing the width of the emitter.

Keywords: Two-diode model · Back-contact back-junction
Dark current-voltage · Solar cells

1 Introduction

The BC-BJ silicon solar cells have high efficiencies and low-cost manufacturing technology. Moreover, the parameters extracted of the two-diode model and the investigation of the parameters provides information on the mechanisms operating without illumination, and their effects on the device performance parameters.

The BC-BJ silicon solar cells are integrated by electrical bars of both polarities, and situated on the backside [1]. Therefore, the contact and the contacted diffused regions

© Springer Nature Switzerland AG 2019
T. Antipova and A. Rocha (Eds.): DSIC 2018, AISC 850, pp. 200–207, 2019.
https://doi.org/10.1007/978-3-030-02351-5_24

are situated on the backside [2]. Then, it was created with the emitter, the back surface field and the contacts designed as an integrated finger pattern.

Analysis of a solar cell without illumination has become important because its features are represented in the model of a diode. On the other hand, under illumination, analysis shows that small fluctuations in the light intensity add considerable amounts of noise, so that dark I-V measurements use inject carriers into the circuit with electrical means instead of light generated carriers.

Dark I-V measurements give extra information about the solar cell characteristics, even in the absence of noise. Within the most important parameter to analyze we can name, the quality of the junction, antireflection layers, grid and contact resistances. Series resistance influences the fill factor losses, also the maximum power and the efficiency of the solar cell [3]. This resistance has the following components: a contact resistance metal-semiconductor, ohmic resistance in the metal contacts, and ohmic resistance in the semiconductor material [4].

The present article is divided in the following sections: Sect. 2 describes the dark analysis of the two-diode model and BC-BJ silicon solar cells. The methodology of simulation used is detailed in Sect. 3 through of Schockley Read Hall recombination, recombination Auger and Surface recombination. Section 4 describes the results and simulation tools as fitting and optimization. Finally, conclusions are explained in detail in Sect. 5.

2   Physics of Solar Cells

2.1   Two-Diode Model – Dark Analysis

Under dark conditions in the two-diode model, diode 1 and diode 2 represent the diffusion and recombination processes, respectively [5]. However, this model adds parasitic resistances that can be analyzed through (Eq. 1).

$$J(V) = J_{01}\left(e^{(q/n_1 kT)(V-JR_s)} - 1\right) + J_{02}\left(e^{(q/n_2 kT)(V-JR_s)} - 1\right) + \frac{V - JR_s}{R_{sh}} \quad (1)$$

Where, J is the current density through an applied load, V is the voltage drop across this load, J_{01} is the generation current density, J_{02} the recombination current density, n_1 is the ideality factor approximate to 1, and n_2 is the ideality factor in ideal cases it should vary between 1 and 2, but general case of silicon solar cells, it could take values higher than 2, R_s are the series resistances, and R_{sh} is the shunt resistance.

In the dark the "-1" term is typically ignored to make the analysis simpler. Moreover, given that R_{sh} has big values, the $V - JR_s/R_{sh}$ term tend to zero, then Eq. 1 simplifies into Eq. 2.

$$J(V) = J_{01}\left(e^{(q/n_1 kT)(V-JR_s)}\right) + J_{02}\left(e^{(q/n_2 kT)(V-JR_s)}\right) \quad (2)$$

The diffusion current density is influenced by the properties of the quasi-neutral regions of the PN junction; instead, the recombination current density is influenced by the density of defect states in the energy bandgap, which is formed by the energy released from recombination of electron-hole pairs [6].

2.2   Back Contact Back Junction Solar Cells

The BC-BJ silicon solar cells have the contact and the diffuse active regions on the backside. The solar cell features a phosphorus-doped front-surface field and is novel cell structure with locally overcompensated boron emitters.

Advantages are the contact movement at the backside giving solution at the front surface contact reflection and may improve cell interconnection in modules allowing an increase in the packing density of solar cells. Hence, the front and rear sides are optimized for optical and electrical performance, respectively.

Due to the high recombination that occurs in the front side of the emitter, the optical part is optimized to obtain greater set of wavelengths [7–9].

3   Solar Cells Simulation

3.1   Carrier Statistics

For carrier concentration in Si lower than $1 \times 10^{19} \, cm^{-3}$ Boltzmann statistics can be used, but if the carrier concentrations in Si are higher than $1 \times 10^{19} \, cm^{-3}$ Fermi-Dirac statistics should be used [10]. In TCAD, the carrier densities can be calculated from the electron and hole quasi-Fermi potentials, and vice versa, using Boltzmann statistics.

3.2   Schockley Read Hall Recombination

In the solar cells, recombination through states in the forbidden bandgap, is explained by Schocley Read Hall (SRH) recombination theory. In TCAD, the following form is calculated (Eq. 3):

$$R_{net}^{SRH} = \frac{pn - n_{i,eff}^2}{\tau_{SRH,n}(p + p_1) + \tau_{SRH,p}(n + n_1)} \tag{3}$$

Where, $\tau_{SRH,n}$ and $\tau_{SRH,p}$ are the carrier lifetimes.

The BC-BJ solar cells are produced from n-type float-zone silicon wafers, with a doping dependence of the SRH lifetimes. Therefore, the lifetime parameters $\tau_{SRH,n}$ and $\tau_{SRH,p}$ can be taken as doping independent in the equation (Eq. 4) [11]. In aluminum-doped silicon, the low lifetimes are attributed to a highly recombination-active defect complex integrated of aluminum-oxygen [12]. Hence, increases J_0, which is measured of Al-alloyed back surface field [13–15].

$$\tau_{SRH,dop} = \tau_{min} + \frac{\tau_{max} - \tau_{min}}{1 + \left(\frac{N_A + N_D}{N_{ref}}\right)^{\gamma}} \tag{4}$$

Where, γ and N_{ref} are fit parameters, τ_{max} and τ_{min} are the best and worst case carrier lifetimes, and N_A and N_D are the bulk acceptor and donor levels.

3.3 Recombination Auger

The band-to-band Auger recombination rate is determined using TCAD application [16]. This equation is calculated by the electron and hole densities together with the temperature dependent coefficients.

$$R_{Auger} = \left(C_n n + C_p p \right) \left(pn - n_{i,eff}^2 \right) \tag{5}$$

Where, $n_{i,eff}$ is the effective intrinsic carrier concentration, and C_n and C_p are temperature dependent coefficients.

3.4 Surface Recombination

When the surface recombination is established by surface recombination velocity, expressed in TCAD by equation (Eq. 6)

$$s = s_0 \left[1 + s_{ref} \left(\frac{N_i}{N_{ref}} \right)^{\gamma} \right] \tag{6}$$

Where, s_0, s_{ref}, N_{ref} and γ are tunable parameters of the model and N_i denotes the doping concentration at interface.

The surface recombination is limited by holes in n-type emitters, that is, by the recombination velocity parameter for holes (S_p). The S_p increase progressively at low than at high dopant densities. The equation which calculates S_p is (Eq. 7):

$$s_p = s_{p1} \left(\frac{N_i}{N_{p1}} \right)^{\gamma_{p1}} + s_{p2} \left(\frac{N_i}{N_{p2}} \right)^{\gamma_{p2}} \tag{7}$$

Where, γ_{p1} and γ_{p2} are a measure for the slope, s_{p1}, s_{p2}, N_{p1} and N_{p2} shift s_p vertically.

The s_p increase dependently on the dopant density of the surface, on the surface (textured or planar), e.g. the pyramidal surface texturing increases s_p approximately by factor of 5. The parameterization of s_p at highly doped Si:P required in TCAD.

4 Results

4.1 Fitting Method

For extraction parameters was used two different ways that are the curve fitting and interpolation standard. The curve-fitting method is used when the parameters can be described by an equation (Eq. 2) and a dark I-V curve. The curve fitting algorithm determines the unknown variables and the unknown constants, but exist a minimal error between the measured output and the estimated output. The interpolation standard is used when the parameters cannot be described by a single formula, therefore is divided in regions [17]. The extraction method used in this study has made use of both approximation and iteration techniques. The limits in the iteration ranges assure that the extraction of the parameters is approximate to the measurement parameters with minimal errors. Also, this is a portion of the total difference between the measured current density (J_m) and the generated current density (J_g) which is accredited to the supposed two-diode model.

The equivalent circuit for a solar cell is determined by equation (Eq. 2), when J is an explicit function of V involving the two-diode model parameters. By reducing the number of parameters, the expression for J only depends of the parameters such as J_{01} and J_{02} where are extracted from both the J and V, also assuming that $n_1 = 1$ that represents the diffusion process within the solar cell at room temperature and $n_2 = 2$ that theoretically expresses the recombination process [18].

First are calculated the parameters with the emitter-Al at a width of 100 nm in the different position and before change the width at 50 nm in the survival position esti-mated (Figs. 1 and 2).

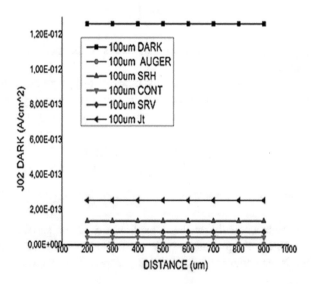

Fig. 1. Specific comparison of the parameters J_{02} with the width 100 μm of the Emitter-Al at survival distances

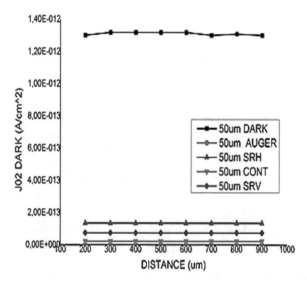

Fig. 2. Specific comparison of the parameters J_{02} with the width 50 μm of the Emitter-Al at survival distances

4.2   Optimization Method

To calculate R_s is only valid if a "good" voltage region is selected, using the optimization algorithm. However, to obtain the most R_s accurate values, one must fit the solar cell parameters and then introduce them in the Eq. 2, given in the optimization algorithm through the Levenberg-Marquardt method (Figs. 3 and 4) [19].

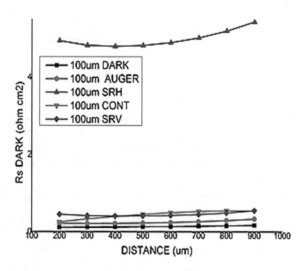

Fig. 3. Specific comparison of the parameters R_s with the width 100 μm of the Emitter-Al at survival distances

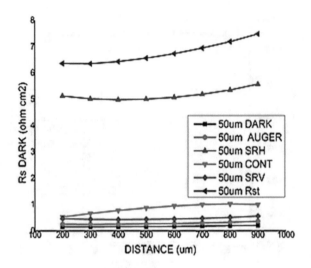

Fig. 4. Specific comparison of the parameter R_s with the width 50 μm of the Emitter-Al at survival distances

5 Conclusions

We designed the TCAD application for numerical simulation of photovoltaic modules that contain BC-BJ silicon solar cells. This two-diode model describes the dark I-V characteristics of parameters solar cells as compared to the one-diode model existing that have been analyzed previous articles. Moreover, although the dark saturation current density is smaller with a width in the emitter-Al of 100 μm than at 50 μm, the decrease of the dark saturation current density increases the open circuit voltage. This means a higher efficiency in the case of study of silicon solar cells BC-BJ. In addition, increasing the distance between the contacts produces an increase in the dark series resistance, which decreases the efficiency of the BC-BJ silicon solar cell.

References

1. Kerschaver, E., Beaucarne, G.: Back-contact solar cells: a review. Prog. Photovoltaics Res. Appl. **14**(2), 107–123 (2006)
2. Desa, M., Sapeai, S., Azhari, A., Sopian, K., Sulaiman, M., Amin, N., Zaidi, S.H.: Silicon back contact solar cell configuration: a pathway towards higher efficiency. Renew. Sustain. Energy Rev. **60**, 1516–1532 (2016)
3. Luque, A., Hegedus, S.: Handbook of Photovoltaic Science and Engineering. Wiley, Hoboken (2011)
4. Agrawal, A., Lin, J., Barth, M., White, R., Zheng, B., Chopra, S., Datta, S.: Fermi level depinning and contact resistivity reduction using a reduced titania interlayer in n-silicon metal-insulator-semiconductor ohmic contacts. Appl. Phys. Lett. **104**(11), 112101 (2014)
5. Khanna, V., Das, B., Bisht, D., Singh, P.: A three diode model for industrial solar cells and estimation of solar cell parameters using PSO algorithm. Renew. Energy **78**, 105–113 (2015)

6. Fell, A., Schön, J., Müller, M., Wöhrle, N., Schubert, M., Glunz, S.: Modeling edge recombination in silicon solar cells. IEEE J. Photovolt. **8**(2), 428–434 (2018)
7. Ceuster, D., Cousins, P., Rose, D., Vicente, D., Tipones, P., Mulligan, W.: Low cost, high volume production of >22% efficiency silicon solar cells. In: Proceedings of the 22nd European Photovoltaic Solar Energy Conference, pp. 816–819 (2007)
8. Köntges, M., Gast, M., Brendel, R., Meyer, R., Giegerich, A., Merz, P.: A novel photovoltaic-module assembly system for back contact solar cells using laser soldering technique. In: 23st European Photovoltaic Solar Energy Conference, pp. 1–5. Spain (2008)
9. Spath, M., De Jong, P.C., Bennett, I.J., Visser, T.P., Bakker, J.: A novel module assembly line using back contact solar cells. In: Photovoltaic Specialists Conference, PVSC'08. 33rd IEEE pp. 1–6. IEEE (2008)
10. Haug, H., Kimmerle, A., Greulich, J., Wolf, A., Marstein, E.: Implementation of Fermi-Dirac statistics and advanced models in PC1D for precise simulations of silicon solar cells. Sol. Energy Mater. Sol. Cells **131**, 30–36 (2014)
11. Altermatt, P.P.: Models for numerical device simulations of crystalline silicon solar cells—a review. J. Comput. Electron. **10**(3), 314 (2011)
12. Schmidt, J., Thiemann, N., Bock, R., Brendel, R.: Recombination lifetimes in highly aluminum-doped silicon. J. Appl. Phys. **106**(9), 093707 (2009)
13. Altermatt, P.P., Steingrube, S., Yang, Y., Sprodowski, C., Dezhdar, T., Koc, S., Schmidt, J.: Highly predictive modelling of entire Si solar cells for industrial applications. In: 24th European Photovoltaic Solar Energy Conference, pp. 901–906 (2009)
14. Bock, R., Altermatt, P.P., Schmidt, J., Brendel, R.: Formation of aluminum–oxygen complexes in highly aluminum-doped silicon. Semicond. Sci. Technol. **25**(10), 105007 (2010)
15. Rüdiger, M., Rauer, M., Schmiga, C., Hermle, M.: Effect of incomplete ionization for the description of highly aluminum-doped silicon. J. Appl. Phys. **110**(2), 024508 (2011)
16. Limpert, S., Ghosh, K., Wagner, H., Bowden, S., Honsberg, C., Goodnick, S., Green, M.: Results from coupled optical and electrical sentaurus TCAD models of a gallium phosphide on silicon electron carrier selective contact solar cell. In: IEEE 40th. Photovoltaic Specialist Conference 2014, PVSC, pp. 0836–0840. IEEE (2014)
17. Haouari-Merbah, M., Belhamel, M., Tobias, I., Ruiz, J.M.: Method of extraction and analysis of solar cell parameters from the dark current-voltage curve. In: Spanish Conference on Electron Devices 2005, pp. 275–277. IEEE (2005)
18. Hussein, R., Borchert, D., Grabosch, G., Fahrner, W.R.: Dark I-V-T measurements and characteristics of (n) a-Si/(p) c-Si heterojunction solar cells. Sol. Energy Mater. Sol. Cells **69**(2), 123–129 (2001)
19. Dkhichi, F., Oukarfi, B., Fakkar, A., Belbounaguia, N.: Parameter identification of solar cell model using Levenberg–Marquardt algorithm combined with simulated annealing. Sol. Energy **110**, 781–788 (2014)

Efficiency of Back Contact-Back Junction Solar Cells with Variable Contact in the Emitter

Alfonso Gunsha-Morales[1]($\boxtimes$) ⓘ, Ximena Acaro[2]($\boxtimes$) ⓘ,
Luis Chuquimarca[3]($\boxtimes$) ⓘ, Luis Villamagua[4]($\boxtimes$) ⓘ,
and David Sánchez[3]($\boxtimes$) ⓘ

[1] Universidad Nacional de Loja, 110103 Loja, Ecuador
alfonso.gunsha@unl.edu.ec
[2] Universidad de Guayaquil, 090514 Guayaquil, Ecuador
ximena.acaroc@ug.edu.ec
[3] Universidad Estatal Península de Santa Elena, 240204 La Libertad, Ecuador
{lchuquimarca,dsanchez}@upse.edu.ec
[4] Grupo de Fisicoquímica de Materiales, Sección de Fisicoquímica y
Matemáticas, Departamento de Química y Ciencias Exactas,
Universidad Técnica Particular de Loja, Apartado, 1101608 Loja, Ecuador
lmvillamagua@utpl.edu.ec

Abstract. As renewable energy becomes more sustainable and more viable than it has ever been, more and more countries are turning 'green' and reverting to sustainable ways. We have carried out the study of a cell Back Contact - Back Junction (BC-BJ) solar cell with an alone contact in the issuer who has as efficiency of 22.027%. This work presents a simulation of BC-BJ, with variation (50 µm, 100 µm) and mobility of a second contact along the emitter, the inclusion of a second contact variable and being moved through the emitter presents an improvement in efficiency at a midpoint of the cell. Increasing the second contact of 50 µm in the emitter with a position of 500 µm has as efficiency of 22.152% in the same position with the contact of 100 µm has as efficiency of 22.109%, On having increased the second issuer in the cell BC-BJ achieved an increase in the efficiency of 0.042% between the two contacts and between the efficiency of a single contact in the cell.

Keywords: Back Contact (BC) · Back Junction (BJ) · Fill factor (FF)
Efficiency (η) · Open-circuit voltage (V_{OC}) · Short-circuit current (J_{SC})

1 Introduction

Nowadays electrical power is essential to most modern-world activities. Besides that, the world has been supplied with energy harvested from non-renewable sources (e.g., crude oil, natural gas, coal, uranium, etc.) for over 140 years, it has been causing irreversible damage to some parts of the planet (e.g. global warming, pollution of air and soil, etc.) since its beginning [1]. In the hope of remedying this, some countries have been focusing their attention towards renewable sources in the last decades [2];

© Springer Nature Switzerland AG 2019
T. Antipova and A. Rocha (Eds.): DSIC 2018, AISC 850, pp. 208–215, 2019.
https://doi.org/10.1007/978-3-030-02351-5_25

renewable energies are ways to obtain energy that can be naturally replenished on a human timescale, such as sunlight, wind, tides, rain, waves, and geothermal heat [3].

While renewable energy is greater for the environment in every sense, they still face major obstacles. Some are inherent with all new technologies, while others are the result of a skewed regulatory framework and marketplace [4]. The most obvious barriers are cost, viability and efficiency [5]. That has brought about a poor progress of these technologies; for instance, still in March 2017, wind and solar accounted only for 10% of all US electricity generation.[1]

Photovoltaics (PV) is the science behind the most popular form of harnessing solar energy [6]. It is the process of converting sunlight directly into electricity. Though this technology has proven to be one of the technologies to replace non-renewables, there is still a long way to go this goal can be fulfilled. Efficiency of PV are still under 22.5%,[2] and therefore it must be improved. It is well-known that efficiency can be improved by analyzing the dark zone of a photocell [7]. With that in mind, we propose to carry out the increase of a second mobile contact in the emitter each 100 μm.

2 Device Solar Cells

2.1 Solar Cells Structure - Back Contact - Back Junction

Solar cell is an electronic tool which converts sunlight into electricity. There are different kinds of solar cells (examples: *GaAs* Thin film, *InP* cell, *CdTe* cell [8]) but silicon based solar cells are used commonly, because the less complicated technology used for their production (BC-BJ). Standard solar cell consists on a large *p-n* junction, with a uniformly doped base or bulk region and a diffused opposite doped region called emitter [9] (Table 1 and Fig. 1).

Table 1. All parameters for cell BC-BJ with emitter contact

Name	Pitch	Ni thickness	SiO$_2$ thick1	FSF thick	Emitter width	Al thick	Em. Al width	Emitter width	BSF width	BSF Al width	Bulk thick	SiO$_2$ thick
Value (μm)	1000	0.069	0.027	1	2	0.5	50/100	770	120	50	300	0.1

2.2 Operation

a. Photoelectric Effect General
When light illuminates a piece of metal, the light causes electrons to the metal surface begin to move and these electrons are detected as a change in electric charge of the metal or as an electrical current. For this reason, they are called photo light and electricity for power (Fig. 2).

---

[1] https://www.ucsusa.org/clean-energy/renewable-energy/barriers-to-renewable-energy#.W0Y0UtL0l PY.

[2] https://news.energysage.com/what-are-the-most-efficient-solar-panels-on-the-market/.

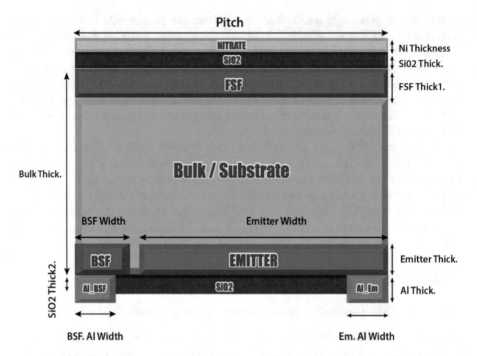

Fig. 1. Architecture of a typical solar cell Back contact - Back Junction

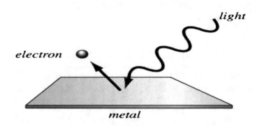

Fig. 2. Photoelectric effect.

There is a simple explanation for this phenomenon where the revolutionary physical concepts that exist today and are nowadays used. In terms of composition of quantum mechanics, the light is known as the "wave-particle duality" of nature [10].

b. Photoelectric Effect Semiconductors

Metalloids or commonly known as semiconductors are interconnected by covalent bonds, thus electrons are more difficult time in the movement, compared with metals. This makes the conduction band of metalloids (semiconductors) is significantly different from the filled valence band of metallic elements.

The electron emission silicon bonds involve removing electrons silicon-silicon. As in the vast majority of semiconductors, high purity silicon conducts current only when a high voltage is applied. This energy is used to jump electrons valence band to the conduction band [11] (Fig. 3).

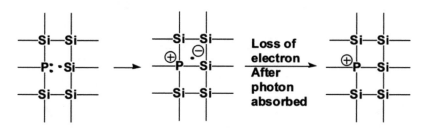

Fig. 3. The n-type semiconductor.

2.3   Figures of Merits

a. Short-circuit current
(J_{sc}) of a solar cell is a function of incoming photon absorption within the solar cell. In a solar cell ideal, the technologies related to optical effects are not known. These effects are reflected in the front surface metallization grid shading, transmission through the silicon wafer and the parasitic absorption in the dielectric layers or silicon regions with high doping [12] (Fig. 4).

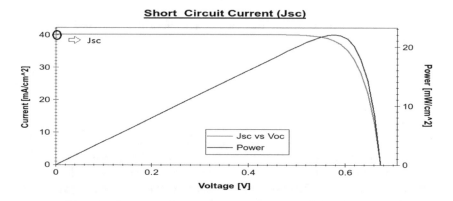

Fig. 4. $J_{SC}-V_{OC}$ curve of a solar cell showing the short-circuit current.

b. Open-circuit voltage
V_{OC} for draw power from a solar cell there has to be a finite voltage drop across it. Theory and experimental data suggest that as this voltage increases the electron hole product and recombination rate throughout the cell increase.

The maximum open circuit voltage is obtained when the recombination process throughout the cell is kept at a minimum; it is possible to eliminate recombination via bulk defects by having defect free substrates of silicon.

The open-circuit voltage is exposed on the IV curve be (Fig. 5).

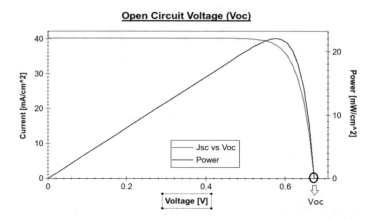

Fig. 5. J_{SC}–V_{OC} curve of a solar cell showing the open-circuit voltage

c. Fill Factor

FF is a measure of maximum power output that can be obtained from a solar cell. It is a very important parameter for a cell. Fill factor is heavily dependent on the open circuit voltage of the cell.

This dependency is somewhat monotonic. Therefore this particular parameter is limited by the limitation of open circuit.

$$FF = \frac{V_{OC} - \ln(V_{OC} + 0.72)}{V_{OC} + 1} \tag{1}$$

d. Efficiency

According to [11], the efficiency of the cell quantifies the fraction of solar energy that it converts into electrical energy. In principle it depends on two parameters.

The generation of current by absorbed incident illumination and the loss of charge carriers via so-called recombination mechanisms. A built in p–n diode separates charge carriers of opposite polarity, and drives the light-generated current through the cell and to the terminals.

Finally, several solar cells will be electrically connected and encapsulated as a module. In mathematical terms the efficiency η of a solar cell is given as

$$\eta = \frac{P_{out}}{P_{in}} = \frac{FF * V_{OC} * J_{SC}}{P_{SUN}} \tag{2}$$

3   Results

The work present to increase contact to the issuer's and with this will see if the efficiency is improved. First, it increases a contact 50 μm and then 100 μm to see which the comparison of these two cells is (Table 2 and Fig. 6).

Table 2. Results first

Name	Result
Short-circuit current (J_{sc})	40.25 mA/cm^2
Open-circuit voltage (V_{oc})	0.6754 V
Fill factor (*FF*)	81.032%
Efficiency (η)	22.027%

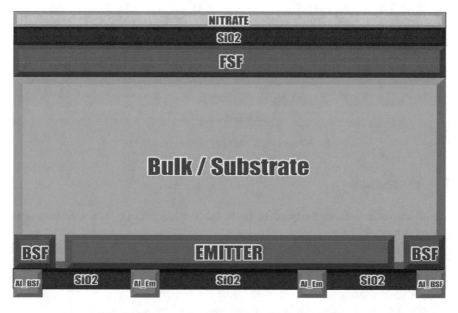

Fig. 6. Element of symmetry – two contact of 50 μm

Symmetrical element or the device simulator to find a solution, it needs to solve all equations at each mesh point. Since solar cells are large area devices with relatively smaller thickness, to model the entire solar cell, it would require a huge number of mesh points to be solved. Therefore simulating the entire device is not feasible, even in two dimensions (Table 3 and Fig. 7).

Table 3. Results with two contacts in the emitter 50 μm and 100 μm

Name	Result to 50 μm	Result to 100 μm
Position of the second contact	500 μm	500 μm
Short-circuit current (J_{sc})	40.26 mA/cm^2	40.25 mA/cm^2
Open-circuit voltage (V_{oc})	0.6754 V	0.6730 V
Fill Factor (*FF*)	81.476%	81.618%
Efficiency (η)	22.152%	22.109%

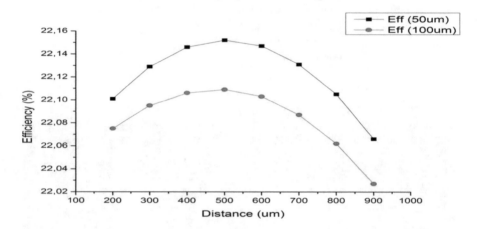

Fig. 7. Efficiency (%) contact 50 μm versus 100 μm

4 Conclusions

The highest solar cell efficiency of 22.109% for distance 500 μm, was achieved, in the case of contact 100 μm.

The highest solar cell efficiency of 22.152% for distance 500 μm, was achieved, in the case of contact 50 μm.

Fill Factor (FF) trend as a function of distance between contacts exhibits 0.142% between the 2 curves.

Efficiency trend as a function of distance between contacts exhibits a 0.042% between the 2 curves this behaviour is explained for taking the trend of fill factor.

References

1. Goldemberg, J.: Energy: What Everyone Needs to Know®. Oxford University Press, Oxford (2012)
2. Johansson, T.B., Reddy, A.K., Kelly, H., Williams, R.H., Burnham, L. (eds.): Renewable Energy: Sources for Fuels and Electricity. Island Press, Washington, DC (1993)

3. Zohuri, B.: Types of renewable energy. In: Hybrid Energy Systems, pp. 105–133. Springer, Cham (2018)
4. Foster, R., Ghassemi, M., Cota, A.: Solar Energy: Renewable Energy and the Environment. CRC Press, Cambridge (2009)
5. Hirst, E., Brown, M.: Closing the efficiency gap: barriers to the efficient use of energy. Resour. Conserv. Recycl. **3**(4), 267–281 (1990)
6. Luque, A., Hegedus, S. (eds.): Handbook of Photovoltaic Science and Engineering. Wiley, Hoboken (2011)
7. Shah, A.V., Meier, J., Vallat-Sauvain, E., Wyrsch, N., Kroll, U., Droz, C., Graf, U.: Material and solar cell research in microcrystalline silicon. Sol. Energy Mater. Sol. Cells **78**(1–4), 469–491 (2003)
8. Green, M.A., Emery, K., Hishikawa, Y., Warta, W.: Solar cell efficiency tables (version 37). Prog. Photovolt. Res. Appl. **19**(1), 84–92 (2011)
9. Fonash, S.: Solar Cell Device physics. Elsevier, New York (2012)
10. Dodd, J.N.: The photoelectric effect. In: Atoms and Light: Interactions, pp. 141–160. Springer, Boston (1991)
11. Kane, E.O.: Theory of photoelectric emission from semiconductors. Phys. Rev. **127**(1), 131 (1962)
12. Tiedje, T., Yablonovitch, E., Cody, G.D., Brooks, B.G.: Limiting efficiency of silicon solar cells. IEEE Trans. Electron Devices **31**(5), 711–716 (1984)

Recommendation System for Material Scientists Based on Deep Learn Neural Network

Andrei Kliuev[✉], Roman Klestov, Maria Bartolomey,
and Aleksei Rogozhnikov

Perm National Research Polytechnic University, Komsomolsky Avenue, 29,
Perm Region, 614990 Perm, Russia
kav-l@bk.ru

Abstract. The paper considers the possibilities of artificial neural networks and deep machine learning in the problem of predicting the physicomechanical properties of functional materials. It is shown that the popular deep neural network VGG with high accuracy solves the problem of hardness classification of metal alloy on the basis of iron. The prospects of building a generative adversarial network that is able to predict the structure of the alloy with predetermined physicomechanical characteristics are discussed.

Keywords: Deep networks · Machine learning · Functional materials

1 Introduction

Works on creation a new functional materials that can withstand loads under certain conditions, dictated by the design element, where they are supposed to be used, are conducted in the laboratories of materials technology at present time. In this work it is extremely important to analyze the microstructure of materials which in most cases is reduced to the analysis of microsections. A microsection – it is a sample with a specially treated surface to detect the microstructure of the material/metal. Digital images of the microsection surface are obtained by an optical microscope in the visible range, an electron microscope or, for example, by atomic force microscopy.

The analysis of the microsection provides a more complete picture of the state of the material and its properties at the macro and micro levels, and also allows you to control a range of physicomechanical properties without separate tests for each of them. A serious drawback of this approach is that the analysis requires the involvement of highly qualified experts in the field of materials technology and solid state physics. It should also be taken into account that the result of the analysis is subjective. At present, it becomes obvious that to improve the efficiency and objectivity of identification of material properties it is necessary to use modern mathematical methods of data processing and artificial intelligence algorithms to solve the problems of classification and identification of the material microstructure.

Functional materials are built for a specific design of the machine or mechanism. Therefore the volume of production of such materials unlike, for example, structural

© Springer Nature Switzerland AG 2019
T. Antipova and A. Rocha (Eds.): DSIC 2018, AISC 850, pp. 216–223, 2019.
https://doi.org/10.1007/978-3-030-02351-5_26

steels is not so large. In such conditions it is very important to consider the time and financial costs of research and preparation of the production technology of such materials. Automated intelligent analysis of microsections allows to reduce these costs, but cannot help in the process of selection of chemical composition and methods of thermomechanical processing, which lead to the creation of functional material with the desired properties. The purpose of this work consists in the construction and training of convolutional neural networks that can predict physicomechanical properties of the material at the images of the microstructure.

Within the framework of the developed approach, the task is to build an intelligent tool capable of generating the structure of the material with predetermined physical and mechanical characteristics. Such a tool can become the core of a materials scientist's recommendation system, which will help to carry out a purposeful search for a method of manufacturing a functional material, using a minimum number of experiments.

2 Related Work

There are successes in the recognition and spatial localization of objects by machine learning methods based on convolutional networks, first proposed by LeCun [1]. With millions of pre-classified images collected in bulky databases of annotated images, such as ImageNet or COCO, and the back propagation method the results were really impressive [2–4].

Modern methods are also used in metallurgy. The problems of classification of microstructures are solved mainly. For example, in [5] the problem of classification of high-carbon steels by their microstructure using the VGG-16 network is studied. For training, testing and verification the base of sections' images of the CMU-UHCS (Carnegie Mellon University Ultrahigh Carbon Steel) taken by an electron microscope (the base contains 961 marked images divided into 7 categories) is used. In [6] an attempt to classify low-carbon steels is made. The training set of Material Engineering Center Saarland (MECS) was used. In [7] recognition of dendritic microstructures on the digital images of the microsections is performed. The training set was formed on the basis of microimages of the DoITPoMS project (Dissemination of it for the promotion of Materials Science) of the University of Cambridge.

Convolutional networks are successfully used in similar problems in the sense of determining surface defects. In the work [8] with the help of MPCNN (CNN with max-polling of subsampling layers) the problem of classification of surface defects of steel pipes is set and solved. Later, the authors [9] solved a similar problem of classification of surface defects of steel sheets. In [10, 11] the method of detection of defects on the surface of rails is investigated.

The purpose of the work consists in the construction and training of a convolutional neural network able to predict the physicomechanical properties of the material at the images of the microstructure. Within the framework of the developed approach the problem of constructing a generative-adversarial network (GAN) [12, 13] able to represent the structure of the material with predetermined physicomechanical characteristics is posed. In [14] is shown that with the help of trained on textures convolutional network it is possible to calculate quite accurate characteristic of texture recorded in the form of Gram's matrix. Components of the matrix are obtained from the cores of convolutions belonging to different levels of the network. In the work a well-known pre-trained VGG-19 is used. The author argues that his approach is much more accurate and flexible than the description of the texture by a two-point correlation function. By applying noise and other transformations, it is able to generate textures similar to the original using a convolutional network. The author [15] used this technique to analyze and generate the microstructure of metals and alloys (Fig. 1). Noise, scaling and rotation were used during generation.

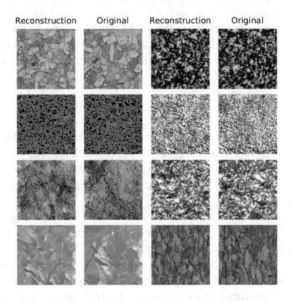

Fig. 1. The original structures and their reconstruction analogues [15]

The network constructed and trained in the work will be the basis for the construction of generative and discriminant parts of GAN.

3   Training Dataset and Learning Process

To solve the problem the VGG-16 architecture was chosen as the most proven convolutional neural network in this area. A pre-trained version of the network with an input image of size 256×256 was selected. The process of preparation the training set and the learning process of the network are described below.

3.1   Training Dataset

The primary data for the training set was prepared by the staff of the Institute of Nanosteels of Magnitogorsk State Technical University in the frame of international project Horizon 2020. The data was represented by 758 digital annotated images of metal alloy sections based on iron of different chemical composition and grain-phase structure. The images were made using an electron microscope with different powers from x1000 to x20,000. Examples of images are shown in Fig. 2. The images were annotated with physicomechanical characteristics measured during the tests.

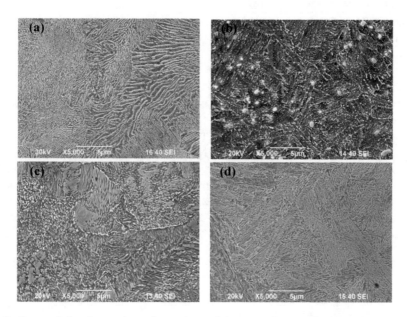

Fig. 2. Images of microsections from the training sample: (a) ferritic-carbide mixture, (b) secondary troostite-sorbite + chromium carbides, (c) lamellar pearlite + granular pearlite, (d) lower bainite

The samples were tested for hardness and strength to determine the physicomechanical properties and the following characteristics were established: microhardness, conditional yield strength, time resistance, elongation, relative contraction. Tests were carried out at room temperature. The manufacture of the sections was performed before the test for the unloaded sample.

During the process of preparing the training set of available images with the help of various transformations, learn and verification sets of 8200 and 2000 images were formed, respectively. This was done in order to increase the learning set and the accuracy of recognition. The learning set was divided into classes on microhardness. Classes of microhardness and distribution of the learning set on them are presented in Table 1.

3.2 Learning Process

The Keras framework in conjunction with the TensorFlow machine learning library was chosen as a software platform for the implementation of the network.

Table 1. Distribution of the training set by microhardness classes

Microhardness classes	Number of images in the learning set	Number of images in the verification set
I class (0–2600 MPa)	940	240
II class (2600–2900 MPa)	1420	260
III class (2900–3200 MPa)	2620	640
IV class (3200–3500 MPa)	1480	360
V class (3500–5000 MPa)	1980	500

The learning was conducted on a hardware platform with the following characteristics: CPU Intel Core i7, 1x GPU GeForce GTX 1080. We used a VGG16 pretrained network with ImageNet weights. Tuning was conducted during 200 epochs. One computational experiment took 9 h. Network learning lasted no more than 525 iterations. The calculation of accuracy was performed every epoch and is equal to the percentage of correctly recognized images from the testing set formed from the learning set using random selection. Figure 3 shows a graph of the accuracy changes in the learning process. At the end of learning, the network recognition accuracy reached 62.5%. Figure 3b shows the dependence of the change of the error function on the epoch number.

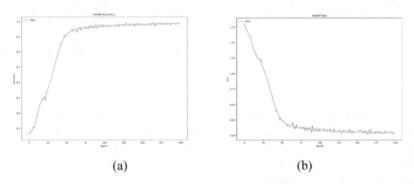

(a)   (b)

Fig. 3. Learning process: (a) accuracy changes, (b) changes of the error function

From Fig. 3 it can be seen that the learning process is typical for this kind of problems and does not have any features.

4   Results and Discussion

On the whole it is shown that the chosen approach to the study of the problem of predicting the physicomechanical properties of materials gives positive results. The resulting accuracy of the method according to different estimates is shown in Table 2.

Table 2. The classification results for the microhardness classes of the learned network

Microhardness classes	Accuracy on the verification set
I class	0.75
II class	0.5
III class	0.8125
IV class	0.5625
V class	0.4375
Average	0.625

Further development of the approach is in the use of generative-adversarial network in the problem of forecasting the microstructure of the material with the given physicomechanical properties. Networks of this kind consist of generative and discriminant parts. The purpose of the generative part is to create various modifications of the microstructure $G(z)$, where z – vector of undetected variables, which is a vector of small dimension, the values of which are random variables that evenly distributed in a certain interval. The discriminator $D(x)$ assesses the proximity of the microstructure x to the real data (to the data on which it was learned). In the process of finding a solution it is necessary to find a minimum of joint error $V(G, D)$ in the problem

$$\min_G \max_D V(G, D) = E_{x \sim d(x)}(\log(D(x))) - E_{x \sim p(z)}(\log(1 - D(G(z)))), \quad (1)$$

where E – loss function, e.g. cross-entropy.

The network learned in this work which belongs to the class of convolutional networks is used as a discriminator. Generative-adversarial network in which is used a convolutional network is called DCGAN (Deep Convolutional Generative Adversarial Networks).

The question of the formation of the generative part of the GAN should be referred to the discussion issues. First, the strategy of generation of new microstructures should be directed to the given values of target indicators, for example, on microhardness. It remains unclear how to ensure this now. Second, on the one hand, following the authors of [15] for the generation of the microstructure Gram's matrix is used taken from the network's convolution layers. This approach has shown high accuracy and that is main it has been tested in the problem of microstructure synthesis. However, there is a different approach. It consists in using the deconvolution network [13]. This approach is simpler in implementation and has a lot of applications in solving various problems. Further studies will compare these two approaches.

5   Conclusions

In the work the convolutional neural network that can analyze images of the microsections of alloys based on iron and carry out the classification of the microstructure on the microhardness with a precision of 93% on a test and 62.5% on the verification set for the evaluation of top-1 error was built and learned.

The problem of creating a recommendation system based on the learned network and GAN architecture is set in the work. A generative model that is able to generate variants of microstructures with the desired production capacity and utility was discussed. In the future, it is expected to implement and compare different approaches to the construction of the generative part of the GAN.

Acknowledgements. The reported study was funded by the Ministry of Education and Science of the Russian Federation (the unique identifier RFMEFI58617X0055) and by the EC Horizon 2020 is MSCA-RISE-2016 FRAMED Fracture across Scales and Materials, Processes and Disciplines. The authors express their thanks to the staff of the Institute of Nanosteels of Nosov Magnitogorsk State Technical University for providing the experimental data allowed to learn the neural network with a given accuracy.

References

1. LeCun, Y., Boser, B., Denker, J.S., Henderson, D., Howard, R.E., Hubbard, W., Jackel, L. D.: Backpropagation applied to handwritten zip code recognition. Neural Comput. **1**(4), 541–551 (1989)
2. Szegedy, C., Liu, W., Jia, Y., Sermanet, P., Reed, S., Anguelov, D., Erhan, D., Vanhoucke, V., Rabinovich, A.: Going deeper with convolutions. In: 2015 IEEE Conference on Computer Vision and Pattern Recognition (CVPR), pp. 1–9 (2015)
3. He, K., Zhang, X., Ren, S., Sun, J.: Deep residual learning for image recognition. In: 2016 IEEE Conference on Computer Vision and Pattern Recognition (CVPR), pp. 770–778 (2016)
4. Hu, J., Shen, L., Sun, G.: Squeeze-and-excitation networks. In: ILSVRC 2017 Image Classification Winner. arXiv:1709.01507 (2017)
5. DeCost, B.L., Francis, T., Holm, E.A.: Exploring the microstructure manifold: image texture representations applied to ultrahigh carbon steel microstructures. Acta Mater. **133**, 20–40 (2017)
6. Azimi, S.M., Britz, D., Engstler, M., Fritz, M., Mücklich, F.: Advanced steel microstructure classification by deep learning methods. arXiv:1706.06480 (2017)
7. Aristov, G.V., Klyuev, A.V.: Recognition and classification of the microstructure of metals and alloys using deep neural networks. In: Proceedings of the 27th International Conference on Computer Graphics and Machine Vision GraphiCon, pp. 180–183 (2017)
8. Masci, J., Meier, U., Ciresan, D., Schmidhuber, J., Fricout, G.: Steel defect classification with max-pooling convolutional neural networks. In: The 2012 International Joint Conference on Neural Networks (IJCNN), pp. 1–6 (2012)
9. Zhou, S., Chen, Y., Zhang, D., Xie, J., Zhou, Y.: Classification of surface defects on steel sheet using convolutional neural networks. Mater. Tehnol. **51**, 123 (2017)
10. Soukup, D., Huber-Mork, R.: Convolutional neural networks for steel surface defect detection from photometric stereo images. In: ISVC 2014: Advances in Visual Computing, pp. 668–677 (2014)
11. Faghih-Roohi, S., Hajizadeh, S., Nu´nez, A., Babuska, R., De Schutter, B.: Deep convolutional neural networks for detection of rail surface defects. In: Proceedings of the 2016 International Joint Conference on Neural Networks (IJCNN 2016), pp. 2584–2589 (2016)
12. Goodfellow, I.J., Pouget-Abadie, J., Mirza, M., Xu, B., Warde-Farley, D., Ozair, S., Courville, A., Bengio, Y.: Generative adversarial networks. arXiv:1406.2661 (2014)

13. Radford, A., Metz, L., Chintala, S.: Unsupervised representation learning with deep convolutional generative adversarial networks. arXiv:1511.06434 (2015)
14. Gatys, L.A., Ecker, A.S., Bethge, M.: Texture synthesis and the controlled generation of natural stimuli using convolutional neural networks. arXiv:1505.07376 (2015)
15. Lubbers, N., Lookman, T., Barros, K.: Inferring low-dimensional microstructure representations using convolutional neural networks. arXiv:1611.02764v1 (2016)

Digital Food Product Traceability: Using Blockchain in the International Commerce

Borja Bordel[1(✉)], Pierre Lebigot[2], Ramón Alcarria[1],
and Tomás Robles[1]

[1] Universidad Politécnica de Madrid, Madrid, Spain
bbordel@dit.upm.es,
{ramon.alcarria, tomas.robles}@upm.es
[2] Ecole nationale supérieure de l'électronique et de ses applications, Cergy,
France
pierre.lebigot@ensea.fr

Abstract. Food products are one of most important elements in the international commerce. They are also one of the most regulated markets, both at national and international level. The potential damage that inadequate food products may cause in the economy, public health or natural environment of any country turns them into a very controlled commercial flow. However, methodologies, processes and instruments to apply these controls and trace the flow of food products through the several countries and states they can pass, are very complex and inefficient. Cultural and administrate problems, language conflicts or lost documents are only some of the main issues future digital food technologies should address. Therefore, in this paper it is proposed a traceability system for food products in the international commerce, based on Blockchain networks and RFID tags. Using REST interfaces, NoSQL databases and JavaScript code it is implemented a distributed solution to collect, sign and store trustworthy information about the food product flow. A security system is also deployed to guarantee the fault-resilience and the access only by authorized users to the system. A first implementation of the proposed system is also described in order to validate the described solution.

Keywords: Blockchain · Traceability · Digital food technologies
RFID · Certification · Security

1 Introduction

Digital traceability solutions have been studied from the early years of twenty-first century, when Radio Frequency Identification techniques [1] were applied to provide with unique identifiers to all kinds of everyday objects, devices, services and products.

In these first digital approaches, traceability systems included different standard SQL databases [2] (one per agent managing the products under control), where relevant information was stored according to the local laws and available technology. However, no centralized data repository was usually considered, and even some agents could still maintain a traditional physical register if no digital solution was available.

© Springer Nature Switzerland AG 2019
T. Antipova and A. Rocha (Eds.): DSIC 2018, AISC 850, pp. 224–231, 2019.
https://doi.org/10.1007/978-3-030-02351-5_27

Food products, which are under more strict sanitary controls than the traditional commercial limitations, are probably the elements to which the most advanced traceability solutions are applied [3]. In this situation two very relevant interests converge: on the one hand governmental administrations need to prevent public sanitary emergencies and, in the event of one, they have to be able to locate the origin in the fastest possible manner; on the other hand, food-related companies desire to guarantee the reputation and quality of their products, to ensure the increase of their markets and their economy.

In that way, many different initiatives to find traceability solutions to solve problems of traditional technologies have been reported: Smart Farming [4], the Internet of Food [5], etc. However, no proposal fulfills all requirements for successful food traceability systems. Namely:

i. The system must be unique and distributed. Several different agents must be able to insert information, but only trustworthy data must be accepted.
ii. The system must be flexible to adapt to the different local regulations and information types that may be part of traceability records.
iii. Information stored in the system cannot be modified under any circumstances. In that way, automatic information generation systems are preferred in order to avoid human errors.

Considering the previous requirements, we argue that Blockchain [6] is a valid solution to develop the future digital food product traceability. Blockchain is a decentralized system composed of an undefined number of geographically sparse nodes, which are able to agree on the global state of the system, through the creation of a chain of linked information blocks which are impossible to modify once stored and consolidated. Although different Blockchain technologies have been reported in the last years, to develop digital traceability systems, Ethereum [7] (a technology to develop, deploy and execute SmartContracts) is the most adequate.

Therefore, the objective of this paper is to design a digital traceability system for food products, based on Ethereum technologies, SmartContract and RFID tags [8]. Information will be stored in centralized databases which are accessed by Ethereum contracts through special connectors named as "oracles" and JavaScript programs.

The rest of the paper is organized as follows: Sect. 2 describes the state of the art on traceability systems and Blockchain solutions; Sect. 3 describes the architecture and implementation of the proposed digital traceability system; Sect. 4 presents an experimental validation using an initial prototype and virtual Blockchain networks; and Sect. 5 concludes the paper.

2   State of the Art on Traceability Systems and Blockchain Solutions

During the last years different proposals to digitalize traceability systems (in particular those for food products) have been reported. These solutions are focused on determining the physical location of any product at any state in the supply chain.

First, innovations in hardware technologies may be found. These techniques include from measuring equipment to new identification tags [9]. Discussions about problems associated to the use of barcodes are also common [10], what justifies the extensive use of RFID tags. Miniaturized tags and specialized readers, to be used in specific markets such as the fresh fish Danish market [11] or the Parmigiano cheese Italian market [3], have been also reported. Second, software tools to create trustworthy distributed traceability systems may be also found. These solutions include electronic identification systems (EID) [12] and specific software programs such as QualTrace, Food Track or Enterprise Quality Management. More scientific approaches such as semantic technologies for food traceability are also available [13].

Finally, most modern innovations on food traceability consist of advanced measuring systems. These systems include mechanisms to obtain information about the product quality and safety [14]; algorithms for environmental monitoring [15], and techniques to attach geospatial information to traceability data [16].

However, all the previously cited proposals maintain the same problems than traditional traceability systems in respect to information losses, trustworthiness, etc. In this work we argue the solution to these problems might be Blockchain. In respect to Blockchain applications, as it is a very innovative technology, only some works have been reported until now. Basically, five different application types may be distinguished: data storage management [17, 18], identity management [19], rating systems [20] and lottery and banking applications [21]. The last group is composed by good and data trading systems (kind of simple traceability solutions) [19, 22].

The proposed technology in this work continues the ideas presented in the last group of the previously referred articles.

3   A New Digital Traceability Systems Based on Blockchain

The general architecture of the proposed solution may be seen on Fig. 1. As can be seen, the proposed architecture is composed by six main components.

- Tagged products: Food products are provided with unique digital identifiers stored in RFID tags. Tags are usually passive, as active technologies are more expensive and potentially toxic due to the use of batteries.
- RFID reader: Composed of two elements, a microcontroller and a RFID antenna or receptor. The microcontroller activates the reader which generates a magnetic field. This field causes the tags included in the products to transmit the stored unique identifier.
- Node.js server: A JavaScript server containing three different scripts. The first script is a web server which obtains information from the database and checks its validity through functions in the SmartContract. This web server generates a web site where information is displayed to be consulted by users. The second script is in charge of receiving the information from the RFID reader, and the third one is responsible for storing the information automatically in the database through the SmartContract and the oracle. Each one of these scripts is executed in a different port.

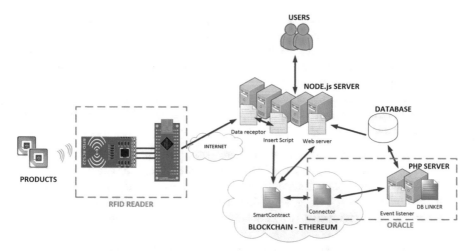

Fig. 1. Architecture of the proposed traceability system

- Blockchain network: It refers to an Ethereum network where two SmartContracts are deployed. The first one contains all functions to store new information in the database and checks the validity of a certain retrieved information from this repository. The second contract is part of the oracle (the connector).
- Oracle: An oracle is a component which is able to insert external information into the Blockchain network. This component includes two modules, a SmartContract acting as connector in the Blockchain network and a PHP server where two scripts are executed (a PHP script and a JavaScript file).
- Database: It is the repository employed to store and to maintain the traceability information. Because of the type of information to be stored and the flexibility required to the system a NoSQL database is the most appropriate technology for this proposal.

The behavior of the proposed system is as described below. The RFID reader obtains the unique identifier of a food product when it is inside its reading area. Then, through the Internet and using wireless communications, the information about this product is sent to the Node.js server, where the "Insert Script" receives the data. These data are enriched with time and/or location information, and are sent to the Ethereum network where a SmartContract is deployed with the appropriate function to receive this information. The smart contract communicates with a second contract which acts as oracle's connector. The information is, then, sent to an external database using an event listener which continuously looks for updates and new information to be stored, and a PHP script which codified the algorithms to interact with the database. Periodically, and using cryptographic hash functions and digital signature techniques, the state of the database is stored in the SmartContract to act as an agent guaranteeing the validity and trustworthiness of stored data.

If a user needs to consult some traceability information, then he connects to a web server running into the Node.js server. This web server retrieves the corresponding information from the database and, using the appropriate checking techniques in the digital signature infrastructure, evaluates the coherence of the obtained data. If all verification processes are successfully finished, the information is presented to the user.

As said, security techniques, in particular digital signature technologies are basic to make the proposed solution work in the adequate manner. These techniques are usually supported by Public Key Infrastructures (PKI), which guarantee a high protection level. In particular, the following security mechanisms are considered in the proposed system to be fault-resilient:

- Communications through the Internet are based on REST interfaces and HTTP messages which are implemented using the secure HTTPS protocol.
- The SmartContract includes modifiers in all functions to make them only reachable by authorized users. This security function is supported by user/password mechanisms and KPI employed to authenticate and identify entities during operation in the Blockchain network.
- Information in the database can only be injected through the SmartContract and the oracle. Following this scheme, the SmartContract must contain information about the entire flow for the food products, but it is guaranteed that no false information is stored in the repository.
- Traceability information stored in the database is "summarized" using anti-collision hash functions and stored in the SmartContract periodically. In that way, the validity of the retrieved information is always guaranteed by the coherence of information in the Blockchain.

4 Experimental Validation: Initial Prototype

In order to evaluate the performance of the propose system, a first initial prototype is built and deployed (see Fig. 2). The proposed prototype is composed of a NodeMCU LoLin V3 (ESP8266) microcontroller and a RFC522 RFID antenna to create the RFID reader. WiFi is selected as wireless communication technology to send the information to the "Data receptor" script running in port 3000 in the Node.js server. This script communicates using HTTP messages and a REST interface with the "Insert script" running in port 1337. The web server is running in port 80 as traditionally. Node.js server is implemented using Express technology.

Contracts are developed using Solidity language. In this first implementation no real Ethereum network is considered, so contracts are deployed in a virtual Blockchain network created using the Ganache software. This software runs a network which can be accessed through the port 7545.

Finally, MongoDB, a NoSQL database, is selected, which can be easily accessed from JavaScript and PHP programs.

Using this prototype an experiment was carried out. Although different parameters could be measured, scalability property is the most relevant in this phase. In real food product traceability systems, thousands of readers may be accessing to store new information at the same time. It is very important to guarantee to all them the access to the database and to investigate the limits to this number in the proposed solution.

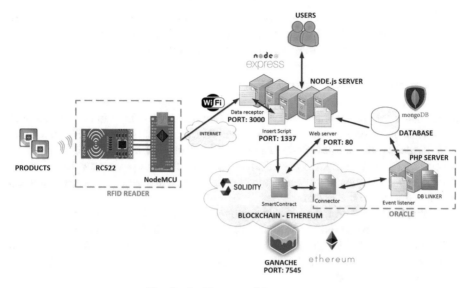

Fig. 2. Architecture of the prototype

To perform the proposed experiment it is employed a 64-bit 1570 Linux Ubuntu 16 operating system, with an Intel i5 processor and 8 GB of RAM memory. As this hardware configuration is not specifically designed to support heavy virtual deployments, the different readers accessing to the Ethereum network are simulated using a unique C ++ program. For each considered number of RFID readers in the system, it is evaluated the number of successful transactions (see Fig. 3).

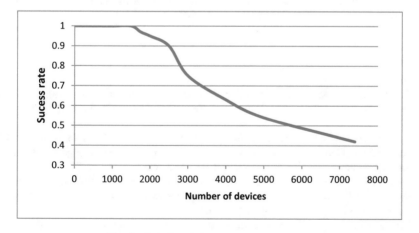

Fig. 3. Results of the proposed experiment

As can be seen, the success rate is 100% until almost reaching the number of 1800 readers in the system. Then, the success rate starts going down, crossing the standard threshold to consider a technological system is working (50%) for (approximately) 6000 readers in the traceability system.

The decreasing speed is stabilizing as the number of readers goes up. For a number of readers around 2000 the decreasing speed is very high, but it is reduced in a 50% (approximately) in the environment of six thousand readers.

5   Conclusions and Future Works

In this paper we present a new traceability system for food products in the international commerce, based on Blockchain networks and RFID tags [23]. Using REST interfaces, NoSQL databases and JavaScript code it is implemented a distributed solution to collect, sign and store trustworthy information about the food product flow. A security system is also deployed to guarantee the fault-resilience and the access only by authorized users to the system.

A first initial real prototype is also deployed using virtual Blockchain networks and hardware nodes based on the Arduino platform and wireless communication technologies (WiFi). Results show that a success rate is 100% until almost reaching the number of 1800 readers in the system, which is a good value for traceability in international commerce.

Future works should consider new proposals about enhanced Blockchain [24] networks such as Algorand [25]. New prototypes and implementations (and experiments) considering real Blockchain networks will be also addressed in future works. Prosumer environments to define SmartContract and all the related logic without requiring programming skill must be also investigated.

Acknowledgments. The research leading to these results has received funding from the Autonomous Region of Madrid through MOSI-AGIL-CM project (Grant P2013/ICE-3019, co-funded by EU Structural Funds FSE and FEDER). Also from the Ministry of Economy and Competitiveness through INPAINK Project (RTC-2016-4881-7).

References

1. Finkenzeller, K.: RFID Handbook: Fundamentals and Applications in Contactless Smart Cards, Radio Frequency Identification and Near-Field Communication. Wiley, New York (2010)
2. Stonebraker, M.: SQL databases v. NoSQL databases. Commun. ACM **53**(4), 10–11 (2010)
3. Regattieri, A., Gamberi, M., Manzini, R.: Traceability of food products: general framework and experimental evidence. J. Food Eng. **81**(2), 347–356 (2007)
4. Tripicchio, P., Satler, M., Dabisias, G., Ruffaldi, E., Avizzano, C.A.: Towards smart farming and sustainable agriculture with drones. In: International Conference on Intelligent Environments (IE), pp. 140–143. IEEE (2015)
5. Kouma, J.P., Liu, L.: Internet of food. In: Internet of Things (iThings/CPSCom), 2011 International Conference on and 4th International Conference on Cyber, Physical and Social Computing, pp. 713–716. IEEE (2011)
6. Zyskind, G., Nathan, O.: Decentralizing privacy: using blockchain to protect personal data. In: Security and Privacy Workshops (SPW), pp. 180–184. IEEE (2015)
7. Wood, G.: Ethereum: A secure decentralised generalised transaction ledger. Ethereum Project Yellow Paper **151**, 1–32 (2014)

8. Pérez-Jiménez, M., Sánchez, B.B., Alcarria, R.: T4AI: a system for monitoring people based on improved wearable devices. Res. Briefs Inf. Commun. Technol. Evol. (ReBICTE) **2**, 1–16 (2016)
9. Jansen-Vullers, M.H., Wortmann, J.C., Beulens, J.A.M.: Application of labels to trace material flows in multi-echelon supply chains. Product. Plan. Control **15**(3), 303–312 (2004)
10. Sahin, E., Dallery, Y., Gershwin, S.: Performance evaluation of a traceability system. In: Proceedings of IEEE International Conference on Systems, Man and Cybernetics, vol. 3, pp. 210–218 (2002). ISSN: 1062-922X
11. Borresen, T.: Traceability in the fishery chain to increase consumer confidence in fish products—application of molecular biology techniques. In: Proceedings of Trans-Atlantic Fisheries Technology Conference—TAFT 2003. 11–14 June, Reykjavik, Iceland (2003)
12. Lima, E., Hopkins, T., Gurney, E., Shortall, O., Lovatt, F., Davies, P., et al.: Drivers for precision livestock technology adoption: a study of factors associated with adoption of electronic identification technology by commercial sheep farmers in England and Wales. PLoS ONE **13**(1), e0190489 (2018)
13. Buchmann, R.A., Karagiannis, D.: Modelling mobile app requirements for semantic traceability. Requir. Eng. **22**(1), 41–75 (2017)
14. Cai, Y., Li, X., Wang, R., Yang, Q., Li, P., Hu, H.: Quality traceability system of traditional Chinese medicine based on two dimensional barcode using mobile intelligent technology. PLoS ONE **11**(10), e0165263 (2016)
15. Bordel, B., Alcarria, R., Manso-Callejo, M.Á., Jara, A.: Building enhanced environmental traceability solutions: from thing-to-thing communications to generalized cyber-physical systems. J. Internet Serv. Inf. Secur. **7**(3), 17–33 (2017)
16. Alcarria, R., Bordel, B., Manso, M.Á., Iturrioz, T., Pérez, M.: Analyzing UAV-based remote sensing and WSN support for data fusion. In: International Conference on Information Theoretic Security, pp. 756–766. Springer, Cham (2018)
17. Zyskind, G., Nathan, O., Pentland, A.: Decentralizing privacy: using blockchain to protect personal data. In: IEEE Symposium on Security and Privacy Workshops, pp. 180–184. IEEE Computer Society (2015)
18. Ateniese, G., Goodrich, M.T., Lekakis, V., Papamanthou, C., Paraskevas, E., Tamassia, R.: Accountable Storage. In: IACR Cryptology ePrint Archive, vol. 2014, p. 886 (2014)
19. Worner, D., von Bomhard, T.: When your sensor earns money: exchanging data for cash with Bitcoin. In: UbiComp Adjunct, pp. 295–298. ACM (2014)
20. Vandervort, D.: Challenges and opportunities associated with a bitcoin-based transaction rating system. In: Financial Cryptography Workshops, Ser. Lecture Notes in Computer Science, vol. 8438, pp. 33–42. Springer (2014)
21. Peters, G.W., Panayi, E.: Understanding modern banking ledgers through blockchain technologies: future of transaction processing and smart contracts on the internet of money. CoRR, vol. abs/1511.05740 (2015)
22. Zhang, Y., Wen, J.: An IoT electric business model based on the protocol of bitcoin. In: ICIN, pp. 184–191. IEEE (2015)
23. Lin, Y.-P., Petway, J.R., Anthony, J., Mukhtar, H., Liao, S.-W., Chou, C.-F., Ho, Y.-F.: Blockchain: the evolutionary next step for ICT E-agriculture. Environments **4**, 50 (2017)
24. Tian, F.: A supply chain traceability system for food safety based on HACCP, blockchain Internet of things. In: 2017 International Conference on Service Systems and Service Management, Dalian, pp. 1–6 (2017). https://doi.org/10.1109/ICSSSM.2017.7996119
25. Gilad, Y., Hemo, R., Micali, S., Vlachos, G., Zeldovich, N.: Algorand: scaling byzantine agreements for cryptocurrencies. In: Proceedings of the 26th Symposium on Operating Systems Principles (SOSP 2017), pp. 51–68. ACM, New York (2017)

Digital Environmental Sciences

A Reliable Wireless Communication System for Hazardous Environments

Sebastian Rosca[✉], Simona Riurean, Monica Leba,
and Andreea Ionica

University of Petrosani, University Street no. 20, 332006 Petrosani, Romania
sebastianrosca91@gmail.com,
{simonariurean,monicaleba,andreeaionica}@upet.ro

Abstract. The hazardous environments such as underground spaces (coal, salt, minerals mines), research laboratory or test centers are dangerous to work in, especially because of air quality, small and difficult to access spaces and in some cases highly explosion hazard due to instant events such as combination of explosive gases or substances. Environmental data acquisition and wireless communication prior to human access in these spaces or sometimes instead of human access, is life savings. The system we propose is low cost, handy, a real time data acquisition and duplex wireless data communication. The first module is mobile data acquisition robot equipped with camera and a network of sensors. It is remotely handled by a person's hand equipped with the second module that consists of a gyroscope and accelerometer sending movements data and receiving video streaming related to space and environmental acquired data.

Keywords: Network sensors · Hazardous environments · Underground spaces
Wireless communication

1 Introduction

There are many spaces dangerous to work in due to environmental health condition and many others difficult or even forbidden to be accessed due to unexpected or fatal events. Acquiring environmental data and sending them from these spaces can be sometimes a reliable solution and more important, human life saving. Environmental data acquisition and wireless communication prior to human access in these spaces or sometimes instead of human access, is life savings. We designed and simulated a duo-module system that is remotely handled by an operator with the support of a gyroscope and accelerometer, acquires environmental data and sends those wirelessly to the operator in real time.

2 Description of the System

As for the functional point of view, the system proposed by us is composed of two modules as shown in the block diagram (Fig. 1).

© Springer Nature Switzerland AG 2019
T. Antipova and A. Rocha (Eds.): DSIC 2018, AISC 850, pp. 235–242, 2019.
https://doi.org/10.1007/978-3-030-02351-5_28

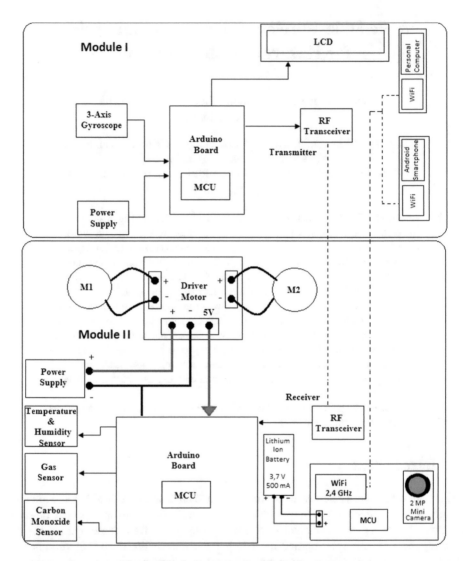

Fig. 1. Block diagram – the monitoring system

The first one, a robotic mobile platform performs a series of movements in the workspace (forward, backward, left or right). It is based on remote human operator control in a manner that depends only by the operator's hand movements through the simple use of a glove that includes a gyroscope.

The other module is responsible for data acquisition based on a network of sensors that are capable of acquiring sensible data of environmental factors from special areas considered to be either hazardous due to emissions of natural gas, or due to the fact of releases of dangerous substances to the workspace.

In order to achieve our goal, we used two Arduino microcontroller boards, both for prototyping the robot remote control system, that is built around an ATmega328P chip and offers the programmer everything needed to achieve an automation system.

It provides a 16 MHz ceramic resonator for dealing with time issues, a series of digital input/output pins can be used to control a wide range of motors such as: DC motor, servo, stepper motor based on PWM modulation or an RF transceiver or an Wi-Fi module for communication purpose and a series of analog pins that can be used to acquire data from sensors such as: environmental parameters or orientation of an object in space using a gyroscope [1], all in a programmable device ready to use that can be programmed based on C language using the USB interface integrated on any ordinary computer to load the program developed by user into embedded microcontroller [2].

The mobile robot used by us provides a relatively simple mechanical implementation with only 3 wheels, of which only two are driven by DC gear servomotors, the third wheel being useful when turning left or right because it can also move on the horizontal axis. Research in the field shows that in order to maintain optimum contact with the soil, especially in the case of uneven surfaces, the required number of wheels is three greater, requiring a suspension system to achieve the same performance [3].

In order to control the direction and speed of the robot we used a driver motors based on a H-bridge Dual Motor Controller that can drive two servomotors in the same time based on the PWM signal received from the Arduino board microcontroller. The driver motors based on H-bridge is useful because it is a current amplifier that converts a low current control signal into a higher current signal [4] since the Arduino board microcontroller can supply a maximum output current of 40 mA per pin [5] and a current between VCC and GND pins limited to 200 mA [6], which proves to be insufficient to control the servomotors.

The supply voltage required for Arduino-based robot operation can be provided directly from an external power source based on a battery that ensure a voltage of 9 V because the Arduino board has a built-in voltage regulator that limits the voltage to 5 V required for most modules compatible with Arduino [7]. In order to have a single power supply for the robot, we have powered the Arduino Board from the driver motors by connecting the voltage input pin - VIN on the Arduino board to the +5 V pin supplied from the driver motors [8].

2.1   Arduino Software – IDE

The Arduino Integrated Development Environment or Arduino Software (IDE) contains a code-writing editor, a message area, a text console, a toolbar with common function buttons and a series of menus.

The Arduino IDE is a cross-platform application written in Java which is derived from the IDE made for the Processing programming language and the Wiring project. It is designed to introduce programming to any new programmer who is unfamiliar with software development. It includes a code editor with features such as syntax highlighting, brace matching, and automatic indentation, and is also capable of compiling and uploading programs to the board with a single click.

The Arduino IDE comes with a C/C++ library called "Wiring", which makes many common input/output operations much easier. Arduino programs are written in C/C++ [9].

2.2   The Sensors Network

Real-time acquisition of sensible data of environmental factors that may be present at a particular time in the workspace from special underground areas is a priority to be considered in risk management to be able to act properly in accordance with the field acquiring data to prevent the errors that can may be arise in taking the decisional measures in case of rescue operations. Therefore, we propose to implementing within our mobile robot a network of wireless low cost environmental sensors compatible to Arduino able to monitor: temperature and humidity, presence of carbon monoxide, and flammable gases, especially methane, the presence of sources of ignition that can generate a fire.

The sensor used to detect temperature and humidity is a hybrid sensor that operates at low energy (3 V–5 V) like most Arduino compatible modules, it can measure the humidity based on the resistor and the temperature based on the NTC component, offers a fast response in time, is immune to interferences, and the fact that it is provided with a single analog data pin for the acquisition of both environment parameters represent an advantage for Arduino board pins management [10]. With respect to the range of measurement values, the sensor can measure temperatures in a range of 0 to +50 °C and humidity in a range of ± 5.0% RH [11].

In order to detect gas leakage in special in underground spaces we implement into the robot low cost gas sensor that is designed for industrial or domestic environment monitoring that can be usefully to detect gasses such as: propane, butane, methane, alcohol and hydrogen. This sensor is also usefully in our monitoring process because it is sensible at smoke [12]. From the constructive point of view, this sensor is based on sensitive component that contains an adjustable resistor and a protective resistor integrated on board that can detect target gas leakage based on the variation of the resistance of the sensitive component [13]. The gas sensor used by us operate at +5 V with a low current consumption of just 40 mA which makes it ideal for use with an Arduino board as a digital or analog data input and as an important advantage it can detect methane emissions with high accuracy between 300 and 10000 ppm according to the datasheet provided by the manufacturer [14].

Because of lethally potential represented by the presence in the air of the carbon monoxide that can present at a time in particular due to an accident in special underground spaces that evolving an incomplete combustion of any organic materials such as: wood, butane, propane and other natural gasses or even by underground machinery malfunction such as air compressor that supply fresh air in workspace, was also necessary to implemented on the robot a carbon monoxide detection specialized sensor that can accurately measure the concentration of this gas in the air which if it remain undetected it present potential to harm the human health because it's particularity that it present: carbon monoxide gas is odorless, colorless, tasteless and nonirritating and is lethally after 1–3 min at a concentration in the air between 12–13,000 ppm or after an hour at a concentration of 1600 ppm [15].

With regard to the construction details of this type of sensor, the sensing component of the measuring circuit consists of two parts: the heating circuit on one side providing the time control function and the output signal circuit that responds to changes of resistance detected on the surface of the sensor [16]. As same in the case of the gas sensor, the carbon monoxide sensor can be connected for supplying data output to any digital or analogue pin from Arduino board and it operates at a +5 V with a low current consumption at only 40 mA and has a high sensitivity for carbon monoxide detection ranging from 20 ppm to 2000 ppm [17].

2.3   The Module with Gyroscope

In order to be able to control the robot into environment that we want to monitoring based on a method that to be able to replicate the hand orientation in space and to transpose it into control commands that we can use in robotics for locomotion purposes in accordance with visual data acquired in real time from the camera kit, we implemented on a glove an intelligent sensor, that can be integrated into our application developed with Arduino, that is based on an I2C motion processor with 6-axis that incorporates a 3-axis gyroscope and a 3-axis accelerometer along with a Digital Motion Processor (DMP) all on one system-on-chip device. The MEMS motion tracking device features programmable gyroscope and accelerometer designed for fast and slow movement's precision tracking.

The motion processing unit incorporates MotionFusion algorithms which will also access external sensors and magnetometers through the auxiliary master I2C bus. The possible applications of this type of intelligent component can to include: development of device that based on wearable sensors or development of smart applications in case of tablets or smartphones in specially for counting steps operations by a mobile processor to display the numbers of calories burned or the quality of sleep and up to could play intelligent games that is dependent for acquired data from sensors. The Platform will extract the motions that are associated and unload the sensor management from the operating system to provide an application program interface (API).

2.4   Video Streaming and Communication

In order to remote control the robot we need to purchase real-time video streaming from the workspace to be monitored and for this purpose we used a low cost special kit based on Arduino microcontroller that can be mounted directly on the robot frame and that offers video streaming capabilities based on a mini camera that can be connected as a daughter card on the board using GPIO pins header and that can capture 2MP full resolution JPEG still image, even stream low resolution at fairly framed video over network via WIFI module embedded on board which operates at a frequency of 2.4 GHz or the data can be directly save to local SD/TF card.

The kit is suitable for portable application, it can be powered from micro-USB or using battery and has built in lithium battery charging circuits with of capacity of +3.7 V and 0.5 A maximum current.

The special kit based on Arduino present a series of key features: 32 bits microcontroller with low power consumption and RISC type architecture [18]; operate at a high frequency clock speed of 80 MHz and can be boost at a frequency of 160 MHz when Real Time Operation System (RTOS) is enabled [17]; supports Arduino sketch script to be programmed and is suitable for android application [19].

The following block diagram shows the connection diagram of our kit based on Arduino microcontroller used for the acquisition of streaming video for real-time monitoring purposes either on a personal computer or on an Android-based device.

2.5 NRF Module

Transceiver NRF uses the 2.4 GHz band and it can operate with baud rates from 250 kbps up to 2 Mbps. If used in open space and with lower baud rate its range can reach up to 100 meters. The radio modules include a 2.4 GHz RF transceiver and a logic that supports a high-speed SPI interface for data connection and exchange.

The module can use 125 different channels which gives a possibility to have a network of 125 independently working modems in one place. Each channel can have up to 6 addresses, or each unit can communicate with up to 6 other units at the same time. Power consumption of this module is just around 12 mA during transmission, which is even lower than a single LED. The operating voltage of the module is from 1.9 to 3.6 V, but the other pins tolerate 5 V logic, so easily we can connect it to an Arduino without using any logic level converters.

So, once we connect the NRF modules to the Arduino boards we are ready to make the codes for both the transmitter and the receiver [20].

3 Software Description and System's Simulation

In order to achieve our goal of monitoring in the real-time the potentially hazardous environment to prevent unpredictable or fatal events that may occur at a time in the workspace we implemented all that was necessary for a good integration powered by C/C++ programming language of both modules to be able to supply the one hand a bidirectional communication required for control operations and the other hand for data acquisition related to existent environments parameters.

From the need to get visual stimuli from the environment we implemented a software solution that allows us to capture video streaming based on a WiFi network using the HTTP protocol to be able to control the robot from a safe distance so that we are not exposed to the hazards unanticipated.

Regarding radio-frequency based wireless communication between modules we developed a software solution by implementing the SPI communication protocol that can check the status of connection between of the modules and which can transmit commands from the transmitter module based on gyroscope that will generate data depending on its position in space and that can give certain commands to DC gear servomotors that acting the robot to change the direction of movement or to perform a turn. Due to bidirectional communication benefit our solution can receive environmental data from sensors network that are placed on the receiver module directly on the transmitter module to analyze and display them. To demonstrate the functionality of the proposed system we also performed a simulation into an environment specialized in design and testing of the embedded system of the two modules handled in this paper for monitoring of a special underground environment (Fig. 2).

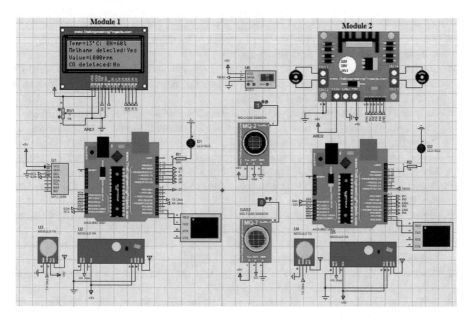

Fig. 2. Simulation of the monitoring system applied in a salt mine

4 Conclusions

There are many workspaces that may endanger the health of workers or can be areas where personnel rescue is necessary to be send in missions during the event of a fire outbreak, for example. Therefore, the danger of harming rescuers during saving operations due to lack of information that otherwise could not be directly obtained from the work environment, is eliminated.

This system we propose is a low cost, handy, a real time data acquisition and duplex wireless data communication duo-module, one of which is managed with one hand by a remote operator. The other one sends video streaming of spaces, acquire and sends environmental data as well. It consists of a sensors networks acquiring data such as temperature and humidity, gasses (propane, butane, methane, alcohol and hydrogen) and carbon monoxide, as well.

References

1. Pan, T., Zhu, Y.: Designing Embedded Systems with Arduino: A Fundamental Technology for Makers. Springer (2017)

2. Sweatt, M., et al.: WiFi based communication and localization of an autonomous mobile robot for refinery inspection. In: IEEE International Conference on Robotics and Automation (ICRA). IEEE, pp. 4490–4495 (2015)
3. Taha, I.A., Marhoon, H.M.: Implementation of controlled robot for fire detection and extinguish to closed areas based on Arduino. Telkomnika 16(2), 654–664 (2018)
4. Behera, S., Muduli, P.K.: Remote speed control of brushless DC motor with display. Int. J. Autom. Smart Technol. 8(2), 65–71 (2018)
5. Nayyar, A., Puri, V.: A review of Arduino board's, Lilypad's and Arduino shields. In: 2016 3rd International Conference on Computing for Sustainable Global Development (INDIACom). IEEE, pp. 1485–1492 (2016)
6. Arduino Playground. https://playground.arduino.cc/Main/ArduinoPinCurrentLimitations
7. Ollukaren, N., Mcfall, K.: Low-cost platform for autonomous ground vehicle research. In: Proceedings of the 14th Early Career Technical Conference (2014)
8. Luiz Jr., A., et al.: A low-cost and simple Arduino-based educational robotics kit. Cyber J.: Multidiscip. J. Sci. Technol. J. Sel. Areas Robot. Control (JSRC) 3(12), 1–7 (2013)
9. Pandya, V., Shukla, D.: GSM modem based data acquisition system. Int. J. Comput. Eng. Res. 2(5), 1662–1667 (2012)
10. Sipani, J.P., et al.: Wireless sensor network for monitoring and control of environmental factors using Arduino. Int. J. Interact. Mob. Technol. (iJIM) 12(2), 15–26 (2018)
11. Ünsal, E., Milli, M., Çebi, Y.: Low cost wireless sensor networks for environment monitoring. Online J. Sci. Technol. 6(2), 61–67 (2016)
12. Sowparanika, E.L., et al.: Wireless communication system for coal mining worker using Arduino. J. Chem. Pharm. Sci. 974, 2115 (2017)
13. Al-Dahoud, A., Jannoud, I., Al-Rawashdeh, T.: Monitoring Metropolitan City Air-quality Using Wireless Sensor Nodes based on ARDUINO and XBEE (2011)
14. Olimex. https://www.olimex.com/Products/Components/Sensors/SNS-MQ2/resources/MQ2.pdf
15. Srivastava, S.K.: Real time monitoring system for mine safety using wireless sensor network (multi-gas detector). Ph.D. thesis (2015)
16. Ramesh, M., et al.: Solid Waste Management Using IoT (2018)
17. Olimex. https://www.olimex.com/Products/Components/Sensors/SNS-MQ7/resources/SNS-MQ7.pdf
18. Gour, G.B., et al.: Helmet sensing speed controller device. Int. J. Curr. Trends Eng. Res. (IJCTER) 2(5), 291–296 (2016)
19. Manikandan, J.: Design and evaluation of wireless home automation systems. In: IEEE International Conference on Power Electronics, Intelligent Control and Energy Systems (ICPEICES). IEEE, pp. 1–5 (2016)
20. How to Mechatronics. https://howtomechatronics.com/tutorials/arduino/arduino-wireless-communication-nrf24l01-tutorial/

Digital Finance, Business and Banking

Approach of Estimation of the Fair Value of Assets on a Cryptocurrency Market

Olga Romanchenko[1], Olga Shemetkova[2], Victoria Piatanova[3], and Denis Kornienko[4(✉)]

[1] Faculty of Finance Department, Plekhanov Russian University of Economics, Moscow, Russia
[2] Faculty of Finance, Plekhanov Russian University of Economics, Moscow, Russia
[3] Department of Financial Management, Plekhanov Russian University of Economics, Moscow, Russia
[4] National Research University Higher School of Economics, Moscow, Russia
dokuniversity@gmail.com

Abstract. Cryptocurrency appears to be a trending payment as well as an investment asset. The substantial number of investors are willing to allocate their wealth in such specific asset in 2018. That led to the increase of the demand on cryptocurrency, which subsequently influenced on its prices a lot. Investors are considering such instrument as an alternative investment, an acceptable substitution to ordinary stocks, bonds or commodities. In that case it becomes necessary to clearly determine the approach of estimation of the intrinsic value of such kind of assets. The article examines what type of asset is closer to the cryptocurrency. Moreover, the extent to which traditional methods of assessing financial assets applicable to the cryptocurrencies is also estimated.

Keywords: Cryptocurrency · Digital currency · Intrinsic value
Alternative investments

1 Introduction

The cryptocurrency market is without doubt one of the more disruptive elements of the Finance industry in 2017. A couple of years ago, cryptocurrencies seemed like a hi-tech story existing only within a Silicon Valley bubble. But now that the cryptocurrency market is generating fresh headlines every week as the key currencies rise and fall.

Basically, cryptocurrencies are a digital money, created from code. The high level of cryptography lies upon the nature of that virtual payment through the consensus-keeping procedure. The cryptography is a core basis of each particular cryptocurrency. The computer code provides encryption of the unit of currency, which means that strong rules of mathematics significantly raise the level of security within the cryptocurrency's system. Moreover, the whole system is supposed to be fully independent from government control. The peer-to-peer internet protocol is the core structure that is underlying the entire economy of cryptocurrency market.

T. Antipova and A. Rocha (Eds.): DSIC 2018, AISC 850, pp. 245–253, 2019.
https://doi.org/10.1007/978-3-030-02351-5_29

2 Literature Review

2.1 Brief Overview of Main Cryptocurrencies

Currently, there exist almost 1381 types of cryptocurrencies with total market capitalization more than $550 bln.

Bitcoin is the first decentralized digital currency. Nowadays it is most famous cryptocurrency and has the largest market share, which is more than 40%. It appeared in 2009 as an open-source software. The Bitcoin protocol uses a trending distributed ledger technology acknowledged as "the blockchain", which provides users with an opportunity to create peer-to-peer connections with the usage of digital currency. There is no central authority, which is responsible for verifying each particular transaction, so that determination the legitimacy of a conducted transaction is always held by the decentralized system.

Litecoin was an early alternative to Bitcoin and has been launched in 2011. Litecoin offered an innovative algorithm in order to try to simplify the mining scheme to allow every person with basic computer hardware to become a part of a cryptocurrency network. Further with the development of the cryptocurrency market the share of other altcoins has captured a significant part of Litecoin's capitalization, nevertheless it still has a large market share and some strong network effects.

Ripple significantly differentiates from other cryptocurrencies. The reason is that it is fundamentally a global settlement network that serves the circulation of other currencies such as Ethereum, GBP, EUR, USD, commodities or another asset. The transaction of any settlement requires a fixed amount of fee to be paid in XRP (the basic tokens in Ripple's system), which are traded on digital currencies' markets. There is no "mining" on the Ripple network – the permanent supply of 100 bn ripples, that is mostly concentrated within the company serves all goals of system.

Ethereum serves mostly as an open software platform and uses blockchain technology. The system is suitable for program developers as it allows to create and dispose decentralized applications. In the system of Ethereum blockchain, as an alternative to ordinary mining of bitcoin, participants work in order to gain a type of digital token called «ether», which is subsequently used in the network. «Ether» is not only a tradeable digital currency, as it is servicing application developers by paying transaction fees during their operational activity in Ethereum system.

2.2 Survey of Relevant Literature

The amount of academic research literature regarding cryptocurrencies is mostly considered bitcoin as a primary asset, which mostly represent the whole conjuncture of a cryptocurrency market. There is a significant amount of scientific surveys and studies, that were mostly focused on the description of a "Bitcoin" as a kind of a fiat currency rather than a commodity. Another group of researchers has identified and considered its' commodity-based specifics and named such asset as a "digital gold" which is definitely accurate due to Bitcoins' origin.

Yermack (2013) deeply analyzed currency-based specifics of the Bitcoin and identified its disadvantages of serving as a fiat currency. Even though I accept such opinion that all digital currencies (and Bitcoin as well) are not supposed to be an absolute substitution to credit money, it is likely to be possible for them to act as a commodity-based money. I assume that it is necessary to investigate more precisely the intrinsic value of a cryptocurrency in terms of its' proxies of production. The value which underneath the concept of cryptocurrency could be described from the point of view of the labor and computational power employed in order to mine Bitcoin so that it makes sensible to use a cost-of-production based approach.

Polasik et al. (2014) revealed the intrinsic value of Bitcoin to be a consequence of its acceptance within the investors and the transactional needs of such instrument for its holders. To obtain the model, authors built a regression based on the Google-searches of the word "bitcoin" and number of transactions as independent variables. The price of Bitcoin in the model was a dependent variable. However, their hypothesis according to the number of transactions was rejected due to low value of t-statistics and, hence, the insignificance of the coefficient. Nevertheless, Google-searches variable appeared to be statistically significant with the probability of 95%. Although such results of approve the hypothesis concerning Google-searches significance, I suppose that use of Google-searches results the word "bitcoin" found might lead to the spurious correlation. The drastic increase in prices of bitcoin caused an increase in media curiosity introducing it to large number of potential investors, who subsequently gathered all available information through the internet-sources. I would consider the variable of difference of Google trends in «buy bitcoin» and «sell bitcoin» to reveal the proxy of people behavior that reflects the level of demand on bitcoin.

Kim et al. (2016) analyzed the relationship between activity in the forum with the number of transactions from 3 Cryptocurrency such as Bitcoin, Ethereum and Ripple. The results show that remarks and responses in forums and in network societies are influential in the amount of virtual currency deals and affect the opinions of investors.

3   Methodology

3.1   Determination of Variables and Sample Selection

The determination of intrinsic value of a digital currency is an unexplored area in modern economy. Nowadays it serves not only for purposes of an alternative investments, but also could be used in some countries for ordinary everyday transactions for buying or selling goods and services. Nevertheless, Bitcoin gives also an opportunity for speculators to try their best in exploiting its volatility and high growing pace to earn additional profit. In that case several factors should influence prices of Bitcoin.

These variables used include:

Coefficient of Complexity of Mining. "Complexity of mining" is one of the important aspects of mining the cryptocurrency. Although the complexity does not affect the production process itself (the need to use equipment, electricity, software and the mining pool), this process becomes more labor-intensive and complex. With the increasing complexity of mining new coins, the miners need more and more powerful equipment to adapt to such changes. The increase in exploiting computational labor to mine digital currency leads to the growth of its intrinsic value.

The Volume of Transactions and the Number of Active Addresses. As it was discovered by previous researchers, the number of transactions was not significant. Nevertheless, to check the assumption of a currency-based origin of cryptocurrency, two alternative indicators were identified, which can serve as a proxy for velocity. In that case, both were included in the model.

Number of Tweets. This indicator reflects the volume of a specific source of information: Twitter. Cryptocurrency is an innovative tool for alternative investments. Only a limited number of people had full information about it before the period of its rapid take-off. The decision on the purchase and sale of cryptocurrency is made by potential investors based on data obtained from various sources of information, including Twitter. In that case, the number of tweets partly forms the scale of demand for cryptocurrencies. Nevertheless, fundamentally the investor after the detection of information spends a certain amount of time on its processing, and therefore in the model should be used use the lagged number of tweets.

Buy/Sell Difference in Google Trends. Despite the fact that the number of tweets a day reflects the scale of the "hype" around the cryptocurrency, it is also necessary to take into account the desire of investors to buy and sell the cryptocurrency. To do this, an indicator that reflects the mood of investors in the "depth of market" is needed. This indicator was identified by taking the difference between "buy cryptocurrency" and "sell cryptocurrency" searches in Google trends. The result will give an opportunity to determine whether the investor moods push the price up (if the indicator is positive) or down (if the indicator is negative). That indicator should be also taken with a lag in the model, but the lag should be less than the one that was taken in number of tweets. The fundamental reason behind that is that potential investors firstly find an information about cryptocurrency, process it and only after that they google how to buy or to sell cryptocurrency.

The span, that is considered in the research paper, includes the time series data from 11/04/2014 to 11/03/2018. The range combines different periods of Bitcoin's history, including the rather stable period of 2014 as well as the phase of its tremendous growth in 2017. Most important, investigated variables are fundamentally disconnected, which means, that they are fully independent from each other.

3.2   Hypothesis

H1. Cryptocurrency might be treated as a currency-based commodity.

The participants of Bitcoin system are evaluating such asset themselves through analysis of all available data, gathered from various sources: web-sites, newspapers, articles etc. The opinion of such investors pushes the market prices of Bitcoin. The increase or decrease in intrinsic value of a digital currency in that case becomes a result of the opinion of buyers, who create demand in Bitcoin and quote prices that are acceptable for them, and sellers, who determine the supply and selling prices.

In addition, commodity-based origin of cryptocurrency also need to be examined as the increase in exploiting computational labor to mine digital currency leads to the growth of its intrinsic value.

H2. Behavioral finance principles can be used in predicting future value of a cryptocurrency.

The increase or decrease in intrinsic value of a digital currency becomes a result of the opinion of buyers, who create demand in Bitcoin and quote prices that are acceptable for them, and sellers, who determine the supply and selling prices.

H3. The regression model could be used in forecasting future value of bitcoin.

The effectiveness of the proposed method for estimating the intrinsic value of a crypto currency (Bitcoin) needs to be determined based on its predictive power. In that case, it is necessary to form a test dataset (excluding a number of historical data) and on its basis to determine the forecast. The resulting forecast should be compared with historical data by calculating RMSE, MAE and MAPE parameters. They will allow to compare the obtained model with previous ones. Nevertheless, it is also necessary to determine the confidence intervals for the proposed model that are also able to reflect its quality.

4 Analysis and Discussion

4.1 Building a Regression Model

According to fundamental view on the origin of cryptocurrency, five factors regression model was built for Bitcoin prices. The market price of Bitcoin in USD was used as a dependent variable. The independent variables are:

- Coefficient of complexity of mining
- The volume of transactions
- Number of active addresses
- Number of tweets
- Buy/Sell difference in Google trends

$$
\begin{aligned}
Bitcoin\ Price = \ & \beta\_0 + \beta\_1 * Mining\ Difficulty + \beta\_2 * Transactions\ Volume \\
& + \beta\_3 * Active\ Addresses + \beta\_4 * Number\ of\ Tweets + \beta\_5 * Buy/Sell\ Difference
\end{aligned}
\tag{1}
$$

The adjusted R^2 from the obtained regression is high. That means that almost 94.4% of variation in Bitcoin values expressed in USD is appeared to be a result of influence of independent variables of the model.

Although the whole model is good, the coefficient of "Number of tweets" is not significant at $\alpha = 5\%$ according to t-statistic. In that case the model was adjusted and formulate in following way:

$$
\begin{aligned}
Bitcoin\ Price = \ & \beta\_0 + \beta\_1 * Mining\ Difficulty + \beta\_2 * Transactions\ Volume \\
& + \beta\_3 * Active\ Addresses + \beta\_4 * Buy/Sell\ Difference\ in\ Trends
\end{aligned}
\tag{2}
$$

The adjusted R^2 from the obtained regression is high. That means that almost 94.1% of variation in Bitcoin values expressed in USD is appeared to be a result of influence of independent variables of the model (Fig. 1).

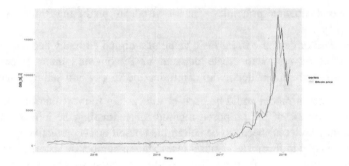

Fig. 1. Multivariable linear regression model (Compiled by the author by using lm() and predict() functions in R)

4.2   Discussion of Results

RMSE, MAE and MAPE are chosen as key parameters for comparison between models. They allow to choose the best models among most appropriate. These parameters are supposed to be obtained on a forecasted test-dataset, which includes 12 variables (data for 3 months). The comparison of measures of quality of models proves, that quality of Linear regression model is higher so that it provides better results (Table 1).

Table 1.   Measures of quality of models (Compiled by the author by using accuracy() function in R on the test set of data)

Type of model	RMSE	MAE	MAPE
Mean forecast	12243.02	11826.22	90.23
Naive forecast	10747.12	9847.46	88.21
Random walk forecast	4757.15	3927.57	37.18
Splines forecast	19272.41	15240.29	133.03
Exponential smoothing	9928.33	8371.50	78.52
ARIMA forecast	19545.58	15509.31	135.37
Neural network #1	18502.51	14310.34	112.81
Neural network #2	25428.93	21432.22	152.43
Linear regression model	4251.34	3543.75	33.36

According to the obtained results of analysis, the model has several parameters, which helps to answer the main queries of the research by testing the formulated hypothesis.

H1. Cryptocurrency might be treated as a currency-based commodity.

In order to check that hypothesis, it was necessary to determine t-statistics of following factors and F-statistics for the groups of those factors.

Commodity-based origin: variable, that serves as a proxy of cost of production:

- Coefficient of mining difficulty.

Currency-based origin: parameters, that together serve as a proxy of velocity:

- Volume of transactions
- Amount of active addresses.

According to the model, those variables have following parameters of t-statistic and F-statistic for the whole group of them (Table 2):

Table 2. Parameters of t-statistic and F-statistic (Compiled by the author in R)

Coeff. of mining difficulty	t-critical = 1.972	t-observed = 14.986
Volume of transactions	t-critical = 1.972	t-observed = 19.992
Amount of active addresses	t-critical = 1.972	t-observed = 4.987
Group of variables	F-critical = 2.417	F-observed = 811.200

Observed values of statistics are higher in absolute value than critical values, so hypothesis that those factors individually or combined are insignificant is rejected at $\alpha = 5\%$.

In that case, a justified confirmation that cryptocurrency might be treated as a currency-based commodity is obtained. It means that cryptocurrency (in case of Bitcoin) has intrinsic value, which is definitely consists of two main parts:

- Commodity-based part: coefficient of proxy of production determines the fact that the increase in exploiting computational labor to mine digital currency leads to the growth of its intrinsic value.
- Currency-based part: coefficients of proxy of velocity determine the fact that the intention of people to buy goods or services with Bitcoin or make an investment in it increase transaction volumes and amount additional active users.

H2. Behavioral finance principles can be used in predicting future value of a cryptocurrency.

In order to check that hypothesis, it was necessary to determine t-statistics of behavior-based factor. In the basic model two factors were considered:

- Number of tweets
- Buy/Sell difference in Google trends.

However, due to the insignificance of "number of tweets" factor, the main variable to be considered is "Buy/Sell difference in Google trends". According to the model, that variable have following parameters of t-stat: t-critical = 1.972, t-observed = 2.150.

Observed values of statistics are higher in absolute value than critical values, so hypothesis that behavior-based factor is insignificant is rejected at $\alpha = 5\%$.

In that case, a justified confirmation that the price of cryptocurrency substantially depends on the willingness of investors to buy or to sell it, so principles of behavior finance are applicable in determining of its intrinsic value.

H3. The model could be used in forecasting future value of bitcoin.

In order to check that hypothesis, it was necessary to conduct two steps of analysis. Firstly, according to the measures of quality of all examined models the linear regression model is better in comparison with others, which means that it provides better and more reliable results. Secondly, the confidence intervals were build and shown following results with the probability of 95% for last 1 month of data (4 periods) (Table 3).

Table 3. Parameters of confidence intervals (Compiled by the author in R)

	Lower bound ($\alpha = 5\%$)	Upper bound ($\alpha = 5\%$)
3 week of February 2018	7 669.55	10 782.40
4 week of February 2018	8 516.38	11 629.23
1 week of March 2018	9 808.14	12 920.99
2 week of March 2018	8 543.73	11 656.58

The confidence intervals are too broad, which leads to substantial difficulties in forecasting of prices of the Bitcoin, which is also increased by significant volatility of most variables included in model. In that case the hypothesis is rejected as the model is not suitable for explicit forecasting.

Nevertheless, model can explain volatility of bitcoin price and allows to check its relative value (whether the Bitcoin is overvalued or undervalued) so that it could be useful for potential investors.

5   Conclusion and Recommendations

This analysis could be applicable in practice as well as in theory. Bitcoin mining, unlike fiat currencies, has the unique feature of being produced as a commodity asset by being mined trough using computational power. However, unlike most produced commodities, Bitcoin has an opportunity to be used as a fiat currency, instrument of payment for goods or services.

The obtained model could be used by potential investors in order to make a decision of including Bitcoin in their portfolio. Nevertheless, there are several areas for future researches, that could improve and expand the scientific knowledge of this topic.

First of all, considering changes in cost of sources of production will allow to explore the variable "mining difficulty" more precisely.

Second of all, there are several variables that were not included in the model and could be tested in another researches in order to attempt to obtain better and more precise results.

Third of all, it is necessary to consider other cryptocurrencies as they have different nature than Bitcoin and, in that case, could be treated differently.

References

Blanchard, O.: Macroeconomics. Prentice Hall, New Jersey (2003)

Breealey, R.A., Myers, S.C., Allen, F.: Principles of Corporate Finance, 11th edn. McGraw-Hill, Irwin (2013)

Hayes, A.: Cryptocurrency value formation: an empirical analysis leading to a cost of production model for valuing Bitcoin (2016)

Baur, D.G., Hong, K., Lee, A.D.: Bitcoin: currency or asset? (2016)

Bouoiyour, J., Selmi, R.: What does crypto-currency look like? Gaining insight into Bitcoin phenomenon (2014)

Caginalp, C., Caginalp, G.: Opinion: valuation, liquidity price, and stability of cryptocurrencies (2018)

Yermack, D.: Is bitcoin a real currency? (2013)

Glaser, F., Zimmermann, K., Haferkorn, M., Weber, M.C., Siering, M.: Bitcoin - asset or currency? Revealing users' hidden intentions (2014)

Chiu, J., Koeppl, T.: The economics of cryptocurrencies – Bitcoin and Beyond (2017)

Kim, Y.B., et al.: Predicting fluctuations in cryptocurrency transactions based on user comments and replies (2016)

Malyshko, M.V.: Cryptocurrency: Bitcoin and its legal status (2016)

Poyser, O.: Exploring the determinants of Bitcoin's price: an application of Bayesian Structural Time Series (2017)

Polasik, M., et al.: Price fluctuations and the use of Bitcoin: an empirical inquiry (2014)

Torbjørn, B.J.: Why Bitcoins have value, and why governments are sceptical (2014)

Quotes, charts and expert opinions on cryptocurrency markets (Electronic resource). https://ru.tradingview.com

Statistics on crypto-currencies key metrics (Electronic resource). https://bitinfocharts.com

Website of crypto-currencies key metrics (Electronic resource). https://coinmarketcap.com

Optimizing Automated Trading Systems

Alessandro Bigiotti[✉] and Alfredo Navarra

Department of Mathematics and Computer Science,
University of Perugia, Perugia, Italy
alessandro.bigiotti@gmail.com,
alfredo.navarra@unipg.it

Abstract. In 2016, more than the 80% of transactions in the Forex market (where the world's currencies trade) have been directed by robots. The design of profitable automatic trading systems is becoming a challenging process. This requires a strong synergy of economists and computer scientists. Our aim is to provide an optimization framework for trading systems that starting from a generic strategy, enhances its performances by exploiting mathematical constraints. Moreover, the growth of new markets requires suitable solutions integrating computer science tools with economic analysis. In this work, we mainly refer to an emerging market known as Binary Options. Starting from basic strategies used every day in the stock markets by professional traders, we show how optimization issues enhance the outcoming performances. Tests on the optimized algorithms are conducted on both historical and real time data.

Keywords: Trading system · Algorithmic optimization · Binary Options

1 Introduction

The increasing of computing capacity and the development of sophisticated hardware, together with the development of the economic science called *Technical Analysis* [15, 17], are the main ingredients for the birth of *High Frequency Trading* (HFT) [4]. This refers to the current trend in the stock markets to program robots that automatically interact with the market itself in order to sell and buy stock options or commodities. The frequency of operations is extremely high, impossible to achieve by human beings, as well as they operate 24 h a day. It implies that even extremely small variations to the prices of the contracts involved can be remarkable due to the high number of operations performed. In 2016, more than the 80% of transactions in the *Foreign Exchange* (FX, [10]) market - where the world's currencies trade - have been performed by robots (see the white paper by the US Commodity Futures Trading Commission [11]). This phenomenon has increasingly become a central factor in the management of market fluctuations. The players involved are users, brokers and financial institutions. Users are anyone who wants to invest. The brokers are responsible for buying the quotations of some *assets* and making them available to users. An asset is intended as each thing, material or immaterial, likely to be evaluated. In the FX market, for instance, an asset is any currency that can be traded. The financial institutions provide the payments and the prices of the various assets.

© Springer Nature Switzerland AG 2019
T. Antipova and A. Rocha (Eds.): DSIC 2018, AISC 850, pp. 254–261, 2019.
https://doi.org/10.1007/978-3-030-02351-5_30

Fig. 1. Japanese candle definition (left). A possible sequence of candles (right).

Over each asset there are different ways to invest, here we consider the *Binary Options* [9, 16]. This is a very young and particular market where who invests does not buy or sell a contract of an asset, but rather bets on the fluctuations that will occur. The bets can be *CALL* or *PUT*: within a fixed time, CALL (PUT) means that one bets on *the price to increase* (*decrease*). After the bet is placed, one has to wait the fixed amount of time in order to achieve a gain or a loss. The success of a Binary Option is thus based on a YES/NO proposition, hence binary. If the bet is successful then the gain is given by a certain percentage called *payout*; otherwise the entire investment is lost. Actually, also nil trades can occur when the price remains unchanged. The payout is not fixed, but might vary during the price fluctuations. It depends on various factors: the current volume of trading; the time of the day; the type of betting (1 min, 3 min, 5 min, 30 min). However, the payout variations are closely related to the broker by which one operates. Reasonable values for the payout are on average around 70%–80%. There are plenty of strategies [4, 6, 12] to try to predict fluctuations, that have been studied and experimented. Some are based on the *Fundamental Analysis* [7], that measures the quality and the potential growth of a company/asset. However, from the computer science point of view, more interesting is the *Technical Analysis* [15, 17] that aims to predict fluctuations on the basis of historical behaviors. The main rule that governs the markets is that to some extent *history repeats*. Starting from a well-known strategy (by economists), we show how algorithmic optimization can be effective.

2 Notations and Terminology

In this section, we introduce some essential financial resources that we need in order to describe the trading strategy. The technical analysis is mainly based on bar charts, here we consider the so-called *Japanese candles* [14], see Fig. 1.

Given a generic asset A, candles are formed observing the fluctuation of A within a fixed time period $t = [t_1, t_2]$. A candle is defined by four prices: *opening* (price at time t_1), *closure* (price at time t_2), *maximum* (maximum price reached in the interval) and *minimum* (minimum price reached in the interval). If a candle satisfies the condition opening < closure (closure < opening) then it is named a *bull* (*bear*) candle. We represent a bull (bear) candle in white (black), a nil candle has no body. A sequence of white (black) candles is named a *bullish* (*bearish*) trend. Figure 1 (right) shows a bar chart as a sequence of candles. In the example we suppose the time period defining candle $i + 1$ has not yet terminated. All candles before $i + 1$ compose the set X. In order to obtain forecasts on candle $i + 1$ we may refer to n candles X_i, suffix of X, for some constant n. Each candle x_i in X_i is identified by a vector with the four prices that define the candle itself: opening, closure, maximum, minimum. These prices are used by particular functions, named *indicators* and *oscillators*, to forecast fluctuations of the chosen asset A.

3   Trading Systems

In general, strategies differ between those that follow the current trend (bearish or bullish), or those that operate counter the current trend. Both approaches make use of indicators and oscillators. Leaving aside the financial aspects and focusing on the mathematical ones, we can see indicators and oscillators as particular functions with several variables defined on the price fluctuations. Indicators aim is to predict the future trend of the prices. Oscillators are an important category of indicators. Their aim is to detect points of excess of bearish or bullish, or the weakening of the current trend. They are named oscillator because their value oscillates between a fixed range. Generally, the domain of these functions is a price in the set X_i, instead the codomain of these functions is closely related to the particular indicator or oscillator. In this section, we need to introduce an indicator and an oscillator that will be used in the strategy we design for Binary Options. The known strategy presented, is based on the *Bollinger Bands* (BB, see [5]) indicator, and on the *Relative Strength Index* (RSI, see [13]) oscillator.

Bollinger Bands (BB). This indicator provides the calculation of a simple moving average (*SMA*) and of two standard deviations σ_+ and σ_- from the *SMA*. We formalize the indicator as a function: $BB_{\{p,w\}}(X_i) := \{ma_i, \sigma_{i_+}, \sigma_{i_-}\}$, where: i is the index of the last candle of a bar chart one refers to; $p = |X_i|$ is the period for the moving average calculation, generally 20 candles; w is a factor for the calculation of the standard deviations, generally 2. The domain of the indicator is the set of closure prices c_j of the candles x_i in X_i, the codomain is a vector $\{ma_i, \sigma_{i_+}, \sigma_{i_-}\}$, where: ma_i is the moving average relative to the i-th candle; $\sigma_{i_+} := ma_i + (w \cdot \sigma_i), \sigma_{i_-} := ma_i - (w \cdot \sigma_i)$, and σ_i is the square root of the fraction between the squared differences of the closing prices of candles j and i, with $j = i, \ldots, i - p$, and $p - 1$. The idea behind BB is: *if the closure price of the last candle goes over σ_{i_+} or under σ_{i_-}, then the market will get in a tense situation, from which it should reverse the current trend.*

Relative Strength Index (RSI). This oscillator provides the calculation of two moving averages over the differences between the opening and the closure prices of a given set of bearish candles X_b, and bullish candles X_w. We could formalize the oscillator as a function $RSI_{\{p,os,ob\}}(X_w \cup X_b) = s_i$, where: i is the index of the last candle of a bar chart one refers to; X_w and X_b are two sets of the same cardinality of white and black candles, respectively, before candle x_{i+1}; $p = |X_b| = |X_w|$: is the period for the moving average calculation, generally 14 candles; *os* is the so-called *over-sold* limit, that is a bound for considering the current trend dictated by 'too many' selling operations; ob is the so-called *over-bought* limit, that is a bound for considering the current trend dictated by 'too many' buying operations. The domain of the oscillator is the difference between opening and closure prices of the candles in the sets X_w and X_b, the codomain is a value s_i ranging in [0, 100]. Value s_i is the result of the calculus $100 - (100/(1 + RS))$, and $RS := m_{i_+}/m_{i_-}$; where: m_{i_+} is the moving average of the absolute value of the difference from opening and closure prices of the last p white candles, x_j in X_w; m_{i_-} is the moving average of the difference from opening and closure prices of the last p black candles, x_j in X_b. Finally, we need to fix the over-sold and over-bought limits, generally 30 and 70, respectively. The idea behind RSI is: *if the oscillator breaks over (under) the over-bought (over-sold) limit, then the market should reverse the current trend.*

Algorithm 1 BB-RSI Reversal Strategy

1:    Set the asset $\mathcal{A}$ and the candles formation time t
2:    Set values (p, w) for BB and (p', os, ob) for RSI
3:    Set the amount b_i and the *expire time* t_e
4: **while** $\mathcal{A}$ is open **do**
5:      **while** a new candle $x_i = \{o_{x_i}, c_{x_i}, l_{x_i}, h_{x_i}\}$ on $\mathcal{A}$ is closed **do**
6:         Calculate $BB_{\{p,w\}}(x_i) = \{ma_i, \sigma_{i_+}, \sigma_{i_-}\}$
7:         Calculate $RSI_{\{p',os,ob\}}(x_i) = s_i$
8:         **if** $c_{x_i} \geq \sigma_{i_+}$ **then**   ▷ if the closure price $c_{x_i} \geq$ upp. standard deviation σ_{i_+}
9:           **if** $s_i \geq ob$ **then**   ▷ if RSI value s_i is over the over-bought limit ob
10:            $result = $ SendOrder(PUT, b_i, t_e)
11:         **if** $c_{x_i} \leq \sigma_{i_-}$ **then**   ▷ if the closure price $c_{x_i} \leq$ low. standard deviation σ_{i_-}
12:           **if** $s_i \leq os$ **then**   ▷ if RSI value s_i is under the over-sold limit os
13:            $result = $ SendOrder($CALL$, b_i, t_e)

3.1 BB-RSI Strategy

Algorithm 1 shows the pseudo-code of the strategy we consider for our optimization intents. At the beginning (Lines 1–3) we need to set some parameters that will be used by the algorithm. These settings will condition the behavior of the strategy, and our purpose is to check if it is possible to discover some *optimal* configurations to enhance the strategy performance. In Line 1 we set the asset A and the candles time t. In Line 2, we set the parameters for the BB indicator and the RSI oscillator. In Line 3 we set the *amount* b_i to invest, and the *expire time* t_e of the investment (i.e. the time during which the order remains on the market). The indicator and oscillator settings will determine the number of operations that the strategy could make. Lines 4–13 implement the reversal strategy. The While cycle at Line 4 can run 'forever'. The While cycle at Line 5 is executed until a new candle is ready. In Lines 6–7, the BB and RSI values are calculated, respectively. Lines 8 and 11 are devoted to check 'when send the signal for the PUT or the CALL orders, respectively.

The If control at Line 8 checks if the price of the closure of a candle (x_i) is over the upper standard deviation (σ_{i_+}), that is a tense situation. The If control in Line 11 checks if the price of the closure of a candle (x_i) is under the lower standard deviation (σ_{i_-}), that is another tense situation. Lines 9 and 12 are devoted to a confirmation for the PUT or the CALL signals, respectively. The If control in Line 9 checks if the value (s_i) of the RSI is over the over-bought limit (ob), this is used as a confirmation of the PUT signal sent by the engine. Similarly, as a confirmation of the CALL signal, there is the If control in Line 12. Lines 10 and 13 are designed to send an order; *result* will contain the outcome of the operation, after the expire time t_e.

3.2   Evaluation Parameters for a Trading System

In the evaluation of a trading system we are interested in the profitability of the strategy. This is a central factor, but other parameters must be considered. Let $n = w + l + nil$ be the sum of winning, losing, and null trades, respectively.

Net Profit (NP): indicates the profit accrued by the strategy. Suppose that each trade can gain $g_\uparrow$ or lose $g_\downarrow$. As in Binary Options we can assume $g_\uparrow$ and $g_\downarrow$ as constants, the net profit is given by NP:= $(w \cdot g_\uparrow) - (l \cdot g_\downarrow)$;

Max Draw Down (MDD): indicates the maximal loss of the strategy. Given an initial *balance* b_1, if the set $B = \{b_1, b_2, \ldots, b_n\}$ maintains the evolution of the balance during the strategy execution, then we need to calculate MDD := $\{max_{b_i, b_j} in\, B(b_i - b_j)$ $|b_i > b_j, i < j\}$;

Profit Factor (PF): indicates the risk of the strategy. It is expressed by the ratio between the average of the gains and the average of the losses. Since in Binary Options we assume $g_\uparrow$ and $g_\downarrow$ as constants, PF: = $(w \cdot g_\uparrow)/(l \cdot g_\downarrow)$;

Winning (W%) and Losing (L%): these parameters are strictly related to Binary Options. W% and L% are simply given by w/n and l/n, respectively.

4   Test and Optimization

Our purpose is to enhance the performance of the strategy provided by Algorithm 1 by means of optimization issues. In particular, considering W% or L%, we look for suitable settings of the indicator and oscillator variables. Maximizing W% (or similarly minimizing L%) may provide reasonable thresholds for judging the strategy as profitable. During our tests we make the following assumptions:

(i) *Average Payout* $(p_{\bar{a}})$: suppose an ideal broker pays the trades with some different payouts $p_1, p_2, \ldots, p_k$ that vary during the fluctuations. We could consider as weights the frequencies of the payouts $w_1, w_2, \ldots, w_k$. So the *average payout* is a good approximation of the payout;

(ii) *Breakeven Level*: there is a relation between the average payout $p_{\bar{a}}$ and the number of the winning trades w. With an investment i, each winning (losing) trade produces a gain $g_\uparrow = i \cdot_{\bar{a}}$ (a loss $g_\downarrow = i$). To find the minimum level for which the strategy is profitable, we need to guarantee $w \cdot g_\uparrow - l \cdot g_\downarrow > 0$. If $p_{\bar{a}} = 75\%$, we obtain $\$ (w \cdot i \cdot 75/100) - (l \cdot i) > 0$, that holds for $W\% > 0.571$.

The testing phase is a simulation of the strategy over a given historical period. We formalize the test as $S_{\{H\}}(BB_{\{p,w\}}, RSI_{\{p',os,ob\}}) = \delta$, where: S is a simulation function, H is the historical data period $(BB_{\{p,w\}}, RSI_{\{p',os,ob\}})$ are the indicator and the oscillator used by the strategy, δ is an evaluation parameter. The optimization process can be seen as a system of simulations $O(BB, RSI)$, in which we fix the asset A, the time for the candles formation t and the historical data period H; then we iteratively execute the strategy tuning the involved variables. Each simulation $S_{\{H\}}^i$ in the system $O(BB, RSI)$ produces a different δ_i that as first approximation relates to $W\%$. Each execution of the system generates a set of configurations of type $c_i = \{(p_i, w_i), (p_i', os_i, ob_i)\}$, concerning BB and RSI settings. Now we are interested in the elements c_i maximizing W %. Note that the search space of the system $O(BB, RSI)$ is rather extensive. Each variable involved by the strategy varies in a bounded interval. Suppose that p in P, w in W, p' in P', os in OS, ob in OB, then the number of simulations in the system are exhaustively given by: $|P| \cdot |W| \cdot |P'| \cdot |OS| \cdot |OB|$. If more indicators or oscillators are introduced, the behavior can become drastically expensive. Moreover, the larger is H the more trustable are the obtained results. A period H of 5 years is reasonably extended. However, not only we want to be profitable along the whole period H, but we aim to gain each year composing H.

To overcome all such requirements, we proceed in two phases. First we choose $H = \{h_1: 2013, h_2: 2014, h_3: 2015, h_4: 2016, h_5: 2017\}$. We consider each year h_i, instead of the entire period H. In this way, the execution of $O(BB, RSI)$ is faster and we obtain a differentiation about the tested configurations. In particular, we obtain a set $C_H := \{C_{h_1}, C_{h_2}, C_{h_3}, C_{h_4}, C_{h_5}\}$. Each element $c_j := \{(p, w)_i, (p', os, ob)_i\}$ contained in a generic C_{h_i} is a configuration produced by the optimization for the i-th year. In the second phase, we look for configurations with 'good' performances for the entire period H. To this respect we select from each C_{h_i}, the best elements c_j that satisfy two properties during H:

- P1 (*Over Breakeven*): c_j must be profitable, that is $g_\uparrow - g_\downarrow > 0$;
- P2 (*Irrelevant Gains*): the number of trades must be greater than some fixed parameter ω to guarantee that the gain of c_j is relevant.

On each configuration satisfying P1 and P2 we execute a final simulation by considering the entire period H, and as evaluation parameters $\{NP, MDD\}$.

5 Experimental Results

In this section, we experimentally show how our optimization can enhance the performance of the BB-RSI strategy over extended historical data. Also, we provide the effectiveness of our results, showing the performance of our optimized strategy on a real time trade system, currently operating. Tests on historical data are made on an in intel core i7-5500U @ 2.4 GHz with 4 GB DDR3, Windows 10 Education x64.

Table 1. Best settings achieved for the entire period H.

Period	p	w	p'	os	ob
h_1: 2013	{19–26, 32, 33, 35, 40, 41}	{2.0–2.5}	{18–20}	{25–27}	{69–75}
h_2: 2014	{25, 30, 31, 34, 36–38, 40, 41, 47, 50}	{1.8–2.5}	{17–20}	{25–28}	{72–75}
h_3: 2015	{10, 15, 26–29, 30, 33, 36, 50}	{1.3, 1.9– 2.5}	{16–20}	{25–28}	{70–75}
h_4: 2016	{22, 25, 38–41, 45, 47}	{2.0–2.5}	{17–20}	{27–30}	{72–75}
h_5: 2017	{18, 19, 24, 25, 28, 30, 33, 35, 48, 49}	{1.9–2.5}	{19, 20}	{25–28}	{72–75}

Table 2. Restrictions on ranges for the involved variables headings.

Configuration $c_i := \{(p,w),(p\prime, os, ob)\}$	NP $	MDD $	# trades	W%
$c_1 = \{(39, 2.2), (17, 27, 73)\}$	68275	2975	14180	59.89
$c_2 = \{(31, 2.3), (18, 28, 74)\}$	56950	3575	10753	60.16

The software used is Meta Trader 4 [2]; Algorithm 1 was implemented in Meta Quotes Language 4 (*MQL4*, see [1]). We execute *O(BB, RSI)* on the last five years H, considering the asset A = EUR/USD (the price of euros in US dollars) and $t = 1$ min for the candles formation. The variables involved by the strategy vary in the following discrete intervals: p in [10, 50], w in [0.3, 2.6], p' in [5, 20], os in [25, 35], ob in [65, 75].

The interpretation of the data is based on an initial balance of 1000 $, an average payout $p_{\bar{a}}$ of 75%, and an investment of 100 $ for each bet. The optimized configurations obtained by the first step of our approach provide the restricted ranges for the involved variables shown in Table 1. The outcome is shown in Table 2 along with the number of trades performed. The best setting is thus c_1 where NP is maximized and MDD is minimized, even though W% is slightly smaller than that achieved in c_2. In

Fig. 2. Balance evolution in the period 2 January–15 June 2018 of the BB-RSI strategy under configuration c_1, starting with 1000 $ and bets of 100 $ each.

Fig. 2 we show the effectiveness of our approach by reporting the balance evolution of a real-time trading system currently operating on the basis of c_1.

6   Conclusion and Future Work

We have shown how optimization issues are effective for the setup of a trading system for Binary Options. It might be reasonable to think about other markets and/or other assets for which multi-criteria optimization is required. Other approaches and methodologies could be applied to further enhance the behavior of a strategy or to provide completely new strategies. For instance, we are thinking about applying *martingale* approaches [3] that could be effective in Binary Options due to the relation of such a market with gambling. Other emerging markets require deep investigations. An example is about crypto-currencies [8], even though the lack of historical data might be critical.

Acknowledgments. Special thanks go to Simone Ciancone for his expertise and useful discussions. The work has been supported in part by the GNCS-INdAM project 2018 "Anti-Social Networks".

References

1. mql4 documentation. https://docs.mql4.com
2. mt4. https://www.metatrader4.com/en
3. Le Gall, J.: Brownian Motion, Martingales, and Stochastic Calculus. Graduate Texts in Mathematics. Springer, New York (2016)
4. Aldridge, I.: High-Frequency Trading: A practical Guide to Algorithmic Strategies and Trading System. Wiley Trading Series. Wiley, New York (2013)
5. Bollinger, J.: Bollinger on Bollinger Bands. McGraw-Hill Education, New York (2001)
6. Bonenkamp, U.: Combining Technical and Fundamental Trading Strategies. Gabler Verlag, Wiesbaden (2010)
7. Bulkowsk, T.: Fundamental Analysis and Position Trading: Evolution of a Trader. Wiley, New York (2013)
8. Clarkson, M.: Cryptocurrency: Best Strategies for Investing and Profiting from Cryptocurrency. CreateSpace Independent Publishing Platform (2018)
9. Coulling, A.: Binary Options Unmasked. Anna Coulling (2015)
10. Guy, D.: Forex Trading Money Management System. Roulette Trader (2016)
11. Haynes, R., Roberts, J.: Automated Trading in Futures Markets Update (2017)
12. Laamb, D.: Automated Trading Strategies by Example. Philip Lim (2017)
13. Perchanok, K., Hrytsyuk, I.: The Encyclopedia of the Indicator RSI (Relative Strength Index). CreateSpace Independent Publishing Platform (2011)
14. Person, J.: Candlestick and Pivot Point Trading Triggers. Wiley Trading Series. Wiley, New York (2018)
15. Pring, M.: Study Guide for Technical Analysis Explained. McGraw-Hill, New York (2002)
16. Schwager, J.: A Complete Guide to the Future Markets. Wiley Trading Series. Wiley, New York (2017)
17. Stevens, L.: Essential Technical Analysis: Tools and Techniques to Spot Market Trends. Wiley Trading Series. Wiley, New York (2002)

Russian Banks Credit Risk Stress-Testing Based on the Publicly Available Data

Davit Bidzhoyan[1,2]([⊠]) and Tatyana Bogdanova[1,2]([⊠])

[1] National Research University "Higher School of Economics",
Myasnitskaya st. 20, Moscow, Russian Federation
bidzhoyan_david@mail.ru, tanbog@hse.ru
[2] Springer Heidelberg, Tiergartenstr. 17, 69121 Heidelberg, Germany

Abstract. This paper suggests an algorithm for stress testing of the credit risk of a Russian commercial bank, intended for use by investors and bank customers to assess the bank's financial stability under stressful scenarios. Indicator of bank losses in this work is the indicator "loan loss provision". An algorithm is proposed that describes the bank's cash flows in stressful situations, taking into account the demand function for the loans of the analyzed bank, the bank's availability of the necessary capital to increase the loan portfolio, and the availability of a sufficient amount of liquid funds to cover losses.

Keywords: Stress-testing · Loan loss provision · Cash flow
Financial stability · Credit risk · Investors

1 Introduction

There are many definitions of the concept of stress testing. According to the Bank of International Settlement (BIS), stress testing is "a variety of methods that are used by financial institutions to assess their vulnerability to exceptional but plausible events." The Bank of Russia defines stress testing as an assessment of "the potential impact on the financial stability of a credit institution of a number of specified changes in risk factors that correspond to exceptional but plausible events." According to the International Monetary Fund, stress testing is nothing more than "methods for assessing the sensitivity of a portfolio to significant changes in macroeconomic indicators or to exceptional but possible events" [1]. Stress testing tools have been successfully integrated into the risk management systems of individual banks (microprudential stress testing) and the banking system as a whole in Central Banks (macroprudential stress testing). The final distribution of stress testing was after the financial crisis of 2008, during which many large banks collapsed (for example Lehman Brothers among many others). At now, many international organizations, as well as central banks of many countries carry out stress testing on a regular basis. Thus, European Banking Authority (EBA) on an annual basis, starting in 2011, conducts stress testing of risks of the largest European banks, for example 2011 EU-wide stress test results [2], according to the developed methodology [3]. Basel Committee of Banking Supervision (BCBS) developed the principles of stress testing for individual financial institutions and for supervisory authorities. The US Federal Reserve, since 2009, has been implementing

T. Antipova and A. Rocha (Eds.): DSIC 2018, AISC 850, pp. 262–271, 2019.
https://doi.org/10.1007/978-3-030-02351-5_31

the Supervisory Capital Assessment Program (SCAR) annually. In addition, at the legislative level, the Dodd-Frank Act of Stress-Testing (DFAST) according to which stress testing of banks with total capital of more than $10 billion conducted, and the Comprehensive Capital Assessment and Review (CCAR) program, which analyzes the financial condition of banks whose total assets exceed $50 billion. Stress testing is one of the key components of the IMF carrying Financial Stability Analysis Program (FSAP) program, which examines the financial systems of countries [4]. The actual issue is the publication of information on the results of stress testing. A number of authors are advocates of disclosure, while other authors actively oppose [5, 6].

As a rule, stress testing is conducted by the supervisory authorities and the banks themselves to identify the bank's ability to absorb losses in stressful situations. However, the results of stress testing can be useful to investors and clients when choosing a bank for investment, as they give an information of the financial statement of the bank in adverse situations. Thus, for conducting independent stress testing, investors may use only published bank reports. The purpose of this study is to develop an integrated approach to stress testing of the credit risk of a Russian commercial bank on the basis of public bank reporting. Within the framework of this work, the process of developing stressful scenarios is omitted.

The rest of the article consists of the following paragraphs: in the second paragraph gives a brief review of the literature on the methodology of stress testing. In the third paragraph, the data source on the basis of which credit risk stress testing of a Russian commercial bank is described, as well as the limitations in which the results of stress testing are consistent. In the fourth paragraph, a method for modeling indicators is provided. In the fifth paragraph we propose an algorithm for calculating the capital adequacy ratio of a bank. In paragraph 6, the areas for future developments are designated, and in conclusion the main conclusions on the work are indicated.

2   Relation to Literature

In the context of stress testing, one of the fundamental issues is the choice of risk factors, which, if worsened, will check the financial state of the bank. In most cases, under stress testing of credit and market risks, macroeconomic variables act as risk factors [4]. However, a more detailed analysis of the structure of assets and liabilities allows you to identify really stressful indicators. It should take into account the sectoral, regional structure of assets, urgency and sources of liabilities.

For the most part, the methodology of scenario stress testing is based on econometric models, with the help of which the predicted values of the analyzed variables are obtained under certain specified scenarios. Regression models of time series, binary choice models, vector autoregressive model and error correction model, simulation modeling, etc. are applied [7, 8].

Particular attention in the stress-testing is given to the development of scenarios. In work [4], current stress testing methodologies used by international organizations and central banks of developed countries that use only 2–3 stress scenarios and do not evaluate the plausibility of the scenarios criticized. A limited number of scenarios can lead to false positive conclusions, since with the same degree of plausibility of different scenarios, the results of stress testing can vary significantly. Authors classify such methodologies as first-generation stress tests, while second-generation stress tests apply many different scenarios and estimate their plausibility, for example, using the Mahalanobis distance or the Kullback-Leibler divergence. However, this approach can lead to computational difficulties. Therefore, the authors of the article propose an approach for selecting scenarios that satisfy the "Worst Plausible Stress Scenario" principle. In [9], in addition to the Mahalanobis distance, the Bregman distance [10] or f-divergence is also proposed, introduced by [11].

Also in the academic literature, indicates the need to take into account the discount rates in the sale of assets. In paper [12] the analysis of the impact of discount factors at stress testing of credit and market risks of small and large banks is given. It is alleged that large banks are forced to sell their assets to the trade book at a discount, in view of large volumes, while small banks are not.

A number of papers [13–15] within the framework of stress testing for the purposes of maintaining capital adequacy, in addition to capital management, also consider the management of the RWA structure necessary. In [12], a hypothesis is proposed that in order to reduce the denominator of the formula for calculating the capital adequacy it is first necessary to sell the assets with the highest risk factor.

3   Data

3.1   Banking Reports

Official public financial reporting is the basis for conducting stress testing of the credit risk of a Russian commercial bank. The data are published on the Bank of Russia website[1]. The following reporting forms are used:

- 0409101 "Balance Sheet";
- 0409102 "Profit and Losses Statement";
- 0409123 "Calculation of own funds (capital) (Basel III)"[2];
- 0409135 "Information on mandatory standards".

Table 1 shows the variables needed to carry out stress testing of credit risk, extracted from the above reporting forms.

The characteristics of credit risk are the portfolios LLPs (hereafter LLP – Loan Loss Provision). There are two types of portfolios: a consumer loans portfolio, which includes all types of loans granted to individuals, and a corporate portfolio, granted to legal entities. As part of stress testing, it is assumed that the LLPs at the time of the beginning of stress testing are formed in full[3]. The peculiarity of the formation of reserves is that a large part of the reserves is formed on the account "45818" of balance sheet. However, the structure of the LLPs on loan portfolios is unknown. In this paper,

---

[1] www.cbr.ru/credit/forms.asp.

[2] Before 01.2014 the form 0409134 "Calculation of own funds (capital)" was adopted.

[3] We also impose other assumptions such as shareholders cannot invest their funds to raise a capital. Securities and bonds are not re-evaluated.

Table 1 Data for credit risk stress testing and their sources

Variable	The reporting form	Designation
Retail loans	0409101	loan_ret
Retail loan loss provision	0409101	llp_ret
Retail overdue loans	0409101	odl_ret
Corporate loans	0409101	loan_corp
Corporate loan loss provision	0409101	llp_corp
Corporate overdue loans	0409101	odl_corp
Overdue loan loss provision	0409101	llp_45818
Corporate deposits	0409101	dep_corp
Funds on corporate accounts	0409101	acc_corp
Individual deposits	0409101	dep_ind
Funds on individual accounts	0409101	acc_ind
Total capital	0409123	cap
Liquid assets	0409135	la
Highly liquid assets	0409135	hla
$H_{1.0}$ capital adequacy ratio	0409135	H1.0
$H_{1.1}$ common equity Tier1 ratio	0409135	H1.1
$H_{1.2}$ common equity Tier2 ratio	0409135	H1.2

it is proposed to divide the *LLP*s formed on the account "45818" by loan portfolios, based on the share of overdue debt for each loan portfolio in the total overdue debt.

$$llp_{i,458} = \frac{llp_{458} * odl_i}{odl_{corp} + odl_{ret}} \qquad (1)$$

where

$llp_{i,458}$ – loan loss provision at the "45818" balance sheet item; *i:* *"corp"* – corporate portfolio, *"ret"* – retail;
res_{458} – loan loss provision, reflected in account "45818";
odl_{corp} – overdue loans of the corporate portfolio;
odl_{ret} – overdue loans on the portfolio of retail loans.

The total volume of *LLP*s for each portfolio is the sum of the *LLP*s reflected in the accounts of each portfolio type and the *LLP* for each portfolio type, calculated by formula (1). The formula for calculating the total amount of provisions for portfolio *i* is presented on the formula (2):

$$llp_{i,all} = llp_i + llp_{i,458} \qquad (2)$$

where

$llp_{i,all}$ – total loan loss provision; *i:* *"corp"* – corporate portfolio, *"ret"* – retail;
$llp_{i,458}$ – loan loss provision, reflected in account "45818" for *i* portfolio;

llp_i – loan loss provision for i portfolio, reflected in correspondent accounts.

3.2   Macroeconomic Variables

Within a proposed framework of credit risk stress-testing macroeconomic variables are chosen as a risk factors. We impose macroeconomic shocks to banks financial stability. To carry out stress testing, the following variables used as variables describing the macroeconomic environment:

- GDP growth;
- consumer price index;
- unemployment rate;
- household income growth;
- mean, standard deviation of RTS index;
- mean, standard deviation of MICEX index;
- mean, standard deviation of USD/RUB currency rate;
- mean, standard deviation of MIACR[4] rate.

The source of macroeconomic data are www.gks.ru, www.finam.ru.

Selected variables well describe a macroeconomic environment from our point of view.

4   Modelling Banking Variables

Modeling of bank indicators within the framework of stress testing is conducted depending on macroeconomic variables. The general specification of the model is as follows (3):

$$y_{j,t} = \beta_0 + \sum_{p=1}^{m} \beta_p x_{p,t} + \varepsilon_{j,t} \qquad (3)$$

where $y_{j,t}$ – j banking variable at time t;

β_0 – constant;
$x_{p,t}$ – p macroeconomic variable at time t;
β_p – coefficient for p macroeconomic variable, $p = \overline{1 \ldots m}$, m – total number of macroeconomic variables;
$\varepsilon_{j,t}$ – error term of j banking variable at time t.

Models are estimates using Least Square Estimator (4):

$$\hat{\beta} = \left(x^T x \right)^{-1} x^T y \qquad (4)$$

---

[4] MIACR – Moscow Interbank Actual Credit Rate.

In the case of heteroscedasticity and/or autocorrelation, robust standard errors are used in the residuals (5):

$$\widehat{Var}(\beta|x) = (x'x)^{-1}x'\hat{\Omega}x(x'x)^{-1}$$
$$\hat{\Omega} = diag\left(\hat{\varepsilon}_1^2 \ldots \hat{\varepsilon}_t^2\right)$$

(5)

Based on the forecast values of bank variables, stress testing is carried out.

It should be noted that at the stage of modeling indicators, methods of machine learning, neural networks can also be applied. However, the lack of the possibility of interpreting the results of the model, and also, in most cases, the dependence of the final result of the model on the initially chosen point does not allow to fully apply this tool for modeling the indicators. In view of this, the choice of linear regression models is justified.

5 Calculating Capital Adequacy Ratio

In each quarter, the capital adequacy ratio is calculated in the horizon. This procedure is carried out in 3 steps.

At the first step, the capital adequacy ratio is calculated taking into account the necessary addition of *LLP*, the volume of which is calculated on the basis of regression models. The calculation formula is as follows (6):

$$H_{1,k} = \frac{C_k - \Delta llp_{corp} - \Delta llp_{cons}}{RWA_k - cr_{corp} \cdot \Delta llp_{corp} - cr_{ret} \cdot \Delta llp_{ret}}$$

(6)

where $H_{1,k} - k$ capital adequacy ratio; $k = 0$ – total capital, $k = 1$ – Core Tier 1; $k = 2$ – Core Tier 2.

RWA_k – risk weighted assets for k capital;
C_k – k capital;
cr_{corp} – risk coefficient for corporate loans portfolio;
cr_{ret} – risk coefficient for retail loans portfolio;
Δllp_{corp} – additional charge of corporate loan loss provisions;
Δllp_{ret} – additional charge of retail loan loss provisions.

In the second step, based on the projected values of the variable resource base, as well as the demand function for the loans of the analyzed bank, the volume of the loan portfolio is calculated. Figure 1 shows the cash flow diagram of the bank that takes into account the inflow (outflow) of the resource base, the demand function for loans, the availability of the necessary capital to meet the capital adequacy requirements and allowances, and the bank's ability to cover the outflows with liquid funds.

The procedure for determining the forecast value of the volume of the loan portfolio is carried out in three stages.

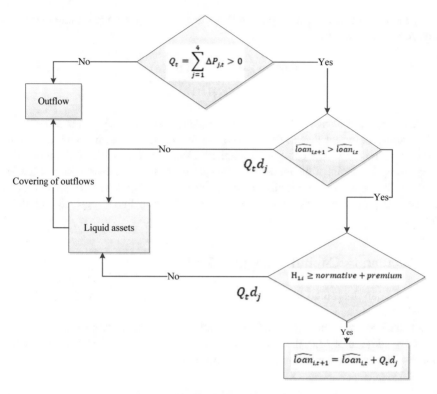

Fig. 1 Cash flow diagram

All investments in assets depend, first of all, on how much the resource base will grow or decrease. At the first stage, the total increment of the variables characterizing the resource base (7) is determined:

$$Q_t = \sum_{j=1}^{4} \Delta P_{j,t} \tag{7}$$

where Q_t – total increase (decrease) of resource base;

$P_{j,t}$ – resource base variable, $j =$ "*dep_corp*", "*acc_corp*", "*dep_ind*", "*acc_ind*" at time t. Total number of resource base is 4.
If $Q_t > 0$ then it is necessary to go to the second stage. Otherwise, there is an outflow of funds, which are covered by liquid funds.

At the second stage, using the estimated function of demand for loans, the predicted values of loans is determined. If the forecast value at time $t + 1$ is less than at time t, the surplus is invested in liquid assets proportionally to their part of total assets d_i. Otherwise, go to step 3.

In the third stage, the ability of the bank's capital to meet capital adequacy standards when investing in loans is tested. If there is not enough capital, the surplus is invested in liquid funds proportionally to their part of total assets d_i. Otherwise, it invests in loans proportionally to their part of total assets d_i

In the third step, the capital adequacy ratio is calculated taking into account newly issued loans (8):

$$H_{1,k,t,new} = \frac{C_{k,t,new} - cp_{ret} \cdot \Delta loan_{ret,t} - cp_{corp} \cdot \Delta loan_{corp,t}}{RWA_{k,t,new} + cr_{ret} \cdot \Delta loan_{ret,t} \cdot (1 - cp_{cons}) + cr_{corp} \cdot \Delta loan_{corp,t} \cdot (1 - cp_{corp})} \tag{8}$$

where $C_{k,t,new}$ – capital k at time t taking into account additional charge of loan loss provision of existing loan portfolios;

cp_{ret} – provision coefficient of retail loans portfolio;
$\Delta loan_{ret,t}$ – retail loans portfolio increase at time t;
cp_{corp} – provision coefficient of corporate loans portfolio;
$\Delta loan_{corp,t}$ – corporate loans portfolio increase at time t;
cr_{ret} – risk coefficient of retail loans;
cp_{corp} – risk coefficient of corporates loans;
$RWA_{k,t,new}$ – risk weighted assets of capital k at time t taking into account additional charge of loan loss provision of existing loan portfolios.

If the obtained values of capital adequacy ratios are higher than the normative value equal to 8% of risk-weighted assets and mark-ups, it is considered that the bank has successfully passed stress testing of credit risk, otherwise the capital deficit is calculated by formula (9):

$$D = \max \left(normH_{1.k} - H_{1.k,new} \right) \cdot RWA_k \tag{9}$$

where D – capital deficit;

$normH_{1.k}$ – normative value of capital adequacy ratio of k capital;
$H_{1.k,new}$ – predicted value of capital adequacy ratio of k capital;
RWA_k – risk weighted assets of k capital.

6 Future Development Areas

The results of stress testing, in many respects, depend on many parameters that the researcher asks as an assumption. These parameters include risk ratios for loan portfolios, as well as the reserve ratio for new loans. As a further prospect of the study, it is proposed to analyze the sensitivity of the results of stress testing to changes above the indicated parameters. This work does not take into account collateral for loans, which can play an important role in the final result of stress testing. Additional parameters may be the discount factor in the sale of assets, as well as the loan impairment factor for loans.

As a supplement to stress testing of credit risk, liquidity stress testing can also be conducted, the purpose of which is the bank's ability to cover outflows from liquid funds. As part of stress testing of liquidity, discount rates are particularly important when selling liquid funds to cover outflows. Thus, carrying out reverse liquidity stress testing and analyzing the sensitivity of test results to discount rates are becoming an urgent task.

It should also be noted that it is necessary to conduct stress testing using a variety of stressful scenarios to avoid the false illusion of a successful stress test. However, the question arises of analyzing the results of the stress test of many scenarios.

7 Conclusion

Thus, this paper presents the developed multi-stage procedure for stress testing of the credit risk of a Russian commercial bank on the basis of public bank reporting. This approach can be used by investors and creditors to assess the financial soundness of banks in stressful situations. A cash flow model has been developed that takes into account the demand function for the loans of the analyzed bank, the availability of capital for investing in the loan portfolio, and liquid assets. However, the process of developing a stressful scenario remained outside the scope of this work.

References

1. Blaschke, W., Jones, M., Majnoni, G., Peria, S.: Stress-testing of financial system: an overview of issues, methodologies and FSAP experiences. IMF working paper WP/01/88 (2001)
2. European Banking Authority 2011 EU-Wide Stress-test Aggregate Report (2011)
3. Methodology EU-Wide Stress-Test (2014)
4. Breuer, T., Summer, M.: Solvency stress testing of banks: current practice and novel options. Report for the Sveriges Riksbank and Finansinspektionen (2017)
5. Goldstein, I., Sapra, H.: Should bank stress test be disclosed? An analysis of costs and benefits. Found. Trends Finance **8**, 1–54 (2015)
6. Latiner, Y.: Should regulators reveal information about banks? Bus. Rev. Fed. Reserve Bank Phila. 1–8 (2014)
7. Melecky, M., Poderiera, A.: Macro prudential stress-testing practices of central bank in central and South Europe. An overview of challenges ahead. Policy research working paper 5434 (2010)
8. Foglia, A.: Stress testing credit risk: a survey of authorities approaches. Bank of Italy occasional paper no. 37 (2008)
9. Breuer, T., Csiczar, I.: Measuring distribution model risk. Math. Finance **26**(2), 395–411 (2016)
10. Bregman, L.: The relaxation method of finding the common point of convex sets and its application to the solution problems in convex programming. USSR Comput. Math. Math. Phys. **7**, 200–217 (1967)
11. Csiszar, I.: Information-type measures of difference of probability distributions and indirect observations. Studia Scientiarum Methematicarum Hungarica **2**, 299–318 (1967)

12. Braouezec, Y., Wagalath, L.: Risk-based capital requirements and optimal liquidation in stress scenario. Rev. Finance **22**(2), 747–782 (2018)
13. Glasserman, P., Kang, W.: Design of risk weights. Oper. Res. **62**(6), 1204–1220 (2014)
14. Mariathasan, M., Merrouche, O.: The manipulation of basel risk-weights. J. Financ. Int. **23** (3), 300–321 (2014)
15. Vallascas, F., Hagenwolf, J.: Risk sensitivity of capital requirements: Evidence from international sample of large banks. Rev. Finance **17**(6), 1947–1988 (2013)

In-System Entries in the Early Memorial Books

Artem Musaelyan, Mikhail Kuter[✉], and Marina Gurskaya

Economy Department, Kuban State University, Stavropol'skaya St., 149,
350040 Krasnodar, Russia
prof.kuter@mail.ru

Abstract. The carried out research is aimed at studying the evolutionary path of the memorial book of Ricordanze, described in 1494 by Luca Pacioli. The peculiarities of its keeping have been considered since 1363. It has been noted that originally Ricordanze served simultaneously for non-system (personal) records, and for in-system (accounting) entries, which in their turn were transferred to the Ledger for the purpose of registering business transactions. In future, the separation of registers reflecting in isolation in-system and non-system entries facilitated the use of Memoriale as books of account and Ricordanze as a register for non-accounting entries.

Keywords: Ricordanze · Avignon-1363 · In-system entries
Non-system entries

1 Introduction

Luca Pacioli (1494) described the triad of books: "Memorial" – "Journal" – "Ledger". In addition, he paid his attention to "Ricordanze", the memorial book of a merchant that was intended for personal non-system entries, which were not related to accounting.

Some authors stated that Francesco Datini's archives in Prato (Tuscany) possess a book that contained both personal (non-system) and in-system (accounting) entries. However, no one has ever referred to this book precisely.

After a long search the Russian scientists have managed to find this book. This "Ricordanze" concerns the company of F. Datini and N. Bernardo in Avignon, established in 1363. The paper describes the course of searches and the study of the book. It has been shown how the book for personal entries gradually turned into a book for dual purpose. The translations of the texts confirming the transformation of entries into in-system ones related to accounting have been introduced. In the diagrams, formed according to the photocopies of Ricordanze and Liber pages, the connection between both accounting registers has been confirmed.

T. Antipova and A. Rocha (Eds.): DSIC 2018, AISC 850, pp. 272–285, 2019.
https://doi.org/10.1007/978-3-030-02351-5_32

2   Prior Literature

Distinguished scientists over the course of the twentieth century argued over the origin sector for double entry and double-entry bookkeeping: banks or merchant trade. Among the factors contributing to the emergence of double-entry bookkeeping R. de Roover distinguished company, credit and mediation.

The role of company is unquestionable, since there is a problem of precise calculation and distribution of the financial result. The role of credit is indisputable as well. The question arises justly though, what kind of credit the conversation is about – bank credit or credit on goods?

One of the most prominent contemporary researchers, Alan Sangster, speaking at the World Congress of Accounting Educators and Researchers (Florence, November 2014) [Sangster] had a hypothesis that a double entry originated in Florence, around 1240, in banks, when constant inspections of the Bankers Guild required immediate delivery of information concerning transactions performed simultaneously in two accounts in one book, that is, the information on intra-bank calculations.

The fact is that bankers, most likely, deliberately did not introduce into the accounting system the accounts for cash accounting or specially designed for those purposes "Entrata e Uscita" books. Such an act was connected with the desire to conceal the charged interest, as the entry in the bank book of account indicated the amount due, and the introduction of a cash account or "Entrata e Uscita" book would make bankers show different amounts when providing and repaying credits, which should lead to the "disclosure" of the facts of interest obtaining.

A. Martinelli holds about the same opinion and indicates the same time, but focuses not on the Florentine banks, but on the banks of Siena, which provided credits abroad, which also required close control (Martinelli 1974). We should mention at once, that no archives possess any archival evidence for the given bank hypothesis.

The trade hypothesis is focused on two types of merchants: travelling and sedentary ones. The former (wholesalers) travelled with the caravans to the East and bought by wholesale consignments of goods. On their return, they sold also by wholesale the consignments of goods to sedentary merchants. In our opinion (Kuter 2012, pp. 32–33), they were interested in two types of results: result from each sale and result from sale of all consignment of goods. Operational result allows to determine the efficiency of buying and selling each particular consignment of goods in order to know the expediency of purchasing a consignment during the following trip. The latter determined the financial result for all the activities (for all the consignments).

Given the relatively small number of consignments during a trip, a travelling merchant did not need any special bookkeeping, it was kept in his memory.

de Roover (1956) considered the sedentary merchant to be a mediator between the travelling merchant and the population. His activity was significantly influenced by the factor of hammered currency deficit. Barter was flourishing for this reason. In those cases, when the seller was not satisfied with the offered goods, the sales were carried out on credit with the settled payment date. In addition, which is very important, in most cases, debts were repaid by means of several payments in separate amounts.

Since the number of buyers was often significant, the merchant had to arrange accounting of settlements with debtors. In addition, the merchant himself acted as a debtor for his suppliers, which determined the existence of accounting for passive liabilities.

And, of course, the mediator merchant as well as the travelling merchant was interested in the result for each consignment of goods and, of course, in the financial result of all activities. Profit was at all times the driving motive of a merchant's activity, although trade and usury were perceived by the church as occupations unworthy of a Christian.

In order to determine the result for each consignment of goods, the merchant had to keep records of the goods separately for each consignment that was received from each supplier, even if there were similar deliveries from several suppliers. It is clear that for record keeping of such a significant number of transactions one person was already not enough, and it was time to form accounting procedures and develop methods for their implementation. There is a question, which is important for every researcher and any practice: how did it all start?

Today in the post-Soviet countries and, first of all, in Russia, there has been a heightened interest in any new findings in the archives of medieval accounting books in Italy. There is a direct explanation for this. The Soviet and Russian scientists have never taken part in them. Moreover, there was no information about the results of our Western colleagues' research. On the other hand, Russian people have entered all known archives today.

In July 2008, the 13th World Congress of Accounting Historians took place in Great Britain. It was already the second Congress, at which two reports of the Russian scientists were presented. Great interest was caused by the report of Professor Alan Sangster from Great Britain, which was devoted to the application of the early memorial book of "Ricordanze". The meaning of the report was in the following: "Double entry bookkeeping emerged by the end of the 13th century and was adopted by, for example, the Datini of Prato during the 1380s. In the transition from the single to the double entry evident in the Datini Archives, the initial accounting records were kept in an ac-count book called a Ricordanze".

According to A. Sangster, "Record books of this name were typical of Tuscany and, when such books were first used in Tuscany, businessmen began to use them also as a form of personal diary and autobiographical record. Others not in business followed suit and maintained purely personal biographical diaries of the same name. For those in business, the Ricordanze thus developed into a hybrid: partly autobiography and personal and, partly, a place to record matters relating to his business, including details of transactions and of other matters he did not wish to forget, such as promises, obligations, and conditional agreements" (Sangster et al. 2012, p. 27).

The author of the report stated: "As revealed in the Datini archives for the 14th and 15th centuries, use of a Ricordanze for this purpose was discontinued in the accounting system and the book was replaced with another called a Memoriale, which contained details of all business transactions. By the time Pacioli wrote the first published description of double entry bookkeeping, the Memoriale was identified as one of the three principal account books of that system. The others were the Giornale (journal) and the Quaderno (ledger)" (Sangster et al. 2012, pp. 27–28).

Pacioli in Chapter 35 of the Treatise gives a description of another book that was used by clever merchants, that was Ricordanze, a book of memorial entries. Many scientists deny this fact. According to Pacioli, the book he described was not a book for personal entries or a hybrid of business and personal entries. Pacioli's variant, moreover, did not represent a certain version of the Memorial. Pacioli's book of entries, which he called Ricordanze, was intended for special purposes: this book contained something which in no case could be forgotten. Consequently, it served as another control tool in managing the merchant's affairs, in addition to those control functions that were offered by the accounting system.

A. Sangster and his colleagues pursue the aim to consider the role of Pacioli's memorial book of "Ricordanze", the specific entries that are entered into the book, to find out the advantages of keeping records in the book. The authors also consider it is necessary to clarify the reasons why such a useful thing did not receive proper recognition, although "Ricordanze" book of entries was introduced in detail in the very Treatise that provided the universal acceptance of double-entry bookkeeping.

3 Research Method

The principal research method adopted in this study is archival. It uses material found in the State Archive of Prato. Generally, this research team has been working with the material in this archive for the past decade and many of the records have been recorded and linked together using logical-analytical reconstruction. This is an approach that we developed for the purpose of enabling entries in the account books to be traced visually between accounts and books and from page to page.

Unfortunately, because the material studied related to an early stage in the development of accounting and was not organised systematically, it was impossible to do this. Instead, the authors had to work from the Ricordanze, entry by entry, searching for each one in the Ledger until they were all traced.

4 The Discovery of the Russian Scientists in F. Datini's Archives in Prato

The presentation aroused the interest of the researchers of our university. Some years earlier (2009), one of the authors of this publication prepared modern interpretation of the Russian translation of Pacioli's Treatise (Pacioli 2009). The previous Russian version of the translation (Paciolo 1893) was performed by E.G. Waldenberg not directly from Pacioli's text, but from the text by E. Jager, which was in German (Jager et al. 1876). Before proceeding to the presentation of our research, let us see how Ricordanze book in Luca Pacioli's Treatise is described. The mentioning of Ricordanze (Memoranda book or Memo) is contained in Chapter 35 and Chapter 36 (subsection C).

L. Pacioli in Chapter 35 notes: "And similarly it is useful to have a book for reminders, a so called memoranda book, in which you daily note matters which you might otherwise forget and easily cause you loss. Every evening before going to bed take a look in this book in case there is something which should have been dispatched or done which has not been dispatched; and cancel with a pen those things which have been done" (Yamey 1994, 88–89).

And, further on, Pacioli enumerates things that are entered into the book of Ricordanze: "And so you will take note of things which you lent to a neighbor or friend for a day or two such as vases for the shop, kettles or other tools. You must follow such instructions together with the other very useful ones given above adjusting them by additions and deletions as you deem fit, according to time and place. For it is not possible in mercantile business to make rules point by point to cover everything; as we have said on other occasions one needs more rules" (Yamey 1994, 89).

In subsection C of the last Chapter 36 Pacioli describes in more detail what a merchant needs to enter for memory: "All the implements and utensils used in the house or shop which you own should be entered in order, that is enter separately all things made of iron, leaving a space to make additions, and leaving a margin to mark those things which have been sold or donated or spoiled, but excluding household implements of little value. And all things made of brass should be entered separately, as mentioned, and similarly all things made of tin, and similarly all things made of wood, and so with all things of copper, of silver, and of gold, etc. always leaving sufficient empty pages to make additions if necessary and also to make note of things which are missing" (Yamey 1994, 89).

"All sureties, or obligations, or promissory notes which you made for a friend, explaining clearly what and how. All merchandise or other things which have been left with you for safekeeping or custody, everything on loan from some friend and also all things which you lent to your friends; all conditional transactions: that is purchases or sales as for example a contract that you send me by the next galley returning from England, so many "*cantari di lane dilimistrī*"; and if they are sound on receipt I will give you so much per "cantaro", or per hundred, or, I will send you so many "cantari" of cotton" (Yamey 1994, 89).

"All the houses, or land, or shops, or jewels, which you rent at so many ducats, or at so many lire annually. And when you collect the rent, enter that cash in the ledger as I mentioned above. If you lend some jewellery or silver or gold vase to a friend for eight or fifteen days, do not enter this in the ledger, make a note in your memoran-dum book since they will be returned to you in a few days. And similarly, per contra such things as are lent to you, do not enter them in the ledger, but make a note of them in your memorandum book, since you will soon have to return them" (Yamey 1994, 89).

The analysis of the subjects proposed by Pacioli for Ricordanze shows that it is not related to double-entry bookkeeping. This reason explains the absence at the initial stage of the Russian researchers' attention to the books of Ricordanze, which are stored in Francesco Datini's archives in Prato. However, the study of our British colleagues aroused heightened interest. Moreover, it is very important that A. Sangster and his co-authors have selected and analyzed substantial theoretical material, the opinions of distinguished scientists from many countries who at different times studied in detail Pacioli's Treatise or examined the archival documents.

Sangster states that the difficulties in understanding the problem are due to the ambiguous interpretation of many provisions of the Treatise performed by various authors of the translations or comments on the Treatise.

This problem was particularly tangible in our country. The Russians translated the Treatise 125 years ago (1893), several years after Jager (1876) and Gitti and Torino (1878) performed their translations. Throughout this period many Russian distinguished scholars were of the opinion that the trial balance described by Pacioli was based on the final sums of the debit and of the credit of each account, and not on the balance of account of closed accounts.

Indeed, in Chapter 34 there is no direct reference to the term "balance". However, in all likelihood, Pacioli saw a real trial balance and worked with the Ledger. If this is the case, "Father of accounting" saw that, as a rule, the totals in the accounts were calculated only on the "strong" side of the account (and not on its both sides), and in these conditions it was impossible to form a trial balance for the sums of the totals at that time. Accordingly, those at whom the Treatise was aimed also saw books, and they could not have a wrong opinion.

Particularly, in the post-Soviet space, all the controversial issues of perusal of the Treatise have always concerned the mistakes contained in the translation.

Sangster makes the reader think that Ricordanze book in the Pacioli's Treatise has nothing to do with "Memorial" book, which is included in the triad "Memorial" – "Journal" – "Ledger".

Sangster's work abounds in numerous references to the significant list of researchers, who, possibly, had to do with the books of Ricordanze in various archives. Sangster states that "a lot has been written, and many scientists have been working in this field today". They and many researchers from other scientific fields drew attention to the fact that among the business documents belonging to the Tuscan companies there are special books called Ricordanze. At the same time, no one has ever paid attention to the fact that the final chapters of Pacioli's Treatise contain a detailed description of the memorial book under the same name. Sangster and his co-authors see the purpose of their research in identifying the origin as well as the purpose of the memorial book of Ricordanze. In addition, they are striving to understand what prompted Pacioli to include this subject in his Treatise.

Here, the attention is drawn to the fact that neither Sangster nor his partners at that time visited the archives of medieval books and, thus, could not see Ricordanze book. It is also interesting that those authors who wrote about the possible use of Ricordanze as a Memorial never gave real examples of such application.

We should note that a century before Pacioli (the era of Francesco Datini) there were no "Memorial" and "Journal" books corresponding to the registers described by Pacioli in the Treatise. If the book called "Memorial" is one of the most common in Datini's accounting systems, then the title "Journal" is generally absent from the lexicon of the accountant. "Memorial" book though is a register of preliminary registration, but in Datini's single entry and double-entry accounting the data in it are not only recorded, but also grouped and they form the in-system entries for the General Ledger or cash book of "Entrata e Uscita". In A. Sangster's article there is no mentioning of "Memorial".

We were really interested in the issue of the register, in which the primary data were fixed before the formation of in-system entries of single-entry and double-entry bookkeeping.

278   A. Musaelyan et al.

The research began in 2012. The first books entitled Ricordanze, which came in view of the Russian researchers when studying the internal balances and postal records of Datini's company in Barcelona, were Ricordanze Prato, AS, D. 824 (for the period of 1397–1399). [4] and Prato, AS, D. 825 (for the period of 1399–1400) [5]. The first book contains 54 completed pages (the last number is 27 Verso). The second book includes 40 pages (the last completed page is 20 Recto). The photocopies of the mentioned books are in full stored in our electronic archives. Figure 1 shows a photocopy of the folio Prato, AS. D. №824, c. 3v-4r, and Fig. 2 - a photocopy of the folio Prato, AS. D. №824, c. 5v-6r.

Fig. 1. Map of Prato, AS. D. No. 824 c. 3v-4r, Ricordanze of the company in Barcelona, 1399

The books of Ricordanze differ from other books that are recognized as book-keepers', they look somewhat "negligent", and, what is just important, they do not contain special signs ("lashes"), which show the facts of transferring data from Ricordanze books to other books or the application of "dotting" with other entries.

We have prepared a translation of one of the pages of Ricordanze book that we found, registered in the archives as Prato, AS, D. 824, c. 3v. The researchers hoped that the translation would help to find out the purpose of Ricordanze books in the accounting system of Francesco Dattini's enterprises and companies. Unfortunately, particular entries are not entirely readable.

Let us look at the translation of the text:
"1398
March, 15. We sent Zanobi Gaddi account to Montpellier.
They are to have 20 shillings
They *still have* 136 pounds 3 shillings 3 pennies.
We sent accounts to Avignon for our partners to speak up today. They are ment for exchange c.5
They *still have* 29 pounds 2 shillings 7 pennies
I remember that we *had* from Andrea di Banco and other cooperation. For Bruno Francesco & Co Genoa date … March Rhubarb ……….. in 1 *barbetto* (…), it weighed 5¼ pounds Meant for Luca de Michele
We weighed the grapes […] Alberto di Bernardo degli Alberti & Co of Bruges on 16 March. It weighed in *rova* XII netto.
We sent balances in Montpellier Giovanni Francesco of May, 26.
They are meant for exchange of money, by 1 pounds c. 2.

Is left to be given 50 pounds 9 shillings 9 pennies in Barcelona currencies.
Meant for Luca de Michele
We weighed the grapes [...] Alberto di Bernardo degli Alberti & Co of Bruges on 16 March.
It weighed in *rova* 12 *netto*.
Still be given 16 pounds 11 shillings 7 pennies, that is why we give more than this
And we owe them for 1 costal of ... 58,5 pounds.
Leaves us with 50.18.2 pounds, giving Andrea ..... April 15".

It follows from the translation that the entries on the page are of fragmentary character, as a rule, there is no connection between the entries. However, the analysis of the text allows to answer the question, for what purposes Ricordanze book was used in Datini's company at the end of the 14th century - for non-system entries that are not related to double-entry bookkeeping.

Later on our plans included the search and study of all the books stored in the Datini's archives entitled Ricordanze, revealing their connection with another memorable register "Memorial". In addition, it was intended to identify the connection of entries in Ricordanze with the entries in other books and, most importantly, to reconstruct the system of books of accounts in Datini's various companies. This will help to reconstruct the historical truth of the accounting profession formation.

Concerning F. Datini's books and the role of Ricordanze and Memorial, Professor Francesco Ammannati of the University of Florence expressed the following opinion: "First of all, it must be borne in mind that the accounting system of Datini's companies did not fully represent an example of a coordinated double-entry bookkeeping, and it should be seen as the system still in its infancy. According to the studied fact, there are some books, the purpose and structure of which underwent changes every year. For example, Ricordanze could be something like "Memorial", but it could also be an ordinary book with various types of information registered in it.

As has been mentioned above, "Memorial", as it was represented in Datini's system, was significantly different from Memorial described by Pacioli. This difference was not in the fact that he supposedly replaced "Journal", which is not used in Datini's system. Some of "Memorials" of Datini's companies contained different sets of entries, somewhat similar to the entries from "Journal", and in other cases resembled the entries of other books.

Fig. 2. Map of Prato, AS. D. No. 824 c. 5v-6r, Ricordanze of the company in Barcelona, 1398.

To assess their role in each particular system, it is reasonable to interpret them in case of researching each individual book.

The Search That Brought Success. Further research began in February 2013. We managed to find the earliest book of Ricordanze, which was used in Datini's companies. This was the book of the First Company in Avignon (1363), which belonged to Nicolo Bernardo and Francesco Datini. According to the information by which the Russian researchers are guided (Nicastro 1914), the accounting of the early companies in Avignon was carried out by Francesco Datini himself.

Here we were lucky. Figure 3 presents the installation based on the photocopies of entries on card 177R in Ricordanze (Prato, AS, D. No. 24) and entries in the accounts on cards 99 V and 227R of the General Ledger (Prato, AS, D. No. 51). In the example studied, the primary data recorded on one page of Ricordanze book are transferred to two accounts in different parts of the General Ledger.

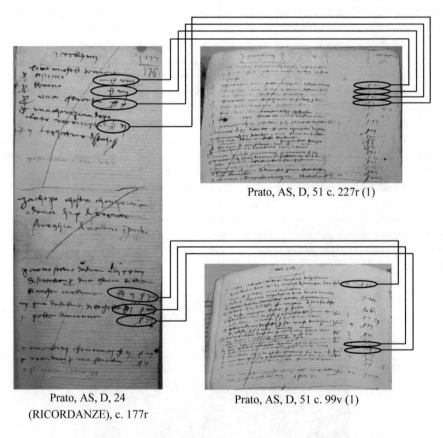

Prato, AS, D, 51 c. 227r (1)

Prato, AS, D, 24
(RICORDANZE), c. 177r

Prato, AS, D, 51 c. 99v (1)

Fig. 3. Transfer of indicators from the Ricordanze to the General Ledger.

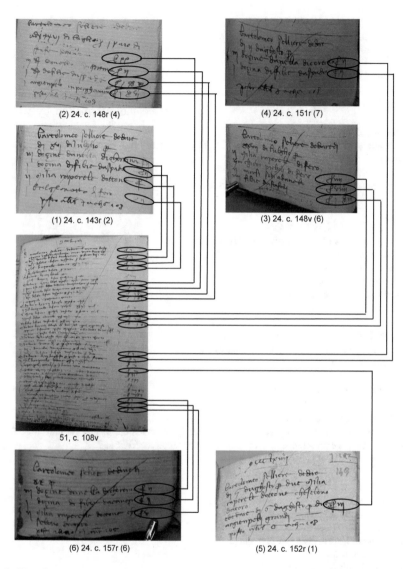

Fig. 4. Forming one account from the data of several Ricordanze cards in the General Ledger (F. Datini and N. Bernardo's company, Avignon, 1363)

The example, in which one account in the General ledger (Prato, AS, D. No. 51, page 108v) is formed from the entries on several pages of Ricordanze book, is shown in Fig. 4.

Of course, it is very important to see the translation of the early in-system entries that occurred in Ricordanze. Let us do this using the example of accounts shown in Fig. 4.

Prato, AS, D. №24, c. 143r(2):
Bartolomeo, a saddler, *must give* on July 16 for:
3 dozens of armoring rings, s. 2.
1 dozen of sword buckles, s. 2.
2 thousands of brass beads, s. 5.
1 iron *brilglonetto*, s. 2.
Put in the Book A at c. 108.

Prato, AS, D. №24, c. 148r(4):
Bartolomeo, a saddler, *must give* on July 26 for a 1 pair of brass stirrups, s. 20.
3 dozens of armoring rings, s. 2.
1 dozen of sword buckles, s. 1 d. 6.
1 *argienpelo* [leather/silver frill] in *prghamino*, s. 1 d. 6.
Put in the Book A at c. 108.

Prato, AS, D. №24, c. 148v(6):
Bartolomeo, a saddler, *must give* on July 27 for:
2 thousands of iron beads, s. 5.
500 small iron nails, s. 4.
2 ring-shaped bits, s. 9.
4 bucklets *da frasali*, s. 1 d. 4.
Put in the Book A at c. 108.

Prato, AS, D. №24, c. 151r(7):
Bartolomeo, a saddler, *must give* on August 2 for:
3 dozens of armoring rings, s. 2.
1 dozen of sword buckles, s. 2.
Put in the Book A at c. 108.

Prato, AS, D. №24, c. 152r(1):
Bartolomeo, a saddler, *must give* on August 5 for 2 thousands brass bead we had from Toro
And he *must give* on August 6th for two large leather/silver frills, s. 3.
Put in the Book A at c. 108.

Prato, AS, D. №24, c. 157r(6):
Bartolomeo, a saddler, *must give* on day 20 for:
3 dozens of armoring rings, s. 2.
1 dozen of sword buckles, s. 2.
2 thousands of brass beads, s. 5.
We got from Toro,
Put in the Book A at c. 108.

And, of course, the reflection of the in-system records registered in Ricordanze in the Ledger is of some interest.

Prato, AS, D. №24, c. 108v:
1364
Bartolomeo, saddler *must give*, as we put away from the Quaderno delle Ricordanze where he *should give* for:
(1) 3 dozens of armor rings at d. 8 for dozen, s. 2.
(2) 1 dozen of iron sword buckets, s. 2, s. 2.
(3) 2 thousands of brass beads at s. 2½ for 1000, s. 5.
(4) 1 iron *brilglonetto*, s. 2, s. 2.

And he *must give* on July 8 for:

(5) 1 thousand small iron nails, s. 8, s. 8.

And he *must give* on July 26 for these goods written here below:

(6) 1 pair of brass knight stirrups, s. 20, s. 20.

(7) 3 dozens of iron armoring, at d. 8 for dozen, s. 2.

(8) 1 dozen of iron sword buckles, at s. 2½ for dozen, s. 1 d. 6.

(9) 1 large leather/silver frill in *perghamino*, s. 1 d. 6.

And he *must give* on July 27 for these goods here below:

(10) 2 thousands of small iron beads, at s. 2½ for 1 thousand, s. 5.

(11) 500 small iron saddle nails, at s. 8 for 1 thousand, s. 4.

(12) 2 ring-shaped bits, at s. 4½ each, s. 9.

(13) 4 iron buckelst *da frasali*, at d. 4 each, s. 1 d. 4.

(14) And he *must give* on July 29 f. 2 of gold he promised us for Peire de Feriere, put that Peire had given where he *should have given* in this book ahead at c. 112, f. 2.

(15) And he *must give* on August 2 f. 1 the aforementioned Bartolomeo *had* in cash, f. 1.

And he *must give* on the same day for these goods written here below:

(16) 3 dozens of armor rings at d. 8 for dozen, s. 2.

(17) 1 dozen of iron sword buckets, s. 2, s. 2.

And he *must give* on August 6 for these goods written here below:

(18) 2 thousands of brass beads at s. 2½ for thousand, s. 5.

(19) 2 large leather/silver frills at d. 18 each, s. 3.

(20) And he *must give* on August 8 for a red Florentine sheepskin brought by Bartolomeo himself, s. 11.

(21) And he *must give* on the same day for two red sheepskins, s. 22.

(22) And he *must give* on August 13 for one *arnese* of Toulouse, s. 10.

(23) 1 iron […] bit of him, in Avignon, s. 4 d. 6.

(24) 1 iron bit […] of Mellino, s. 4 d. 5.

(25) And he *must give* on August 20 for these goods for 3 dozens of armoring rings, s. 2.

(26) 1 dozen of buckles, s. 2.

(27) 2 thousands of brass beads, s. 5.

And he *must give* on August 27 for:

(28) 2 pieces of white ox leather tanned […] at f. 2, f. 2.

(29) And he *gave* on August 31 f. 5 in cash […], f. 5 s. 10.

(30) And he *gave*, and he [……..] *must give* at […], f. 5 s. 10 [….].

The document from Ledger Prato, AS, D. No. 24, p. 108v is a personal account of debtor Bartolomeo, a saddler. The account is performed according to the adjacent system: debit entries at the top of the paragraph, and credit ones at the bottom.

The account includes 30 entries. 18 entries (1–4, 6–13, 16–17, 19, 25–27) are transferred from the above six pages of Ricordanze book (Prato, AS, D. No. 24, p. 143r (2); 148r (4), pp. 148v (6), 151r (7), pp. 152r (1), 157r (6)). 10 entries (5, 14, 15, 19, 20–24, 28) are referred to the accounts that we did not include in Fig. 4 because of format limitations. One entry (29) a partial repayment of the debt in cash is entered directly in the General Ledger without prior registration. The last entry, the account closing and closing balance of account transfer, is performed in the same way.

The above example is very significant, since it represents one of the earliest examples of in-system entries of the registration of primary economic facts in Ricordanze and their reflection in the Ledger.

Here it is possible to make an assumption. It seems that Francesco Datini's accounting in Avignon began by analogy with the accounting of merchants-mediators. Initially, the accounting information for memory was entered in Ricordanze, used in the form of a regular book for entering "ordinary" cases, not related to accounting.

The search for the data in the non-system book presented certain difficulties and the merchant began to distribute the information on each counteragent to separate sheets of paper. Probably, in such a way the first accounts of debtors and creditors appeared that were systematized in the General Ledger.

Initially, a merchant, and then an accountant assessed the need for the systematization of entries in the accounts in the trading books. In the future, the separation of in-system entries from non-system ones occurred. This led to the division of the memorial book into two independent ones. The first of these is Ricordanze intended for non-system notes and which acquired the appearance that Pacioli had described. The second book is "Memorial" containing in-system entries used in the formation of accounting entries in the General Ledger.

The study of accounts in F. Datini's First Company in Avignon (from 1363) is of particular interest, since this accounting system is Datini's personal contribution to the formation of accounting, since it was he who formed the system of accounts and personally kept records.

5   Conclusion

In Luca Pacioli's Treatise the book Ricordanze intended for non-system entries, is described. Parallel to it, there was an accounting book of primary registration of in-system (accounting) data, intended for transfer to the Ledger or Entrata e Uscita. In the Francesco Datini's archives in Prato the early Ricordanze book of the year 1363 was found, in which only non-system (personal) entries were initially kept, and later non-system and in-system (accounting) entries were registered simultaneously.

Such a book was the only one in Datini's archives. The Figures clearly illustrate how the entries from Ricordanze are transferred to the Ledger. This was the very first Ledger, in which the records of Datini's companies were kept. The accounts with parallel debit and credit have not been used in the company yet. Adjacent accounts in the form of a "paragraph" were used in the book.

In the paper it is suggested that over the period of one hundred years before the appearance of Pacioli's Treatise the differentiation of the register into two separate ones occurred: Ricordanze for non-system (personal) entries and Memorial for in-system (accounting) entries.

References

Archivio di Stato di Prato, Prato, AS, D., Fondaco di Barcelona, Company Francesco di Marco Datini e Luca del Sera, Ricordanze e Quadernacci/Quaderni di Ricordi, 1397–1399, No. 824

Archivio di Stato di Prato, Prato, AS, D., Fondaco di Barcelona, Company Francesco di Marco Datini e Luca del Sera, Ricordanze e Quadernacci/Quaderni di Ricordi, 1399–1400, No. 825

Archivio di Stato di Prato, Prato, AS, D., Fondaco di Avignon, Company Francesco di Marco Datini e Niccolò di Bernardo, Ricordanze e Quadernacci/Quaderni di Ricordi, 1363–1364, No. 24

Archivio di Stato di Prato, Prato, AS, D., Fondaco di Avignone, Company Francesco di Marco Datini e Niccolò di Bernardo, Libri contabili/Memoriali, 1363–1364, No. 51

Sangster, A.: The genesis of double entry bookkeeping. In: Sangster, A. (ed.) 12th World Congress of Accounting Educators and Researchers: Abstracts, Florence, Italy (2014). http://prpg.usp.br/dcms/uploads/arquivos/PPGCC/TEXTO_ALAN%20

Martinelli, A.: The origination and evolution of double entry bookkeeping to 1440. In: Martinelli, A. (ed.) ProQuest Dissertations & Theses Global (1974). https://digital.library.unt.edu/ark:/67531/metadc504552/

Kuter, M.: Introduction to Accounting. Prosveshchenie-Yug, Krasnodar (2012)

de Roover, R.: The development of accounting prior to Luca Pacioli according to the account-books of Medieval merchants. In: Yamey, B.S. (ed.) AC Littleton. Studies in the History of Accounting. Sweet & Maxwell, London (1956)

Sangster, A., Stoner, G., De Lange, P., Scataglini-Belghitar, G.: Pacioli's forgotten book: the merchant's Ricordanze. Acc. Hist. J. **39**(2), 27–44 (2012)

Pacioli, L.: Traktat o schetakh i zapisiakh [Treatise on accounts and records] ed. by M.I. Kuter Finansy i statistika, Moscow; Prosveshchenie-Yug, Krasnodar (2009)

Pacioli, L.: Traktat o schetakh i zapisiakh [Treatise on accounts and records], translation and notes by E.G. Valdenberg. Commerc. skoropechatnya preemn. E. Tile, St. Petersburg (1893)

Jager, E.: Lucas Pacioli und Simon Stevin, nebst einigen jungeren Schriftstellern uber Buchhaltung. Skizzen zur Geschichte der kaufmdnnischen, staatlichen und land-wirtschaftlichen Buchfuhrung/E. Jager. Stuttgart (1876)

Gitti, V.: Tratato de' computi e delle scritture. Fru Luca Paciolo/V. Gitti. Torino (1878)

Yamey, B.S.: Luca Pacioli. Exposition of Double Entry Bookkeeping Venice 1494/B.S. Yamey, A. Gebsattel. Albrizzi Editore, Venice (1994)

Nicastro, S.: L'Archivio di Francesco di Marco Datini in Prato, edited by Casciano, R.S., and Cappelli, L. (1914)

Accounts of Household Expenses in the Medieval Companies

Mikhail Kuter[✉] and Marina Gurskaya

Economy Department, Kuban State University,
Stavropol'skaya St., 149, 350040 Krasnodar, Russia
{prof.kuter,marinagurskaya}@mail.ru

Abstract. In Luca Pacioli's Treatise, the account of household expenses of an independent merchant has been described. This paper considers similar accounts of the medieval companies by means of the example of Francesco Datini's companies. It has been proved that unlike sole merchants' businesses, where the given account contained family expenses, the companies kept records of the expenses for covering their personnel needs only, mainly the need for housing and sustenance.

Keywords: Accounts of household expenses · Medieval accounting

1 Introduction

In Luca Pacioli's Treatise the account of household expenses of an independent merchant has been described. This paper considers similar accounts of the medieval companies by means of the example of Francesco Datini's companies. If we compare this account in family trading businesses and companies, then there are no special differences in the composition of the expense items. When analyzing the course of expenditure, then there is an essential difference. In the first case, the income of a store covers both the expenses of business servicing and the full expenses of a merchant's family maintenance. Consequently, the account of household expenses described by the "father of accountancy" is an account of family expenses. In the second variant, this account is intended for the bookkeeping of general household expenses caused by the social security of company employees, who were often seconded to a company and geographically remote from their families. As a rule, these are the expenses related to living, sustenance and creating conditions for a normal work activity, but by no means to the satisfaction of personal needs of owners and employees.

2 Review of Prior Literature

In Chapter 22 "Of the manner in which expenses of each kind, household expenses, extraordinary expenses, merchandise expenses, wages of apprentices and stewards should be entered in the books" Luca Pacioli's Treatise it is said: "… … you must also have in all your books these accounts, that is: expenses of merchandise, ordinary household expenses, extraordinary expenses, and an account of incomings and

© Springer Nature Switzerland AG 2019
T. Antipova and A. Rocha (Eds.): DSIC 2018, AISC 850, pp. 286–295, 2019.
https://doi.org/10.1007/978-3-030-02351-5_33

outgoings and an account of profit and loss. Such accounts are of major importance in every mercantile business so that you may always know the amount of your capital, and when the books are closed how the business has gone; and we will clarify in sufficient detail how they must be kept in the books" (Yamey 1994, c. 70).

Particularly important are household expenses: "One cannot do without an account for household expenses; by which I mean expenses such as grain, wine, wood, oil, salt, meat, shoes, hats, dressmaking, jackets, stockings, and tailoring, beverages, tips, barbers, rakers, water-carriers, laundry, kitchen ware, vases, glasses and glass – are of all kinds, buckets, tubs, and casks. Some keep separate ac-counts for household goods in order to be able quickly to ascertain the position" (Yamey 1994, c. 71).

"Keep the household expenses account in the same way the account for merchandise. If you have major expenses enter them m your books day by day, as for grain and wine and wood, for which many keep separate accounts so that at year end or from time to time they may know how much has been consumed. But for minor expenses, such as meat or fish, barbers or ferries, one needs to keep one or two ducats at a time in a bag for petty spending because it would not be possible to enter such expenses individually" (Yamey 1994, c. 71).

About the account of household expenses Professor Tuyakova Z.S. says: "The procedure for defining the financial result in the Venetian version of accounting was almost the same as that in today's accounting. So, incomes and expenses were registered, as indicated above, on the result accounts and goods accounts. Then, after closing the General Ledger, the final figures (the algebraic sum of income and expenses of a merchant) were transferred to "Profits and Losses" account as a financial result. However, these features of the financial result calculation reflect only certain differences and the specific character of accounting of that time, but the main distinctive difference was in the following – when calculating it the household expenses of a merchant and his family were registered. Thus, in Chapter 22 of the Treatise it is stated that one cannot do without opening an account for ordinary household expenses, which mean…" (Tuyakova 2008, p. 112).

Professor K.Yu. Tsygankov introduces a sort of continuation: "Thus, there was no clear boundary between individual property and the property of a business as an economic entity. *Separate Entity Concept* as one of the basic principles of accounting, was accepted much later. However, the separate maintaining of the book of account for household and for the store was not a rare case" (Tsygankov 2002a, pp. 53–61). This statement seems to be very controversial: for 12 years of working in the medieval archives of Venice, Genoa, Florence and Prato, having more than 60,000 photocopies of early accounts, we have not been able to find any confirmation of it.

In another work, Tsygankov contradicts himself: "The most important of them, in our opinion, is the "joint accounting of these expenses", which was described in the Treatise and, as the archival research showed, was widely used in the Venetian accounting practice. Venetian merchants did not separate property and expenses of their business from their personal (household) property and expenses. The records for the former and the letter were kept jointly, in the same accounting books.

In the process of financial result calculating, the household expenses of the Venetian merchants were added in "Profits and Losses" account to the business expenses of their business. … It is not difficult to guess that in consequence of joint accounting the reporting was distorted so much that it hardly presented any value" (Tsygankov 2002b, pp. 163–174).

According to Tsygankov, the simplest explanation of joint accounting is the mistake conditioned by the initial stage of accounting development, which contradicts the opinion about the almost modern level of medieval accounting. Tsygankov even offers advice to his medieval colleagues: "… It is not difficult to see and eliminate this mistake, new and complex methods are not needed for this. It is enough just to exclude household property and expenses from the number of objects that are subject to accounting. At the accounting level that the Venetian bookkeepers had reached, they had to find such a simple solution" [ibid.].

And, here the same author without any proof summarizes: "Studies have also shown that the Venetian accounting was not the best one in the medieval Italy and was significantly inferior to the accounting of the Florentine companies. Florentine bookkeeping happened to be not just at the higher level, the Venetian accounting anomalies were not characteristic for them. Florentine companies clearly separated commercial property from their household, regularly (mostly annually) formed reporting, confirmed by inventories, used modern methods of assets valuation. This gives, in our opinion, all the grounds for speaking about the identical character of the Florentine and modern accounting and, at the same time, about the Venetian lag behind them" [ibid.].

In this case we should take a note of complete misunderstanding of the problem by K.Yu. Tsygankov. Two factors contribute to this: he is absolutely unfamiliar with the literature concerning archival research, and even more so, he did not take part in the research of that character. Unfortunately, the practice of such publications has been going on to this day.

"Formation of reporting, confirmed by inventories" – this is another fragment of imagination of the Novosibirsk professor. As for the inventory, it really had wide independent application, but not for the confirmation of reporting formation, but for forming inventory analytical balance sheets of the financial result drawing up, as Jacques Savary wrote in 1675 (Savary 1993). The authors described this problem in (Kuter 2018a).

For our research, it is interesting to see how accountants registered household expenses a century before Pacioli had published his Treatise. Moreover, Francesco Datini's accounting in Pisa seems a little more complicated: Pacioli's household account is aggregative; the Pisan accountant applied aggregative and clearing account. The amount of accumulated expenses is distributed between Datini's company and the partners who used the office of Datini's company to carry out their business.

3   Research Method

The principal research method adopted in this study is archival. It uses material found in the State Archive of Prato. This research team has been working with the material in this archive for the past decade and many of the records have been recorded and linked together using logical-analytical reconstruction. This is an approach that we developed for the purpose of enabling entries in the account books to be traced visually between accounts and books and from page to page. It is described in the Appendix. By adopting this approach, we are able to see the entire accounting system electronically, making entries and their sources clear in a way that is not possible if all that you have is the original set of account books. This enables us to consider each transaction in detail, trace its classification, and so explain the bookkeeping and accounting methods adopted without possibility of misinterpretation. This approach represents a new paradigm in how to analyse and interpret accounting practice for periods when there was no concept of either a standard method or a unified approach to either financial recording or financial reporting (Kuter 2017).

4 Statement of Basic Materials

We have stated that in Francesco Datini's company accounting (1392–1394) an intermediate "Profits and Losses" account was formed, which was closed July 12, 1393. The second (closing) account was formed in 1394 at the next closing of the company.

When from the opening to closing there are two "Profits and Losses" accounts in the company, then, consequently, two "Household expenses" accounts are applied. Figure 1 presents a photocopy of "Household Expenses" account of Datini's company in Pisa over the period 1392–1393. (the translation of entries in the account is given in Table 1).

Fig. 1. "Household expenses" account of Datini's company in Pisa over the period 1392–1393. (Prato, AS, D, No. 361, page 279v) [Here and below, reprinted with permission of the Ministry of heritage, culture and tourism of the Republic of Italy. Protocol No. 659/28.13.10 dated 13. 03.2014]

Today it is difficult to define the form of accounting register. It resembles an adjacent account in a paragraph from single-entry bookkeeping, as all the indices are

Table 1. Household expenses in account Prato, AS, D, No. 361, p. 279v (F. Datini's company in Pisa, 1392–1393)

Code	The nature of the transaction	Sum
A	B	1
EH-1	**House expenses** for food and beverage ***must give***, for 1 carriage of wine we had from Luca di Antonio di Pacie, and for Antonio is debtor in the old book we put him in this book at c. 14 Francesco di Marco himself who **must have**	f. 13 s. 6 d. 4
EH-2	For 1 *mezina* of salted meat of Provençe lib. 50 that we bought the 2 of August from Inghilese, we put him in this book at c. 28	f. 1 s. 7 d. 2
EH-3	And they ***must give*** for the amount we calculated from the old account from Francesco di Marco himself in this book at c. 22, for *staia* 12 of Provençal wheat f. 13 s. 50 and lib. 1 of oil s. 35 […]	f. 15
EH-4	For many expenses made for food and beverage from the **1st of July** to **11 of August** and minor expenses at the book Uscita at c. 81	f. 6 s. 11 d. 6
EH-5	For many expenses made for food and beverage from the the **12 of August** to the **14 of September** as the book Uscita at c. 83	f. 6 s. 11 d. 1
EH-6	12 barrels of red wine from Cecco di Vannuccio at the book Uscita at c. 83	f. 10 s. 2 d. 10
EH-7	3 ½ barrels of wine from Cristofano Becarelli at the book Uscita at c. 83	f. 5 s. 6 d. 2
EH-8	For many expenses made for food and beverage from the **15 of September** to the **21 of October** as at the book Uscita at c. 84	f. 10 s. 19 d. 6
EH-9	For gingerbread we had from our partners of Florence as at the Memorial at c. 66	f. 2 s. 1 d. 6
EH-10	For figs and rice they sent us from Genoa our partners for the house in this book at c. 80	s. 16 d. 2
EH-11	For expenses made in the house for food and beverage from the **21 of October** to the **1st of March** as at the book Uscita at c. 87	f. 41 s. 10 d. 10
EH-12	For 6 barrels of red wine from Dino da Palaio the 8 of March at the book Uscita at c. 87	f. 7 s. 4 d. 5
EH-13	For expenses made for the house until the **9 of March** at the book Uscita at c. 88	f. 5 s. 9 d. 6
EH-14	For 40 of candles we had from Pietro and Pagolo of Prato in this book at c. 54	f. 1 s. 12
EH-15	For 1 *schienale* and 1 *sportino* of *zibibbo* and rice for the house and many other things we had from Genoa as in the Memorial at c. 100	f. 2 d. 3
EH-16	For 1 *schienale* we gave to Stefano Guazalotti and Co. of Pistoia of a lot of 6 at the Memorial at c. 123	s. 5 d. 10
EH-17	За 9½ *staia* пшеницы провансальской […], как в Мемориале, на c.130	f. 8 s. 2 d. 10
EH-18	For *staia* 9 ½ of Provençal wheat […] as at the Memorial book at c. 130	f. 44 d. 4
EH-19	For 16 cheeses of Maiorca from Miniato di Nucio in this book at c. 116	s. 16

(*continued*)

Table 1. (*continued*)

Code	The nature of the transaction	Sum
A	B	1
EH-20	For many expenses made for food and beverage from the **21 of June** to the **1st of September 1393** as at the book Uscita at c. 93	f. 13 s. 8 d. 7 **198.12.10**
EH-21	For expenses of the house made from the **2 of Septamber** at the Uscita at c. 93	s. 8 **199.0.10**
EH-22	For loss on furnishings/household goods until this 2 of September in this book at c. 290	f. 20 **219.0.10**
EH-23	They ***gave*** the 31 of August 1393, we calculated for our partners of Florence for [...] come back to us Baldo Vilanuzi in many entries in this book at c. 334	f. 11 s. 15
EH-24	They ***gave*** the 2 of September f. 72, they are for 9 months [...] of Iacopo with Bandino his son in this book at c. 129, messer Niccolò di Pagnozo and Co. ***must give***	f. 72
EH-25	We spent f. 135 s. 5 d. 10 *a oro* put in the account «loss on merchandises» in this book at	*f. 135 s. 5 d. 10*

written in a column. In content, it coincides with the dual account of the old Tuscan form. The top part of the account is a debit, and the bottom one is credit. At the same time, the account is balanced (balance of account - the difference between debit and credit - is transferred to "Profits and Losses" account). True, the debit total (f. 219 d. 10) is not indicated after the balanced credit entries.

Conspicuous is the article "Losses on Furniture", speaking modern language, impairment. In the works (Gurskaya 2016; Kuter et al. 2016; Kuter 2018b) it is proved that Datini's companies in Pisa (1982–1400) never charged deprecation, and periodically estimated the impairment of long-term property.

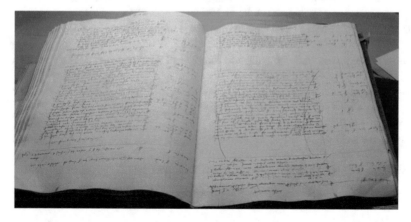

Fig. 2. "Household expenses" account, Datini's company in Pisa over the period 1393–1394. (Prato, AS, D, No. 361, p. 349v-350r (2)

Table 2. Household expenses in the account Prato, AS, D, No. 361, p. 349v (2) – debit (F. Datini's company in Pisa, 1393–1394)

Code	The nature of the transaction	Sum
A	B	1
EHD-1	*Expenses for the house,* for food and beverage *must give* the 3 of September for may expenses made from the 3 until the 15 of September, as in the Uscita book A at c. 93	f. 4 s. 13 d. 2
EHD-2	For *staia* 18 of wheat we had from Pietro da Righaccio at s. LI for *staio*, Lorenzo Ciampolini and Co paid for it the 30 of October as in the Memorial A at c. 179	f. 13 s. 2 d. 3
EHD-3	For a barrel of 7 barrels of Provençal wine from Pietro di Bindo and Co f. 10 we had many days ago and Lorenzo Ciampolini and Co paid for us the 31 of October as in the Memorial A at c. 179	f. 10
EHD-4	For many expenses made for the house on this November for 7 barrels of red wine bought from Giovanni Ventie from Sant [???] as in the Uscita book A at c. 95	f. 5 s. 17 d. 4
EHD-5	For many expenses made for the hous from the 16 of September until the 1st of November as in the Uscita book A c. 95	f. 23 s. 9 d. 8
EHD-6	For lib. 20 of candles from Pietro of Prato the 19 of December at the Memorial A c. 204	s. 14 d. 4
EHD-7	For staia 6 of wheat we had from Antonio di Giovanni Macheroni in two times many days ago as in the Memorial A c. 205	f. 4 s. 9 d. 2
EHD-8	For many expenses made for the hous from the 2 of November until the 12 of January as in the Uscita A c. 97	f. 41 s. 14 d. 11
EHD-9	For many expenses made from the 19 of January until the 28 of February as in the Uscita book A at c. 98	f. 15 s. 11 d. 5
EHD-10	For salted fish from Antonetto Micheli at the Memorial A c. 242	s. 2 d. 2
EHD-11	For salted eels and fish for the Lent as in the Memorial A c. 252	s. 17 d. 9
EHD-12	For 8 *schienali* and 25 lib. Of gelding that came from Genoa for the house, a part was given as in the Memorial A c. 253	f. 4
EHD-13	For 1 1/3 piles of wood we had from Pietro of Sinena some time ago for Niccolaio di Pino in this book at c. 135	f. 2 s. 13 d. 4
EHD-14	For expenses made from the 2 of March until the 7 of June 1394 as in the Uscita book A c. 101	f. 23 s. 12 d. 2
EHD-15	For expenses made from the 8 of June until the 13 of July 1394 as in the Uscita book at c. 102	f. . 7 s. 5 d. 7
Sum of f. 158 s. 3 d. 3 a oro		
To this sum should be subtracted for commission s. 12 of our gain s. 11 in sum, in golden money		f. – s. 6 d. 7
The net sum is f. 16 s. 4 d. 7 of gold as in this book at c. 193		f. 16 s. 4 d. 7

Table 3. Household expenses in the account of Prato, AS, D, No. 361, p. 350r (2) – credit (F. Datini's company in Pisa, 1393–1394)

Code	The nature of the transaction	Sum
A	B	1
EHC-1	**Expenses for the house *must have*** the 12 of November we had back from ser Andrea our *sindaco* for II barrels of red wine of Provence, he had from us as in the Uscita book A at c. 14	f. 2 s. 17 d. 2
EHC-2	For *staia* 4 ½ of flour we gave to our partners of Genoa as in the Memorial A at c. 201	f. 3 s. 9 d 5
EHC-3	For […] we had from messer Niccolò di Pagnozo and Co from the 3 of September until the 1st of January as in this book at c. 169	f. 28
EHC-4	For *staia* 4½ of flour we sold to our partners of Genoa as in the Memorial A c. 219	f. 3 s. 9 d. 5
EHC-5	For ½ *staio* of walnuts we sent to our partners of Genoa in this book at c. 183	s. 5 d. 8
EHC-6	For 6 1/3 months that Antonetto Micheli came back to our house until the 28 of March in this book at c. 186	f. 15 s. 8 d. 1
EHC-7	For 3 *staia* of flour we sent to Genoa as in the Memorial A at c. 238	f. 2 s. 2 d. 2
		55.18.11
EHC-8	For 7 *schienali* and lib. 25 of capers which came from [canceled]	
EHC-9	For 2 days that Nicolò del Barna came back to us in the house as in the Memorial A at c. 374	s. 17
EHC-10	For 2 *schienali* messer Nicolò di Pagnozzo and Co had from us, they were part of a shipment of 8 *schienali* that can be seen in the other page, they were put in the house expenses account at c. 20 we had the 3 of June 1394	f. 1 **57.13.11**
EHC-11	For expenses for the ones of messer Niccolò di Pagnozzo and Co from the 1st of January until the 1st of July for six months between Nutino and Antonio Mancini with Bandino, f. 50 from which we subtract for flour and oil and walnuts we had from them some time ago as in the Quadernaccio A at c. 198 (f 3), f 47 remain in this book at c. 266	f 47 **104.13.11**
	And they *must have* the 8 of July f. 53 s. 9 d. 4 for balance of this account we put in this book at c. 397	f. 53 s. 9 d. 4
Sum of f. 158 s. 3 d. 3 a oro		

It seems that in this account impairment is observed the first and the last time. Further, the accountant, like all his colleagues in the future, will charge the impairment directly to "Profits and Losses" account.

There are 3 entries on the credit side: 2 entries of the distribution of expenses between the partners and the third entry refer to closing the account by transferring the balance of account to "Profits and Losses" account.

Figure 2 (translation in Table 2 (debit) and in Table 3 (credit)) presents a double-page spread c. 349v-350r of the book Prato, AS, D, No. 361, on which "Household expenses" account over the period 1393–1394 is placed. On the folio are 3 accounts, which are kept according to the Venetian form: debit is opposed by a symmetric credit. The account that interests us is located in the center of the folio, the second one.

The second "Household expenses" account in content almost does not differ from the similar account of the previous period. The expenses for food, drinks and, of course, wine still prevail. New items have not appeared.

Conspicuous is the first credit entry of the account. It says: "**Household expenses** *must have* 12, November we received paying back from ser Andrea, our Sindaco (mayor), for 2 barrels of red Provence wine, which he received from us, as in Entrata A on p. 14".

The situation looked something like this. The managers of the company presented the city of Pisa a couple of barrels of red wine from Provence. At the same time, the expenses for buying wine, of course, were included in household expenses. Apparently, the amount of f. 2 s. 17 d. 2 according to the medieval legislation was considered essential and could be recognized as "a bribe", or for ethical reasons, the mayor paid the given amount to the cashier of the company, and there is a corresponding entry in Entrata e Uscita.

Below on the credit side of the account there are entries that reduce the amount of household expenses due to the charge of their part to the partners of the company. Apparently, the accountant did not record purchasing as material inventory, but wrote them off directly as household expenses. Then, in the course of transfer of these reserves to the partners, the sum of expenses decreased. The sum balancing the account (f. 53 s. 9 d. 4) is transferred to the debit of "Profits and Losses" account.

It is very important to see the differences between the household expenses described by Pacioli for an independent merchant and the expenses of a company 100 years before Pacioli's Treatise. In the time of Pacioli, the income of a store owned by an independent merchant covered the expenses for purchasing and marketing of goods as well as expenses for the maintenance of the merchant's family.

In the era of Francesco Datini household expenses (whether that be Datini's individual company or a company with his participation) did not contain personal expenses items, but included only the expenses for office maintenance and the sustenance of its employees, including capital owners, top managers and minor clerks and workers.

5   Conclusion

Luca Pacioli described the account of household expenses of an individual merchant. Expenses included both the expenses for carrying out business (excluding the cost of goods) and expenses for family maintenance.

This paper is aimed at studying the structure of household expenses of Francesco Datini's companies in Pisa during the last decade of the 14th century. If we compare this account with the same account in family trading businesses and companies, then there are no special differences in the composition of the items of expenditure. When analyzing the directions of expenditure, it is clear that the account is aimed at the accounting of general economic expenses caused by the social security of company employees, who were often seconded to the company, geographically remote from their families. As a rule, these are the expenses related to living, sustenance and creating conditions for normal work activity, but in no case to the satisfaction of personal needs of owners and employees.

Moreover, there are significant differences in the classification of accounts by purpose. At family merchant businesses, this account is aggregative, and in companies it could sometimes be aggregative and clearing. The sum of accumulated expenses is distributed among Datini's company and the partners who used Datini's office to carry out their business.

References

Primary Sources: *Archivio di Stato di Prato*

Prato, A.S., Fondaco di Pisa, D.: Company Francesco di Marco Datini e Manno d'Albizo Degli Agli, Libri Grandi, pp. 1392–1394, no. 361

Prato, A.S., Fondaco di Pisa, D.: Company Francesco di Marco Datini e Manno d'Albizo Degli Agli, Libri Grandi, pp. 1394–1395, no. 362

Secondary Sources

Yamey, B.S.: Luca Pacioli: Exposition of Double Entry Bookkeeping Venice 1494. In: Yamey, B.S., Gebsattel, A. (ed.). Albrizzi Editore, Venice (1994)

Tuyakova, Z.: Methodology and the organization of cost measurement of the capital in accounting. Ph.D. dissertation. Saratov (2008)

Tsygankov, K.: Origin of double-entry bookkeeping. Accounting **16**, 53–61 (2002a)

Tsygankov, K.: What the modern accounting began with. EKO **2**(332), 163–174 (2002b)

Savary, J.: Le parfait négociant ou instruction générale pour ce qui regarde le commerce ... et l'application des ordonnances chez Louis Billaire...; avec le privilège du ROY. (Reproduction en fac similé de la 1ère édition par Klassiker der Nationalökonomie, Allemagne) (1993)

Kuter, M., Gurskaya, M.: The unity of the early practieces of financial result formation: the cases of Alberti's, Datini's and Savary's works (1675). In: Théories comptables et sciences économiques du XVe au XXIe siècle: Mélanges en l'honneur du professeur Jean-Guy Degos - Mélanges en l'honneur du Professeur Jean-Guy Degos. Editions L'Harmattan (2018a)

Kuter, M., Gurskaya, M., Andreenkova, A., Bagdasaryan, R.: The early practices of financial statements formation in medieval Italy. Account. Hist. J. **44**(2), 17–25 (2017). https://doi.org/10.2308/aahj-10543

Kuter, M., Gurskaya, M., Andreenkova, A., Bagdasaryan, R.: Depreciation accounting in Francesco Datini's companies. In: 5th International Conference on Accounting, Auditing, and Taxation (ICAAT 2016) (2016). https://doi.org/10.2991/icaat-16.2016.26

Gurskaya, M., Kuter, M., Musaelyan, A.: Specific features of depreciation accounting at the end of the 12th early 13th centuries. In: 5th International Conference on Accounting, Auditing, and Taxation (ICAAT 2016) (2016). https://doi.org/10.2991/icaat-16.2016.10

Kuter, M., Gurskaya, M., Andreenkova, A., Bagdasaryan, R.: Asset impairment and depreciation before the 15th century. Acc. Hist. J. **45**(1), 29–44 (2018b). https://doi.org/10.2308/aahj-10575

Retrospective Analysis of Institute
of Bankruptcy in Russia

Ruslan Tkhagapso

Ruslan Tkhagapso[✉] and Mikhail Kuter

Kuban State University, 149, Stavropolskaya Street,
350040 Krasnodar, Russian Federation
{rusjath, prof.kuter}@mail.ru

Abstract. The political and economic crisis of the end of the 1980th and the change of eras which followed in the 1990th showed extremely sharp crisis phenomena in the Russian economy. Change of public formations was shown in total reorganization of the social and economic relations in society. Many organizations were unable to pay bills of suppliers, to pay wages in time as well as make payments on taxes, etc. The financial position of most of accounting entities and citizens sharply worsened because of the acute shortage of financial resources caused by a hyperinflation.

The whole economic history of mankind is followed by problems of payments and the problems of bankruptcy that coincide with this phenomenon. Even in modern history the problem of bankruptcy remains the sharpest (on average, the statics of bankruptcy cases in Russia is about 30,000–40000 cases per year). In this paper analysis of history of origin and development of institute of bankruptcy in Russia is carried out.

Keywords: Bankruptcy · Insolvency · Bankruptcy Law

1 Introduction

Development of the Russian Bankruptcy Law can be provided three main stages matching social, economic and political transformations in the history of Russia. First stage *pre-revolutionary Bankruptcy Law*; the second—*the Bankruptcy Law of the Soviet period* (existing only in the period of the New Economic Policy); the third—*the modern Bankruptcy Law* (developing since June 1992).

2 Pre-revolutionary Bankruptcy Law

The norms, regulating the problem of insolvency in the prerevolutionary legislation of Russia, have undergone a long evolution, in many respects like the development of similar legislation of the countries of Western Europe.

In ancient Russia there was no separation of society into estates, which in the mutual struggle would advance their privileges. On the contrary, Russian history represents the unity of the entire civil society. Therefore, there was no ground for the emergence of feudal or merchant law. If, at times, trade turnover required for itself

T. Antipova and A. Rocha (Eds.): DSIC 2018, AISC 850, pp. 296–305, 2019.
https://doi.org/10.1007/978-3-030-02351-5_34

some legal features, then they found their place in general legislative monuments, and did not isolate themselves into a special system.

On the other hand, the isolation and mainly land way of life of the state slowed down development of the Bankruptcy relations and in general institute of bankruptcy.

Rudiments of Bankruptcy process can be already found in the most ancient code of laws which arose and existing in Ancient Russia the XI—XII centuries—the Russian Truth of Extensive edition of Yaroslav the Wise. The Russian Truth fixed the differentiated approach to insolvency assessment. For example, art. 54 said: «*Аже которыи купец, кде любо шед с чюжими кунами, истопиться, любо рать возметь, ли огонь, то не насилити ему, ни продати его но како начнет от лета платити, тако же платить, зане же пагуба от бога есть, а не виноват есть; аже ли пропиеться или пробиеться, а в безумьи чюжь товар испортить, то тако любо тем, чии то товар, ждут ли ему, а своя им воля, продадять ли, а своя им воля*» [9, V. 2, p. 68].

The legislation distinguished the unfortunate insolvency that arose not by the debtor's fault, in which he was able to pay the debt by installments, but this privilege did not extend to the merchant who spent on drink the capital or lost it in a fight (careless bankruptcy). In such cases, the fate of the debtor depended on the will of creditors, who at their discretion either agreed to repay the debt by installments or sell the property and the debtor himself in bondage. From the casual and careless Old Russian legislator distinguishes *malicious bankruptcy*: «*Аже кто многимъ долженъ будеть, а пришедъ* гость *из иного города или чюжеземець, а не ведая запустить за нь товаръ, а опять начнеть не дати гости кунъ, а первии должебити начнуть ему запинати, не дадуче ему кунъ, то вести и на торгъ, продати же и отдати же первое гостины куны, а домашнимъ, что ся останеть кунъ, тем же ся поделять; паки ли будуть княжи куны, то княжи куны первое взяти, а прокъ в делъ; аже кто много резаималъ, то тому не имати*» [9, V. 2, p. 68].

According to the Art. 55 of the Russian Truth of Extensive edition, a merchant who was deprived of the credit of his citizens, who took for sale the goods from a merchant, who did not know about his insolvency and did not pay for the received things, should be sold by creditors to bondmen, payment by installments was not provided to him.

The division of insolvency into categories and the establishment of the character of the debtor's guilt are peculiar only to Russian legislation. In any phenomenon, there is an attempt to get to the root of the causes that generate it, in the name of a fair assessment and retribution. Rational and business-like West does not have such a trait in the character of its peoples [19, p. 157].

The special interest is the phenomenon that the more remote development of law from the original era, the weaker become the features of insolvency. If the provisions of the Russian Truth were sufficient for their time and to the extent consistent with the general system of law, it is impossible to tell the same about the newest time when credit relations have developed so much that it was already impossible to do without Bankruptcy legislation. And meanwhile only by the XIX century there is a Bankruptcy charter for us.

The traces of Bankruptcy Law occurred in the contract of Smolensk prince Mstislav Davydovic with Riga, Gotland and German cities in 1229. In the Art. 10 of this agreement, we indicate the order of satisfaction in the event of the existence of several creditors. In this case, both the German with respect to his Russian debtor in Smolensk and the Russian with respect to his German debtor in Germany enjoyed the privilege of preferential satisfaction. This privilege remained according to the Art. 11 even in the case of confiscation of property for a crime—«*аже разгневается князь на своего человека*».

The Russian Truth so well satisfied requirements of princely courts that it was included in legal collections up to the 15th century. Lists of the Extensive Truth actively extended in the XV—XVI centuries. And only in 1497 the Code of Law of Ivan III Vasilevich was published, replacing the Truthful Truth as the main source of law on the territories united in the centralized Russian state [3, p. 24].

Analyzing ancient monuments of institute of bankruptcy, it is possible to draw a conclusion that the level of the economic relations existing in those days didn't reach that mark when there is a need for detailed regulation of the Bankruptcy relations. This period is characterized by formation of the Russian centralized state.

For four centuries up to the Cathedral Code of 1649, no laws regulating Bankruptcy relations in Russia have been found, although in other countries such laws have already existed. It is obvious that in Russia at that time there was no need for such laws. This is also evidenced by the fact that the Cathedral Code practically repeats what was established 400 years before by the "Russian Truth" [10, p. 117].

It should be noted that in the insolvency law of Ancient Russia the priority of satisfaction of claims of creditors was already entered. The prince was the first on a priority, he was followed by foreign and nonresident merchants, and the last-local creditors. By the same, creditors were not allowed to recover a significant amount of interest to satisfy the requirements, because of which the creditor practically recovered the loan from the debtor. The above provisions seemed so important and justified that they were almost completely reproduced by the subsequent legislative acts of Russian law. In later Russian legislation, for example, in Aleksey Mikhailovich's Code of 1649, the priority in creditors was given to the state treasury and foreign creditors.

In Ch. 4 "Brief image of the processes or litigation" (the manuscript version of which is supposed to be edited by Peter the Great himself), published in March 1715, we find Art. 5 of the following content: «*Купеческие книги, из которых видимо, что тот и тот толикое число винен, могут токмо вместо половины доказания служить...*» [9, Vol. 4, p. 41].

Despite almost absent legislative regulation, the problems connected with bankruptcy became more and more relevant. Trade relations considerably became complicated, and over time there were not enough available legislative provisions which determined only the most general operations procedure in case of insolvency of the debtor. Nevertheless, long before the first systematized laws on insolvency, Bankruptcy processes were conducted, especially in large trading cities.

The Bankruptcy Law in Russia developed in three directions which were not interconnected among themselves. *The first* direction consisted in attempt to adapt regulations of foreign legislations for the Russian conditions and features. *The second* direction in the basis of the institution of insolvency saw the development of national Bankruptcy relations. *The third* direction—creation and codification of regulations of the Russian Bankruptcy Law.

In Russia the foreign legislation was generally applied in case of trials within the largest processes. The government not only didn't interfere with application of precepts of law of a foreign state, and moreover, it strongly recommended sometimes, explaining its actions with readiness of the foreign legislation.

For certain reasons, the *Bankrupt charter* accepted on December 15, 1740 was *the first regulating document* in Russia regulating insolvency proceeding. The decree accompanying the Charter it was offered to arrive in all questions concerning bankruptcies according to the Charter. As it was noted above, in many respects this document just copied the Western European legislation, and some provisions couldn't be applied in Russia only for that simple reason that didn't correspond to either national peculiarities, or economic realities of that time.

This explains the constant efforts of the Senate to pass a new law, which were made periodically: in 1753, 1763, 1780, etc. It should be noted that projects in most of provisions were better, than the Charter which was accepted later.

On December 19, 1800 considering national peculiarities and the state of the Russian economy Paul The First adopted the new law—Charter on bankrupts. According to the Charter of 1800 the person was considered as the bankrupt, who "cannot fully pay their debts". Three types of bankruptcy were entered: *from misfortune, from negligence and defects, from forgery*. Various measures of influence were taken with respect to each type of insolvency [12, p. 43].

On June 23, 1832 the new law—"The Charter about Trade Insolvency" acceptable and adequate to the Russian reality which was applied only to the persons belonging to trade estate in the beginning was accepted, and in 1846 it was distributed on all people, who are engaged in trade. This Charter was applied with minor amendments up to 1917.

The history of the Bankruptcy Law both Russian, and foreign, carries out separation of insolvency into types. As criteria of classification it is possible to allocate:

(1) nature of the reasons which entailed insolvency;
(2) nature of activities of the debtor.

The Charter of 1832 distinguished three types of insolvency: simple insolvency, forged or malicious in the presence of intent and forgery; and in the absence of guilt, the debtor was declared insolvent for misfortune and was not subject to arrest.

According to another criterion, depending on the conduct of a person's trading activities, trade and non-trading insolvency are singled out.

The pre-revolutionary legislation in Russia provided three forms of initiation of proceedings of insolvency: according to the statement of the debtor, at the request of creditors and on an initiative of the court. First, the right to ask about opening of Bankruptcy process was provided to creditors, but it was supposed that the debtor before the creditors can find the insolvency. Frank recognition of the debtor was very important circumstance as, first, creditors could be convinced of conscientiousness of such debtor, secondly, in case of personal application the debtor could avoid arrest.

It should be noted that along with traditional procedures of Bankruptcy production the Russian legislation of that time regulated the procedures directed to recovery of solvency of debtors during trade insolvency i.e. what in the modern language is called rehabilitation procedures. For their carrying out administrations on trade affairs were created [16, p. 315].

It is necessary to emphasize that the existing institute of administration differed from the western analogs where for prevention of insolvency the debtor was granted delay in payments, such measures as precautionary peace deal, judicial liquidation was applied. At the same time by the foreign legislation for the debtor management of its property remained, and by the legislation of Russia the debtor was discharged of a property management.

It should be noted that the law on insolvency in Russia was used often. Thus, during the period 1893—1903. 546 joint-stock companies with capital of the owner in 11,420 million rubles were liquidated, and during the same period 1190 joint-stock companies with the owner's capital of 2147.7 million rubles were created [1, p. 46].

Along with the improvement of the insolvency law in the 19th century there was a formation and development of regulation of financial accounting in Russia. Regardless of merchants performed trading activities independently or collectively (in a partnership), the law of July 14, 1834 assigned an obligation of maintaining trade books which number depended on a trade sort. For wholesale trade were provided: "(a) memorial the book for daily record of affairs, (b) the ledger (gross-book) opening separate accounts on all turnovers of trade" [18, p. 82]. In this occasion, the famous Russian civilian P. P. Tsitovich, in his Essay on the Fundamental Concepts of Commercial Law, noted: "The Commercial Charter ... gave the whole management of obligatory accounts department, various, however, by the amount of trade. Bookkeeping is compulsory for everyone, so the banker should keep commodity and invoices books, although he has no goods, receives and sends no invoices" [19, p. 107].

In 1857 regulations about annual creation of accounting balance were entered the Charter. Also the legislation on the public reporting was improved: par. 45 of the Regulations on the companies on shares approved on December 6, 1836 determined the structure of the annual reporting: "Reports are signed by all board members, and should include:

(a) condition of the equity of the company;
(b) total income and expense for the time for which the report is submitted;
(c) detailed account to costs for management;
(d) account of a net profit;
(e) the account of the reserve equity when that is, and
(f) the special report on artificial and other institutions if those are in case of the company" (paragraph 14).

Besides, the Regulation contained the rules relating to the liquidation of the enterprise: the obligatory notification was provided in a seal about the beginning and the termination of a liquidation procedure. In relation to owners of the company the priority of a covering of their requirements according to which they could receive the shares "from the company so far was established the amount necessary on payment of all its liabilities won't be brought in one of the state credit establishments" (paragraph 15).

During the analysis of the legislation of the period 1800—1917. We can distinguish some of its characteristic features:

(1) bankruptcy is considered as insolvency of the dealer, banker because of default of liabilities through his fault; the choice of insolvency proceeding depended on degree of guilt;

(2) a clear terminological distinction is made between insolvency and bankruptcy, where the former is a more voluminous concept, and the latter is a separate case of insolvency;

(3) classification of concepts of insolvency (trade and non-trading) and bankruptcy (mercenary and simple) is conducted.

Besides, in case of determination of insolvency of an accounting entity accounting records were not practically used that, first, is explained by unified accounting methodology and by the linkage of insolvency only to the debtor's non-payment feature.

3 The Bankruptcy Law of the Soviet Period

The next, very short period of the development of the institution of inconsistency can be called *Soviet*. At this stage of development of the Bankruptcy Law the procedure of insolvency as a method of the termination of business activity, wasn't practically applied.

The transition to a peaceful life, implementation of the new economic policy and emergence of free trade turnover have led to the emergence of the problem of non-payments and, accordingly, the need for their legal regulation. The first Civil Code of the RSFSR of 1922 contained norms regulating relations arising in the conditions of insolvency of civil and commercial partnerships and individuals. However, in practice, the application of these regulations was difficult as the Civil Procedural Code of RSFSR accepted in 1923, didn't regulate the procedure of the announcement of insolvency. In the legislation the accurate procedure of insolvency proceedings wasn't established. This shortcoming was compensated only in 1928 when the Civil Procedure Code of RSFSR was added with hl. 37 "On insolvency of individuals, physical and legal". Insolvency signs were provided in Art. 318: "The debtor who stopped payments on debts on the amount over three thousand rubles or forced to stop them on the meant amount on a condition of the affairs can be acknowledged insolvent if the court establishes its inability to complete payment of cash requirements to creditors" [19, p. 46].

Since 1928 in process of folding of the New Economic Policy and entering of the program of industrialization of the country and the first 5-years periods from laws also those insignificant Bankruptcy regulations which existed gradually began to disappear. All events were explained by the fact that the institute of insolvency, unlike some other institutes, was incompatible with state planned economy and undivided state monopoly for property, model of insolvency existed only in regulations of the Soviet State, in practice it wasn't implemented. The problem of non-payments was regulated at the national level by means of the developed system of dating of unprofitable industries of national economy and the separate entities. The unprofitable entities and farms were artificially built in economy and existed as planned-loss-making at the expense of state financing, periodic debt write-off from farms, constantly existing schemes of redistribution of financial flows in the economic systems of the ministries and departments.

Concerning citizens, the question of bankruptcy did not arise at all as they had no right to be engaged in independent economic activity. Non-payments of taxes were collected as civil legal proceedings by the address of collection on the property belonging to the debtor.

Thus, by early 1960, general rules on bankruptcy were excluded from the civil law of the USSR.

Given the fact that the insolvency institution is inherent only in the market economy, then its revival became possible only with the transition of the Russian economy to market relations.

With the beginning of reforms on forming of market economy, the problem of creation of the standard legal basis of insolvency of accounting entities became a pressing practical task. Already in the very first Russian economic reform law "On Enterprises and Entrepreneurship" (adopted by the Supreme Soviet of the RSFSR on December 25, 1990) there was Art. 24, par. 3 of which said: *the entity which isn't fulfilling the liabilities on calculations can be declared judicially insolvent (bankrupt) according to the legislation of RSFSR.*

4   The Modern Bankruptcy Law

The *third, modern stage* of functioning of the bankruptcy mechanism begins with the publication of a detailed, rich in regulatory and legal content of the Decree of the President of the Russian Federation of June 14, 1993, No. 623 "On measures for maintenance and improvement of the insolvent state entities (bankrupts) and application to them special procedures". The decree appeared in an extremely difficult period for the USSR, when the actual procedure for the disintegration of the country began, which led to the breakage of the existing ties between socialist enterprises for years, the vertical of economic management was lost, and the privatization of state and municipal enterprises began.

The decree No. 623 was issued for support of the state entities which appeared insolvent (bankrupts) and ensuring effective use of the state-owned property assigned to the state entities, protection of the rights and the interests of the state, creditors and labor collectives of the insolvent entities. By that time, the need to regulate relations connected with insolvency of the entities based on other patterns of ownership ripened.

For regulation of institute of insolvency, the Law of the Russian Federation of November 19, 1992 No. 3929-1 "On insolvency (bankruptcy) of the entities" [8] was soon adopted which in a decade underwent three absolutely opposite editions. Here the legislator gives a priority to external signs of bankruptcy, thereby narrowing a concept of insolvency and destroying historically developed concept of bankruptcy.

The law of 1992, in comparison with the Presidential decree of the Russian Federation of June 14, 1992, essentially differently resolved the main issue: the insolvency of the entity began to be considered the insolvency taking place after recognition of the fact as arbitration tribunal or after the official announcement of it the debtor in case of his voluntary liquidation.

The implementation of the Law of 1992 [8] didn't begin right after its entry into force. Not everyone, who has to execute it, turned out to be ready to the application of the Law of 1992. Considerable preparatory work on creation of necessary organizational and legal prerequisites was required. Important issues of legal regulation of institute of insolvency (bankruptcy) were resolved in the regulations of the President of the Russian Federation and Government of the Russian Federation accepted according to the Law of 1992.

The practice of applying the Law of 1992 also revealed its weaknesses. First, the problem is connected with the fact that the norms and principles set forth in the Law were "adjusted" to the conceptual apparatus of the then effective legislation of the RSFSR. Secondly, this normative legal act became morally outdated, ceased to answer the developing new economic relations fully. It was explained, first, by the fact that it was adopted in 1992, long before the adoption of the Constitution of the Russian Federation of December 12, 1993 and the new Civil Code of the Russian Federation of 1994.

Work on review of the Law of 1992 was conducted within several years. The Federal law "On Insolvency (Bankruptcy)" of January 8, 1998 No. 6-FZ "On insolvency (bankruptcy)" [6] became the result of work. By the Arbitration Courts, the Law of 1998 was applied in cases of insolvency (bankruptcy) proceedings, the proceedings for which were initiated on March 1, 1998.

The law No. 6-FZ "On Insolvency (Bankruptcy)" [6] was aimed at solving only one task—the protection of the interests of creditors. The procedure for bankruptcy was so simplified that every enterprise that had a 3-month overdue debt in a meager amount— 50 thousand rubles, was threatened with liquidation. But in those days in Russia about 40% of enterprises were unprofitable, and each of them was threatened by liquidation. As a result, the Law on bankruptcy from the means of economic recovery turned into a source of conflict, led to the ruin of many solvent enterprises. In general, it seems that the developers of the Law pursued one purpose, super-fast secondary privatization of state property.

Unfair creditors often showed interest not in paying off debt by debtors, implementing measures for financial recovery, but in their liquidation and seizure of property complexes for nothing [14].

A serious drawback of the former Law was, in fact, the disenfranchised position of the debtor. Of course, he could submit to the court a response to the creditor's statement with his charges in five days (now—ten days), but otherwise his interests were not considered.

The problem of insolvency became particularly relevant in the Russian Federation because of the August 1998 crisis and was identified as a result of the threat of a "sovereign default" caused by the termination by Russia of payments on state bonds of federal loan and short-term bonds (OFZ and GKOs) duty of the USSR. These events regarding instruments placed on foreign markets through commercial banks could not but cause a chain reaction in the sphere of financial transactions of the latter [11, p. 274].

In December 2002, the new Federal Law "On Insolvency (Bankruptcy)" No. 127-FZ [7] came into force, which is still functioning today. It applies to individuals and legal entities, except for state enterprises, institutions, political parties and religious organizations. Bankruptcy procedures help to overcome the crisis of non-payments, stop the growth of overdue debt and prevent negative social consequences associated with crisis processes. The law contains more stringent requirements for arbitration managers, who are entrusted with conducting bankruptcy procedures.

In comparison with previous edition, the last Law regulates bankruptcy process in more detail. All these changes are very positive and, undoubtedly promote balance of interests both debtors, and creditors.

5 Conclusion

Summarizing the above, we conclude that the emergence of bankruptcy as an economic and legal category goes back to the deep past. The bankruptcy procedure in a market economy serves as an important component of public relations in the sphere of economics and law. It is the most effective way to protect violated rights and legitimate interests of economic activity participants, as well as an important tool for self-regulation of the business environment from elements that are not viable.

At the same time, it is necessary to develop and adopt separate rules that establish the possibility for the effective functioning of the institution of bankruptcy in Russia. It is also essential to use special procedural rules within the framework of existing norms, taking into account the peculiarities of cases connected with insolvency.

All these things can be realized by paying attention to the accumulated historical experience of Russian and foreign legislation.

References

1. Gavrilova, V.E.: Bankruptcy in Russia: Issues of History, Theory and Practice: Textbook, p. 207. TEIS, Moscow (2003)
2. Holmsten, A.H.: Historical Outline of the Bankruptcy Process. St. Petersburg (1872)
3. Kuter, M., Tkhagapso, R.A.: Accounting in Insolvency: Textbook, p. 204. KUBAN State University, Krasnodar (2005)
4. Nosov, S.I.: Joint-stock law in Russia (experience of historical and legal research). Application, p. 32. RAGS, Moscow (2000)
5. Oberbrinkmann, F.: Modern understanding of the balance sheet: Trans. with him. Ed. I'M IN. Sokolov, p. 416. Finance and Statistics, Moscow (2003)
6. On Insolvency (Bankruptcy): Federal Law No. 6-FZ of 08 January 1998
7. On Insolvency (Bankruptcy): Federal Law No. 127-FZ of October 26, 2002 (as amended on 31 December 2004)
8. On Insolvency (Bankruptcy) of Enterprises: Federal Law No. 3929-1 of 19 November 1992
9. Chistyakov, O.I. (ed.): Russian Legislation X–XX Centuries, vol. 9, Under the general. Legal Literatureю, Moscow (1984)
10. Stepanov, V.V.: Insolvency (Bankruptcy) in Russia, France, England, Germany, p. 204. Statute, Moscow (1999)
11. Telyukina, M.V.: Comments on the Federal Law "On Insolvency (Bankruptcy)", p. 591. Yurayt-Izdat, Moscow (2003)
12. Tkachev, V.N.: Insolvency (Bankruptcy) in the Russian Federation. Legal Regulation of the Competitive Relations. The 2nd edition Processed and Added, p. 362. The Book World, Moscow (2004)

13. Tikhomirov, M.N.: A Manual for the Study of the Russian Truth. Publishing House of the Moscow University, Moscow (1953)
14. Tkhagapso, R.A.: Formation and development of the institution of insolvency in Russia. Econ. Theory Pract. Sci. Educ. J. **8C**, 18–21 (2004)
15. Tsitovich, P.P.: Essay on the Basic Concepts of Commercial Law, p. 448. Center for JurInfo, Moscow (2001)
16. Shershenevich, G.F.: Competition Law, 2nd edn, p. 509. Imperial Publishing House, Kazan (1898)
17. Shershenevich, G.F.: Textbook of Commercial Law (on the edition of 1914), p. 335. Spark, Moscow (1994)
18. Shershenevich, G.F.: The Bankruptcy Processes, p. 477. Statute, Moscow (2000)
19. Yudin, V.G.: Insolvency (Bankruptcy): Historical Aspect. Bulletin of the Supreme Arbitration Court, no. 1, pp. 155–162 (2002)

Digital Technology in Risk-Based Approach of Continuous Audit

Julia Klimova[(✉)]

Kuban State University, Krasnodar, Russia
ladycat23@mail.ru

Abstract. The formation of an internal control system for the company is an actual and necessary process that indicates the level of maturity of the business. The existence of developed organizational measures, methods and procedures allows in a preventive manner to focus management attention on risky areas of business processes and determine activities to optimize the performance of company to achieve its goals and strategy. In these conditions, a continuous audit becomes a priority for increasing the information openness of business as a part of the company's internal control system. In present work we examine the implementation of digital technology in continuous audit through the automatic processing of data in the company's information systems based on the analysis of the criteria of risk-oriented approach to improve risk-management process.

Keywords: Continuous audit · Risk-oriented approach
Internal control system · Risk-management

1 Introduction

The influence of the uncertainty on business necessitates the timely identification and assessment of risks in order to prevent and reduce the possible negative consequences for the financial position of the company and its performance. In this connection, the formation of an internal control system becomes the necessary process, which helps management to focus attention on risky areas of business and optimize activities to achieve the company's goals [19].

An important factor of the company's competitiveness is the ability of management to build a competent and effective policy of risk management. Studies of recent years [20–22] have noted the increasing role of risk management as an integral part of management functions in the company. Timely identification of risks and the development of a set of measures for preventive exposure make it possible to create a reliable basis for decision-making and planning and improve the management of the company to achieve the objectives of business processes [13, 15–17, 23]. Despite a significant number of works [1, 7, 9, 18] devoted to the application of risk-based approach in the activities of companies, the issue of practical implementation of control procedures in the company's business processes remains relevant. The development of elements of the internal control system as a whole and continuous audit in particular is becoming more urgent for improving the risk-management process [2, 7, 9–12, 14, 18, 24].

© Springer Nature Switzerland AG 2019
T. Antipova and A. Rocha (Eds.): DSIC 2018, AISC 850, pp. 306–312, 2019.
https://doi.org/10.1007/978-3-030-02351-5_35

2 Implementing Continuous Auditing and Continuous Monitoring

Companies have begun to focus their efforts by implementing continuous auditing (CA) and continuous monitoring (CM) disciplines around their business processes, transactions, systems and controls. Implementation of technology-based applications to manage key areas of risk and control has become a practical and necessary alternative to meet the growing ever-changing regulatory, business, and industry requirements and needs of the organization. Together, CA and CM offer a broad range of benefits that can help organizations provide greater transparency into the operations and improve business performance by removing excess costs from operations, improving controls and processes, and preventing and detecting fraud and misconduct [5].

In general, CA/CM strategy is influenced by a variety of drivers [6]:

– heightened demand for faster, better decisions and for improved, but cost-effective risk management;
– rising pressures on internal audit to provide timely assurance to stakeholders;
– increasing complexity and change in regulatory requirements;
– greater efforts to align internal audit activities with management's strategic business goals.

CA/CM seeks to add value by improving compliance and supporting business goals. Potential benefits of CA/CM as a monitoring mechanism include [4, 5]:

– delivering regular insight into the status of controls and transactions across the global enterprise;
– enhancing overall risk and control oversight capability through early detection and monitoring;
– using automation to efficiently test a broader range of transactions and controls leading to cost-reduction opportunities;
– enhanced and more timely oversight of compliance across the enterprise;
– business improvement through reduced errors and improved error remediation, allowing reallocation of resources to value-adding activities;
– the ability to report more comprehensively on compliance with internal and regulatory requirements.

In this regard, the topical task is to improve the quality of the management level, which implies creating an internal control, audit and risk management system that meets the modern needs of corporate governance. The most effective is the construction of a single risk-oriented system of internal audit and control, which should be implemented in three key strategic areas:

(1) formation of a risk-based internal control system based on a strategy of continuous improvement of its quality;
(2) development and improvement of the activity of a functionally structured system of internal audit on the conditions of centralization and strengthening of the vertical control;

(3) establishment of a control mechanism for prevention of fraud in the financial and economic sphere with the development of a single integrated program. The relevant information must be identified, recorded and transmitted in a timely manner in a form that allows auditors to use it effectively.

Although CA/CM are often considered together, they are actually two distinct types of controls. As the name implies, continuous monitoring enables management to continually review business processes for adherence to and deviations from their intended levels of performance and effectiveness. Similarly, continuous auditing enables internal audit to continually gather from processes data that supports auditing activities. Researchers [3, 6, 8] define CM as an automated, ongoing process that enables *management* to:

- assess the effectiveness of controls and detect associated risk issues;
- improve business processes and activities while adhering to ethical and compliance standards;
- execute more timely quantitative and qualitative risk-related decisions;
- increase the cost-effectiveness of controls and monitoring through IT solutions.

CM enables management to determine more quickly and accurately where it should be focusing attention and resources in order to improve processes, implement course corrections, address risks, or launch initiatives to better enable the enterprise to achieve its goals.

CA, in its turn, is defined as an automated, ongoing process that enables *internal audit* to:

- collect from processes, transactions, and accounts data that supports internal and external auditing activities;
- achieve more timely, less costly compliance with policies, procedures and regulations;
- shift from cyclical or episodic reviews with limited focus to continuous, broader, more proactive reviews;
- evolve from a traditional, static annual audit plan to a more dynamic plan based on CA results;
- reduce audit costs while increasing effectiveness through IT solutions.

A company may maximize the value of implementing both CM and CA by:

- integrating management's responsibility for the performance of controls with internal audit's responsibility for assurance regarding management's controls—while preserving audit's independence;
- increasing coordination between management and internal audit in order to minimize duplication of controls and efforts;
- adapting more quickly and effectively to changes in the risk and regulatory climate.

As we can see, the most relevant method of internal audit, which allows to ensure a control procedure at all levels of management, identify the main risks and their acceptable level is the method of continuous audit. It can act as an early warning system to detect control failure on a more timely basis than under traditional approaches. The goal of implementing continuous audit is to enhance the overall visibility of the organization to risk and performance through the effective use of technology.

3 Procedural Steps for Automated Continuous Auditing

With the use of continuous audit, the company can establish automated collection of audit evidence and indicators from an entity's IT systems, processes, transactions, and controls that allow real-time analysis of financial and business processes with a significant reduction in time and resources. This information enhances auditor capabilities and helps to ensure compliance with policies, procedures, and regulations.

To develop a continuous audit, it is necessary to create an automated information system of financial and economic indicators integrated into the IT-system and allowing, with the help of logical compliance settings, to determine the degree of correspondence between the actual parameters and the target criteria. Reports from this automated information system should be generated in accordance with established criteria, which allows to detect deviations from the targets of business processes.

The Implementation of a Continuous Audit Consists of the Following Procedural Steps

(1) establishment of priority areas for continuous audit;
(2) definition of monitoring rules within the continuous audit;
(3) determining the frequency of the audit process;
(4) setting parameters for continuous audit;
(5) management of results and subsequent actions;
(6) reporting results [2].

At the stage of establishing priority directions based on the risk-oriented approach we should:

- identify the most important business processes that have the greatest impact on the company's activities;
- evaluate the projected benefits from the inclusion of the business process in the continuous audit process;
- assess the availability and structure of the required data.

The definition of monitoring rules in continuous audit includes the establishment of rules/scenarios/sequences for processing analytical data for specifying a programming algorithm according to the objectives of a particular business process.

When determining the frequency of the continuous audit process, it is necessary to take into account the natural rhythm of the process being tested, the issue of the cost, risks and benefits of the proposed periodicity, taking into account monitoring objectives such as containment or prevention.

Setting parameters for continuous audit should include the identification of target and limit values of the analyzed indicator criteria for the planning needs in order to set relevant activities in audit and management.

Management of results and subsequent actions of risk owners should be built on the basis of evaluating the results of signal processing and error detection. The results of continuous audit are considered in comparison with internal and external measures to determine their impact on the process and sufficiency to cover risks.

The results of continuous audit should be submitted to management in a timely manner in a consistent official report, including observations and understanding of the risks, control measures and consequences associated with the findings.

In order to prevent significant costs, it is advisable to create automated information system on the basis of existing network resources and information databases in the company. Modern information technologies make it possible to create digital software that aggregates data from information systems with both financial and technical parameters according to given algorithms. Using digital technologies with the function of indicator system will allow to detect inconsistency with the indicator and to receive the detailed information in real time. It is possible that basic data analysis can be performed using a range of tools, including spreadsheets and database query and reporting systems. There are certainly risks from using spreadsheets, apparent to any auditor, because of the difficulty of ensuring data integrity. General purpose analysis tools also have their own limitations. It is clear that the analytics process must be managed in order to be relied upon by audit, which is why audit-specific analysis software should include capabilities such as:

- maintaining security and control over data, applications, and findings
- logging all activities;
- analysis techniques designed to support audit objectives;
- automated creation and execution of tests [8].

For purposes of visualization of the current status of the system of indicators, both reports from the system, IT-enabled dashboards or colored scoreboard with the reflection of business processes for which deviations can be used. The objective is to make the use of data analytics a sustainable, efficient, and repeatable process.

4   Conclusion

The use of digital technologies in continuous audit provides an integrated approach to real-time risk assessment that combines data on various functional tasks in company and efficiently test a broader range of transactions and controls in order to enhance risk and control oversight capability through monitoring and detection. The move to automated testing and continuous audit procedures also changes the traditionally cyclical nature of the audit process. Comprehensive testing of transactions and controls effectiveness, on an ongoing automated basis, enables audit to move to a more risk-based approach. The results of continuous auditing techniques provide visibility into whether risk is increasing in specific areas and warrants additional audit focus. This use of analytics provides continuous insight into control effectiveness and the compliance of transactions. As long as internal audit can depend on the integrity of these testing procedures, it frees up audit resources to address other areas of risk. Reducing the need to commit substantial resources to regular financial and operational audits provides the ability to focus more on areas of higher risk in which professional judgment and expertise are key [8].

References

1. Belousov, S.A.: Risk-oriented internal audit as an element of key risk management of the company. Risk Manag. **2**, 14–18 (2011)
2. de Aquino, C.E., Lopes da Silva, W., Sigolo, N., Vasarhelyi, M.A.: Six steps to an effective continuous audit process. https://www.researchgate.net/publication/266059571_Six_Steps_ to_an_Effective_Continuous_Audit_Process
3. Continuous auditing: is it fantasy or reality? http://m.isaca.org/Groups/Professional-English/ continuous-monitoring-auditing/GroupDocuments/ISACA%20Continuous%20Auditing.pdf
4. Continuous auditing and continuous monitoring: the current status and the road ahead. http:// mcr.doingbusiness.ro/uploads/517518efdd1e7CA%20and%20CM%20WEB%20DE-Survey.pdf
5. Continuous auditing and continuous monitoring: transforming internal audit and management monitoring to create value. https://home.kpmg.com/content/dam/kpmg/kz/pdf/cacm-brochure.pdf
6. Continuous monitoring and continuous auditing: from idea to implementation. https:// www2.deloitte.com/content/dam/Deloitte/uy/Documents/audit/Monitoreo%20continuo% 20y%20auditoria%20continua.pdf
7. Firova, I.P., Bikezina, T.V.: Modern problems of integrated risk management in order to reduce the financial risks of economic subjects. Sci. Bus. Ways Dev. **11**, 35–38 (2016)
8. How data analytics and continuous auditing and monitoring are evolving. https://www.acl. com/pdfs/eBook_CW_Continuous_Monitoring.pdf
9. Grischenko, O.V., Efimenko, A.: The role and place of internal audit in corporate governance system. Bull. Taganrog Inst. Manag. Econ. **1**, 46–53 (2009)
10. Kovalenko, A.I.: Features of financial risk management of international corporations. Econ. Sustain. Dev. **4**(28), 36–44 (2016)
11. Luchakova, E.V., Tuvaeva, A.M., Sergienko, L.V.: Risk management in accounting. Actualscience **12**, 277–279 (2016)
12. Nigrini, M.J.: Continuous auditing. Ernst & Young Center for Auditing Research and Advanced Technology and Advanced Technology University of Kansas. http://aaahq.org/ audit/midyear/01midyear/papers/nigrini_continuous_audit.pdf
13. Mihret, D.G., Khan A.A.: The role of internal auditing in risk management. In: Seventh Asia Pacific Interdisciplinary Research in Accounting Conference, Kobe (2013)
14. Kokemuller, N.: The advantages of continuous auditing. http://small-business.chron.com/ advantages-continuous-auditing-39568.html
15. Paschenko, T.V., Tarasova, K.J.: Methodical approaches to assessing financial investments for the purpose of the financial reporting and expertise of the balance sheet of assets. Probl. Mod. Econ. **4**(64), 82–86 (2017)
16. Pislegina, N.V.: Problems of estimation of financial investments in accounting and financial statements. Proc. Altai State Univ. **2**, 104–106 (2002)
17. Pravkina, E.I.: The role of risk management in the process of business planning. Bull. Univ. (State Univ. Manag.) **7–8**, 247–250 (2016)
18. Selezneva, E.S.: Information support of risk-oriented internal audit. Bull. Saratov State Soc. Econ. Univ. **1**(50), 107–110 (2014)

19. Shtiller, M.V.: Theory of risk management and development of risk management of companies. Stat. Account. Audit **2**(41), 28–31 (2011)
20. Spira, L.F., Page, M.: Risk management: the reinvention of internal control and the changing role of internal audit. https://poseidon01.ssrn.com/delivery.php?ID=2850880950020251210 95064068077103091033020039072045089028087112023101115077118083075049063097 01511202301602706610301408909501808305500007910308508909400612110909506901 40091270920650900080090161081091060200880000730930880761191061261081171 18119026&EXT=pdf
21. The changing role of internal audit. https://www2.deloitte.com/content/dam/Deloitte/in/Documents/audit/in-audit-internal-audit-brochure-noexp.pdf
22. The role of internal audit in enterprise-wide risk management. https://www.ucop.edu/enterprise-risk-management/_files/role_intaudit.pdf
23. Vasile, E., Croitoru, I., Mitran, D.: Risk management in the financial and accounting activity. Intern. Auditing Risk Manag. **1**(25), 13–24 (2012)
24. Stippich, W.W.: Continuous auditing=continuous improvement. http://www.corporatecomplianceinsights.com/continuous-auditing-continuous-improvement

Cryptocurrency in Digital Wallet: Pros and Cons

Tatiana Antipova[1]([✉]) [ID] and Irina Emelyanova[2]

[1] Institute of Certified Specialists, Perm, Russia
antipovatatianav@gmail.com
[2] Sigma Ltd, Perm, Russia

Abstract. This paper considers the birth of cryptocurrencies, their concept and classification, pro and contra cryptocurrencies. The percentage growth analysis makes it possible to draw conclusions about the profitability and reliability of certain types of cryptocurrency. Authors have designed a model of decision-making to transactions cryptocurrency in daily life. Most of the governments have not given their legal status for the use of cryptocurrency in their country yet. But if cryptocurrency will be stable in the future, then it is easily accepted through worldwide and in the long run, people will have more trust to the cryptocurrency and its usability.

Keywords: Cryptocurrency · Digital wallet · Bitcoin · Mining
Forging · Blockchain

1 Introduction

XXI century is inextricably linked with the emergence and implementation of innovative technologies that have a significant impact on society, in which virtual money has been widely developed. Cryptocurrency is a kind of virtual/digital money based on cryptographic methods. For the first time cryptography for the purpose of confidential payments began to be used in 1990 in DigiCash system but the use was short. Theoretically, the term "cryptocurrency" appeared as the Bitcoin payment system, which was developed in 2009. Bitcoin was invented in 2008 by a person or a group of individuals under the pseudonym of Satoshi Nakamoto. According to Satoshi, Bitcoin was created in order that any person could manage their money independently [1, 2].

But the exact name cryptocurrency was appeared in the magazine "Forbes" in 2011. And since that time the name has become firmly in use. Crypto currency is a special kind of electronic means of payment. Strictly speaking, this is a mathematical code. It is called so because the cryptographic elements, namely the electronic signature, are used for the circulation of this digital money.

The units of measurement in this system are "coins". Cryptocurrency has no real expression such as metal coins or paper banknotes. This money exists exclusively in digital form.

The principal feature that distinguishes crypto money from real ones is the way they arise in the digital space. Cryptocurrency is a decentralized register of data on transactions performed on the basis of cryptographic algorithms. This group of relations can

© Springer Nature Switzerland AG 2019
T. Antipova and A. Rocha (Eds.): DSIC 2018, AISC 850, pp. 313–322, 2019.
https://doi.org/10.1007/978-3-030-02351-5_36

be aligned to fiduciary relations (relations based on trust between the trustee and the beneficiary, the founder of the company and the established company, etc. [3]. Based on this definition, we can not call the currency an electronic cash, because the bank account for transactions does not open - all transactions occur through digital wallets [4].

Mark [5] argue that the digital currency is something different from government issued currency as it is issued by the private parties and circulate only through the Internet. Dwyer [6] establishes that current digital currencies such as bitcoin is helpful for double-spending problem and create finality of transactions.

Digital wallets are computer software applications that store and transmit payment authorization data for one or more credit or deposit accounts. Digital wallets differ from traditional plastic cards in that they are potentially smart wallets. Traditional payment cards are "dumb" devices that are capable of doing a single thing and nothing more: transmitting payment authorization data to a merchant. In contrast, a digital wallet can provide two-way communication between a consumer and a merchant. That communication need not be limited to payment authorization data, but could include virtually any type of data (geolocation, coupons, loyalty program information, etc.) about the consumer to the merchant. This means that a digital wallet can potentially integrate payments into a comprehensive digital retail services suite of advertising, search, payment, customer service, and loyalty program features [7].

Consequently, if there is one lesson to be learnt from the evolution of cryptocurrencies, it is that there is nothing constant in the definition of a cryptocurrency. The only constant is the fast evolution of cryptocurrencies and businesses centered on them and the lack of robust legal framework regulating them in many areas [8].

2 Cryptocurrency's Birth and Measurement

Understanding the reasons for the birth of cryptocurrencies gives a complete picture of the regulatory problem embodied in them. It also allows one to appreciate whether there was a realistic vision for the creation of a system of payment or currency. Since the creation of bitcoin coincided with the 2008 global financial crisis, it is commonly suggested that cryptocurrency is the result of the global financial crisis, although there is no concrete evidence for that. The purpose and the core principles of cryptocurrencies do not address the causes or consequences of the global financial crisis. Nevertheless, it cannot be ignored that the lack of trust in government institutions and central banks that ensued during and in the aftermath of the global financial crisis might have been exploited as a marketing tool by the developers and backers of bitcoin and other cryptocurrencies. But that remains the only connection between cryptocurrencies and the global financial crisis [8].

In other hand, the development of computer technologies and networks has affected the change in the form of money. Banknotes and coins are gradually replaced by bank cards. In connection with this, a lot of payment systems operate on the Internet today, which were originally created only for electronic payments, such as PayPal, WebMoney, Yandex.Money. Cryptocurrency's birth has changed the notion about a traditional payment system over the last decade in the world. In many countries cryptocurrencies are considered illegal. At the same time some countries mark positive sign for the use of cryptocurrence (USA, UK etc.) and some did not give the legal permission to use it in their territory (Bangladesh, India, Bolivia, Ecuador, Russia etc.). At present, digital currencies are not issued by national banks. But it should be noted the growth of the world's currency.

Cryptocurrecies are created by programmers basically in three ways [9]:

- ICO (initial coin placement, investment system);
- Mining (maintenance of a special platform for the creation of new cryptogenes);
- Forging (formation of new blocks in existing cryptocurrencies).

Only the Central Bank is entitled to emit real traditional money, but any person can issue cryptocurrency. In order to make transactions using cryptocurrency, you do not need to contact any financial organizations (banks). A further explanation of cryptocurrencies is likely to state that unlike traditional currencies issued by the central banks, a cryptocurrency has no central authority that controls its creation and circulation. The circulation of cryptocurrency occurs in the "Blockchain" system. This system is a database distributed over millions of personal computers around the world. The storage and recording of information on the transfer of crypto currency is carried out in Blockchain system on all devices at once, which guarantees absolute transparency and openness of the transactions being made.

Gikay states [8] that the blockchain is the only infrastructure necessary for the functioning of cryptocurrencies. But that simply is naïve and perhaps a denial of the reality of cryptocurrencies. Cryptocurrencies cannot function without other supporting infrastructures such as exchanges and digital wallets. First time users of cryptocurrencies must necessarily purchase cryptocurrencies from exchanges using traditional currencies and this renders the exchange a necessary part of the cryptocurrency ecosystem, unless the user in question is a miner who earns cryptocurrencies by mining; in the latter case, the user has the option to transact directly from the blockchain. Then users may need to store their cryptocurrencies in third-party administered digital wallets who should keep funds safely but may also use their power to the detriment of users.

Advanced distributed system architecture built to protect against potential threats. More than 98% of digital assets stored in multi-signature, digital wallets. Security protocols are fully aligned and compliant with industry best practices.

The following methods of obtaining cryptocurrency are available:

I. Acquisition through exchanges. Exchange systems flooded the Internet. They allow users to deal with the system of purchases and sales, as well as converting cryptocurrency. The commission on stock exchanges is the lowest, but you have to wait until your buyer finds a product for you. A popular exchange is a great chance to quickly exchange a coin. The most reliable exchanges are those that after registration are asked to make a deposit and undergo verification, and only after that they provide a complete list of operations.

II. Exchangers. The most reliable and proven way by many people. Almost instant payments to the wallets you want. But of course, the exchanger requires a fee for its operations.

III. Forums. This is perhaps the most unsafe method of currency exchange. It is built through a forum on the complete trust of strangers to each other. And here the chance to become a participant in a dishonest transaction increases. At specialized forums, it is possible to find a specific person with whom you will make a transaction, but it will take some time. The advantages of this option are the zero commission and instant exchange.

By now, there are already several thousand varieties of cryptocurrencies. There is a large category (almost 50%) of cryptocurrencies, in fact not provided with any content. These are so-called soap bubbles and we will not take them into account.

The most common types of cryptocurrency are [10]:

1. Bitcoin (BTC, bitcoin, currently one bitcoin costs about 10,000 US dollars). Crypto currency bitcoin in simple words is the very first digital currency on the basis of which all subsequent ones were developed. For this currency, a limit of 21,000,000 is stated, however, it is still not reached at present.
2. Etherium. This is the development of the Russian programmer Vitaly Buterin. This currency appeared relatively recently - in 2015. Now it is quite popular along with bitcoins.
3. Litecoin (LTC is equal to 40 US dollars). The currency is developed by the programmer Charlie Lee and is released from 2011. Lightcoin is considered an analogue of silver among the crypto currency (and bitcoin is analogous to gold). The release of lightcoins, like bitcoins, is also limited and amounts to 84,000,000 units.
4. Z-Cash (Z-cash, ~$ 200).
5. Dash (dash, ~$ 210).
6. Ripple (~$ 0.15 USD).

The most popular of all cryptocurrency is bitcoin. Its name is made up of the words "bit" - the smallest unit of information and "coin". For BTC, or bitcoin, not only the program is created, but also a special digital wallet, in which you can store this

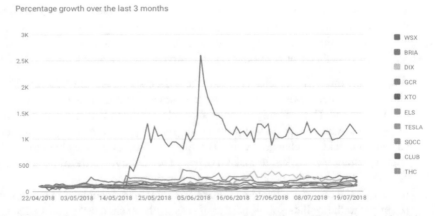

Fig. 1. Top 10 best cryptocurrencies for mining and buying, steadily growing in the price on cryptocurrency markets in the last 3 months 2018. Source: https://www.cryptoratesxe.com/best-cryptocurrency.html.

currency. In addition, now you can withdraw usual paper money in special ATMs for bitcoins exchanging. Also a number of retail chains and stores accept this currency for calculation along with ordinary banknotes and coins. But bitcoin is not among top ten best cryptocurrencies, and we can see it on Fig. 1 and Table 1.

Table 1. Fast growing 10 best cryptocurrencies.

Cryptocurrency	Percentage growth for the quarter
WeAreSatoshi (WSX)	997.21%
BriaCoin (BRIA)	179.99%
Dix Asset (DIX)	117.11%
Global Currency Reserve (GCR)	95.02%
Tao (XTO)	84.28%
Elysium (ELS)	78.24%
TeslaCoilCoin (TESLA)	57.93%
SocialCoin (SOCC)	34.85%
ClubCoin (CLUB)	18.54%
HempCoin (THC)	0.58%

Source: https://www.cryptoratesxe.com/best-cryptocurrency.html

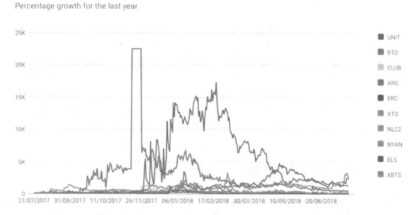

Fig. 2. Top 10 most reliable cryptocurrency with stable growth of profitability in the last year. Source: https://www.cryptoratesxe.com/most-reliable-cryptocurrency.html.

Percentage growth of ten best cryptocurrencies in the last 3 months 2018 is shown in Table 1.

According to Table 1 and Fig. 1 data, we can see that in the last quarter the most dynamically growing cryptocurrency was WeAreSatoshi (WSX). But every quarter the composition of the cryptocurrency is changing. For example, 20.12.2017 - 14.03.2018 the most dynamically growing cryptocurrency was Argentum (ARG).

The first in Reliability also has changed for a long period (for a year). Information about top 10 most reliable and safest cryptocurrency from 21.07.2017 to 20.06.2018 is shown on Fig. 2 and Table 2.

Percentage growth of ten best cryptocurrencies from 21.07.2017 to 20.06.2018 is shown in Table 2.

Table 2. Safest cryptocurrencies.

Cryptocurrency	Percentage growth for a year
Universal Currency (UNIT)	2 460%
Bitcloud (BTD)	399.82%
ClubCoin (CLUB)	42.44%
Argentum (ARG)	1 205%
EuropeCoin (ERC)	28.32%
Tao (XTO)	2 187%
NoLimitCoin (NLC2)	10.11%
Nyancoin (NYAN)	57.99%
Elysium (ELS)	28.76%
Beatcoin (XBTS)	150.66%

Source: https://www.cryptoratesxe.com/most-reliable-crypto
currency.html

According to Table 2 and Fig. 2 data, we can see that 21.07.2017 - 20.06.2018 the most reliable and safest cryptocurrency was Universal Currency (UNIT) with 2 460% growth for a year. But the most reliable and safest cryptocurrency was Bitcoin (BTC) with 747.54% growth for a year (26.03.2017–25.02.2018). So we can see that cryptocurrency is very volatility. With a long-term investment in the crypto currency, it is possible as a price increase, stop, and a decline. That is, the investment has a certain risk. In following part we will take into account another pros and cons.

3   Cryptocurrency: Pros and Cons

Cryptocurrency affects consumers and merchants quite differently. Different cryptocurrency entail different form factors, technologies, and business models. Accordingly, there are different risks and features involved with different cryptocurrency.

The main advantage of cryptocurrency is their independent on official institutions. The cost of ordinary fiscal money directly depends on gold, silver, the rate of the leading international currencies, political events in the world, volatility of the price of oil, etc. But cryptocurrency is not depending on financial institutions and tools, not inflation, it unique code cannot be copied. This allows you to ensure the safety and security of cryptocurrency. By cons can be attributed the speculative nature of this payment instrument, the high volatility of the course and the further complication of the "mining" process due to the growth in demand.

At this moment, there are more than 2000 types of cryptocurrencies, the most popular are Bitcoin and Litecoin, created on its basis, as well as the gaining popularity of Etherium, created by Russian programmer Vitaly Buterin in 2015. Usually, crypto-currencies are created for specific tasks. For example, Ehtereum is provided not only for the implementation of transactions. It already had smart contacts. The role of the analogue of gold in the world of crypto-currency is assigned to bitcoin, but the role of silver is Litecoin. The higher the demand for e-currency, the more expensive it is to pay for it.

Pros:

1. Anyone can get cryptocurrency with the help of a specially organized activity (mining). Since there is no single emission center and no official bodies controlling this process, no one can prohibit obtaining crypto money in the network for ordinary citizens.
2. All transactions with cryptocurrencies occur absolutely anonymously. The only open information in this case is the number of the electronic wallet. And all information about its owner is closed.
3. Decentralized output, in addition to the possibility of extracting money by everyone, determines the lack of control over this process.
4. For each type of cryptocurrency, the release limit is provided. Thus, excessive emission is impossible and, as a consequence, there is no inflation in relation to this money.
5. Cryptocurrency is protected by a unique code like an electronic signature that is copy-protected, and therefore it can not be counterfeited.
6. There are practically no commissions for transactions, since during the operations with the help of cryptocurrency, the role of the third party of relations - banks - is not required. Consequently, such payments are relatively cheaper than using traditional money.

According to their main characteristics, digital money is very different from the usual ones. This entails not only solid pluses, but also some disadvantages for users. Let consider some cons cryptocurrency.

Cons:

1. If the user has lost the password from his electronic wallet, it means for him the loss of all funds in it. For example, one friend of us lost the password from his electronic wallet so he could not withdraw about one Bitcoin, which meant for him a loss about $ 10,000. Since there is no control over the conduct of transactions using digital money, there are no guarantees of their safety.
2. Crypto currency is characterized by high volatility due to the specifics of its circulation (volatility means a frequent change in its value).
3. In relation to the crypto-currency, attempts may be made by various negative influences from the national regulators of monetary circulation (for example, the Central Bank).
4. As over time the process of obtaining cryptonyms becomes more and more complicated, mining with the help of individual equipment becomes less and less profitable.

Each of the existing types of crypto currency has both advantages and disadvantages inherent in all of them in the aggregate. In general, the same features characterize all of cryptocurrency as traditional money, namely: They are universal; They are exchangeable; They can be accumulated; Perform the settlement function.

The cost of digital money varies depending on supply and demand. The rapid growth of Cryptocurrency is provoking start-up investors to invest massively in the crypto industry, however, such investments are adventurous, and they should be treated accordingly. An investor who decided to enter this market should understand that he seems like sitting on a powder keg. Nevertheless, for those who are not afraid to take risks, we have developed an approximate decision-making scheme for cryptocurrency investment.

Fig. 3. Decision-making scheme for cryptocurrency investment. Source: authors' elaboration.

Figure 3 reflects following stages:

1. **Be aware = Stay inform**. In the absence of sufficient training, knowledge and intuition, there is a very high risk of losing your investments. The rate of cryptocurrency is not provided and extremely volatility. It changes almost every hour. This is because its fall and increase depends on two factors only: buying and selling. The basic knowledge for successful investment is laid in the publications of N. Kondratieff, J. Schumpeter, A. Chizhevsky, and others.
2. **Data analysis** (safest & profitability). Study the graphs of cryptocurrency value in relation to USD. Based on this study result, one can conclude: buy or sell.
3. **Choose cryptocurrency** for investment with the highest reliability indicators, for example, with stable growth of profitability for the last quarter.
4. **Buy-selling decision**. The decision to sell could accept, for example, when the price graph has a trend for raising and looks like head-shoulders figure. Otherwise, we

could buy cryptocurrency when the price graph has a downward trend looks like a head-shoulders figure, for example.

5. **Result evaluating (profit or loss).** When you get enough profit, you can try to continue. But at the same time it is necessary to build on the knowledge gained. At a significant loss, it is recommended to stop investing in cryptocurrency.

4   Discuss and Conclusions

Cryptocurrency is a new word in money circulation. Its occurrence is conditioned by the needs of the time. Despite the fact that crypto money does not have a real expression, they can almost equally with traditional currency units participate in various operations on the market. In general, the cryptocurrency in its characteristics is largely similar to traditional money; however, it also has a number of fundamental differences that allow digital money to gain popularity in the modern information space. At present it is difficult to predict the further fate of Cryptocurrency, including Bitcoin. As point Monyakova & Tsoy [7], Warren Buffett shared his opinion about the future of Bitcoin: "You can not evaluate bitcoin, since it is not an asset-based. This is a real bubble".

As a result of this study we have conducted following analysis:

Ways of creating and obtaining; Pros and cons of cryptocurrencies introduction into everyday use; Top ten best cryptocurrencies for mining and buying, steadily growing in the price on cryptocurrency markets in the last 3 months 2018; Top ten most reliable cryptocurrency with stable growth of profitability in the last year. Based on these analysis we have reached a conclusion that cryptocurrency is very volatility. For those who are not afraid to take risks, we have developed an approximate decision-making scheme for cryptocurrency investment.

Mainly two scenarios might be considered for the future of cryptocurrencies. One option is a complete ban on cryptocurrency using. Another scenario is based on the legislative design of the crypto currency market. To successfully implement this path, it is necessary to create "working" laws aimed at creating a transparent market for cryptocurrency, and, most importantly, for controlling transactions in cryptocurrency. In addition, it is necessary to license activities with cryptocurrencies and maintain a corresponding register for monitoring. In modern conditions, "no country in the world has the necessary means, technologies and competencies for qualitative assessment and control over cryptocurrency transactions" [11].

The development of the digital economy will entail radical changes in the way of life of country citizens, not to mention the financial structures, in particular the banking sector, where the introduction of new technologies threatens to reduce the cost of banking services, and a partial waiver of commissions. Therefore, banks will have to focus on lending to specific projects of the real sector of the economy.

References

1. Sofronova, N.S.: Cryptocurrency - as an alternative means of payment. In: Actual Questions of Finance and Insurance of Russia at the Present Stage. Materials of the IV regional scientific and practical conference of university teachers, scientists, specialists, post-graduate students, students. The University of Minin, pp. 289–292 (2017)
2. Tyan, N.G., Zamotayev, E.V.: Crypto currency is a special kind of electronic money. In: Modern Conditions for Interaction of Science and Technology, a Collection of Articles on the Results of the International Scientific and Practical Conference, 3 parts, pp. 121–124 (2017)
3. Pronina, Y.O.: Innovative payment unit - crypto currency?! To be or not to be? In: Youth and Science: A Step to Success. Proceeding of the All-Russian Scientific Conference of Promising Developments of Young Scientists: 3 volumes, pp. 322–326 (2017)
4. Fatkhutdinova, T.F.: Cryptocurrency as an instrument of digital economy. In: Interdisciplinary Approach to the Study of the Economy. Proceeding of the III International Scientific and Practical Conference, Dedicated to the 70th Anniversary of the Department of General Economic Theory of Bashkir State University, 3 parts, pp. 434–442 (2017)
5. Mark, H.: Observations on digital currency industry (2011). http://papers.ssrn.com/sol3/papers.cfm?abstract_id=1721076
6. Dwyer, G.P.: The economics of Bitcoin and similar private digital currencies. J. Financ. Stab. **333**, 1–11 (2014)
7. Levitin, A.J.: Pandora's digital box: the promise and perils of digital wallets. Univ. Pa. Law Rev. **166**(2), 305–376 (2018)
8. Gikay, A.A.: Regulating decentralized cryptocurrencies under payment services law: lessons from European Union Law. J. Law Technol. Internet **9**, 1–36 (2018)
9. Maximova, A., Okhotnik, A., Kolokolnikova, A.: Cryptocurrency as a new payment instrument. Bull. Mod. Stud. **11-1**(14), 233–235 (2017)
10. https://calc.ru. Accessed 18 Mar 2018, 21 July 2018
11. Monyakova, A.S., Tsoy, I.D.: Cryptocurrency as an object of speculative operations on a virtual exchange. In: The Development of Science and Technology: A Mechanism for Selecting and Implementing Priorities. Proceeding of the International Scientific and Practical Conference, pp. 14–17 (2017)

Fuzzy Logic Approach for Evaluation of Sovereign Wealth Funds' Management

Svetlana Tsvirko[(✉)]

Financial University Under the Government of the Russian Federation,
Leningradskiy prospekt 49, Moscow, Russian Federation
s_ts@mail.ru

Abstract. The article reveals the features of a sovereign fund as a management object. Different types of sovereign wealth funds and their aims and tasks are discussed. The results of the investment of the Reserve Fund and the National Wealth Fund of Russia are analyzed. The conclusion about the need to improve management of Russia's sovereign funds is made. It is stressed that the sovereign funds are unique and the management of them is characterized by uncertainty of the non-statistical nature, so the usage of classical methods of probability and statistics for evaluation is difficult. The possibilities of models based on mathematical fuzzy modeling and fuzzy logic are revealed. There is an opportunity of managing sovereign funds on the basis of fuzzy logic, that allows to combine the subjective preferences of the decision maker with quantitative estimates, including approximate, to ensure prompt and effective solutions.

Keywords: Sovereign wealth fund · Management · Investment
Profitability · Liquidity · Evaluation · Fuzzy logic · Fuzzy sets

1 Introduction

The study of issues related to the evaluation of sovereign funds' management is of current interest. In recent years, sovereign funds of many countries have accumulated significant reserves (\$ 7.862 trillion as of April 2018) and strengthened their influence on the global financial market [1].

When managing sovereign funds, the task is the increase of the profitability of investments and stimulation of the growth in the national economy. As for the global practice, as a rule, sovereign funds conduct a fairly active investment policy. They invest in less liquid and more risky assets, providing higher returns than typical gold and foreign exchange reserves of the Central Banks.

For effective management of sovereign funds a system for evaluation of the investment is needed. Characterizing the activities of managing a sovereign fund, Russian financial authorities, as a rule, indicate only the profitability of the fund for a certain period, and in some cases even limit information to the absolute values of the income received. It is obvious that this is not enough. In modern conditions, the ratio of "profitability/risk" is at the center of all investment decisions, as well as assessment of them. Another important criterion in evaluating the management of sovereign funds should be the minimization of the costs of managing funds.

© Springer Nature Switzerland AG 2019
T. Antipova and A. Rocha (Eds.): DSIC 2018, AISC 850, pp. 323–330, 2019.
https://doi.org/10.1007/978-3-030-02351-5_37

To date domestic and foreign practice has accumulated a lot of experience in assessing profitability and risk in their various aspects. However, it seems that there is no holistic approach to an integrated assessment of sovereign fund management, which is really necessary to solve the problems in the relevant sphere.

There is significant amount of publications devoted to the possible directions of investments of sovereign wealth funds – both in western economic literature and in Russia. Among them we can name publications by such groups of authors, as Al-Hassan, Papaioannou, Skancke and Sung [2], Bodie and Briere [3], Brown, Papaioannou, Petrova [4], Hentov and Petrov [5], Kunzel, Lu, Petrova, Pihlman [6], Scherer [7], researchers from Preqin company [8], Kazakevich [9], Tsvirko [10]. But there is a lack of publications devoted to the multifactorial decisions in the sphere of sovereign wealth funds' management under different conditions on the global financial market.

Sovereign funds have a high degree of uniqueness and specificity, their management is characterized by uncertainty of non-statistical nature, therefore the usage of methods of classical probability and statistics is difficult.

The situation is complicated by a high degree of uncertainty in the world economy, which makes it impossible to accurately assess the state of the economic entity.

In this context the aim of this study is to analyze possibilities of application of fuzzy logic for evaluation of sovereign wealth funds management. For achieving the results of the research, it is necessary to reveal the essence of sovereign wealth funds, their classifications, different aims and tasks, clarify the problems of evaluation of management of sovereign wealth funds and to show possible approaches to the applications of fuzzy logic in this sphere.

2 Essence of Sovereign Wealth Funds, Their Tasks and Management

A sovereign fund is a state investment fund with assets that include various financial instruments, for example, shares, bonds, as well as property, precious metals, etc. According to the functional approach, a sovereign fund is defined as a separate fund established by the state and in direct state ownership, the final beneficiary of which is the population of the country as a whole.

International practice shows that sovereign funds pursue an active investment policy, invest in less liquid and more risky assets, which provides higher yields compared to typical investments of central banks. The features of sovereign funds are as follows: long-term investments, tolerance to short-term fluctuations in income, limitation of speculative transactions.

Studying foreign experience in managing sovereign funds has made it possible to identify the main goals of their creation, types of sovereign funds and the specific features of investment strategies. In accordance with the approach of the IMF, there are five types of sovereign funds:

(1) stabilization funds,
(2) savings funds (funds of future generations),

(3) reserve funds (pension reserve funds),
(4) state investment corporations,
(5) development funds [11].

Stabilization funds invest their assets conservatively - mainly in high-quality sovereign obligations. Sovereign fixed income securities account for 69% of investments of stabilization funds; other debt liabilities account for 22% of investments. The objects of investment are highly liquid sovereign securities with low profitability. Stabilization funds need assets that have negative correlation with the source of risk for budget revenues. The directions of investment are explained by the need for a countercyclical fiscal policy. In this case short and medium-term investment horizons and high liquidity requirements for stabilization funds are similar to the investments of official reserves.

Savings funds are designed to accumulate wealth from one generation to another; transforming non-renewable resources into a diversified portfolio of financial assets. Sometimes these funds are called investment funds, future generations' funds, funds with constant income, etc. When investing, they are guided by high income and a long investment period. Requirements for liquidity of assets in the case of savings funds are lower than those of stabilization funds. In the structure of the assets of savings funds shares prevail (55%), significantly exceeding the share of government securities (21%).

Reserve funds (pension reserve funds) are designed to ensure the fulfillment of the state's pension obligations. In the structure of the assets of such funds shares prevail (39%), a significant portion of assets is invested is in other financial instruments (33%). The investments of pension reserve funds are generally long-term; it depends on the expected time of future payments on pension liabilities, in some cases investments are made for decades.

State investment corporations seek to obtain greater returns than official reserves run by the central bank. Funds are placed primarily in shares (66%). The investment horizon is longer than that of the central bank reserves.

Development funds are established to place resources in priority socio-economic projects, usually infrastructure ones. Long-term investment is typical for development funds.

Obviously, criteria for the effectiveness of sovereign fund management should be applied taking into account the type of sovereign fund and the tasks assigned to it.

Managers of official reserves traditionally prioritize investment in the following ways:

(1) safety of investments,
(2) liquidity,
(3) profitability.

Thus, sovereign fund management refers to tasks with several objective functions.

3   Management of Sovereign Wealth Funds in the Russian Federation

In the Russian Federation the first sovereign fund (Stabilization Fund) was established in 2004. It was organized to ensure the balance of the federal budget when the oil prices fall. In February 2008 the Stabilization Fund was separated into the Reserve Fund and the National Wealth Fund. Management of the Reserve Fund aimed at capital preservation and stable level of return in long-term perspective. The National Wealth Fund is dedicated to support pension system of the Russian Federation to guarantee long-term sound functioning of the system.

Creation of the Reserve Fund and the National Wealth Fund made it possible to pass the period of the world financial crisis (in 2008–2009) without reducing the standard of living of citizens, financial resources from these funds supported the financial infrastructure and the real economy. Under the conditions of aggravation of the geopolitical situation and restrictions on access to the international capital market, the Reserve Fund was actively spent and by the end of 2017 it was exhausted. As of July 1, 2018, the total amount of the National Welfare Fund is equivalent to $ 77.11 billion, that is 5% of GDP [12].

For several years both sovereign funds of Russia were managed extremely conservatively, mainly by investing in foreign currency and in low-risk or risk-less but low-yielding debt obligations of foreign countries. The revenues received from the investments of the Reserve Fund and the National Wealth Fund in 2017 amounted to 51.49 billion rubles, including revenues from the investment of the Reserve Fund funds - 0.65 billion rubles and from the investment from the National Wealth Fund - 50.84 billion rubles. The aggregate profitability of the placement of the Reserve Fund's funds on foreign currency accounts with the Bank of Russia amounted to (−)0.10% per annum in 2017, since the foundation (January 30, 2008) - 1.33% per annum; in rubles – 13.49%. As for results of investments in particular currencies, the profitability was as follows:

– on accounts in US dollars - 0.57% per annum (1.01% per annum since the foundation);
– on accounts in euro - (−)0.76% per annum (1.29% per annum from the creation of the fund);
– on accounts in pounds sterling - (−)0.16% per annum (2.91% per annum since the foundation).

The yield of investing assets of the National Wealth Fund for 2017 was as follows:

(1) on accounts in foreign currency with the Bank of Russia - (−)0.10% per annum;
(2) on deposits with Vnesheconombank: (a) deposits in Russian rubles - 6.44% per annum; (b) deposits in US dollars - 0.25% per annum;
(3) on deposits with VTB Bank and Bank GPB in order to finance self-supporting infrastructure projects - 8.27% per annum (10.39% per annum from the beginning of deposit operations) [12].

Thus, the return on investment of sovereign funds in Russia remains extremely low.

In the current situation new tasks arise for sovereign funds. Some of the resources of the National Wealth Fund can be used to finance infrastructure projects and other priority projects for the Russian Federation. In 2019 the Government of the Russian Federation plans to create a special fund for investments in the infrastructure.

During the global financial crisis sovereign funds in Russia, similar to sovereign funds from other emerging markets, changed their investment strategy, reorienting to domestic markets. The inflow of foreign capital into the markets of the countries that own sovereign funds was reduced, so the state funds were forced to replace foreign investors. Thanks to such sovereign funds the governments provided support to banks, the real sector and the national currency, alleviating the effects of the crisis.

It should also be taken into account that in the context of increasing volatility of exchange rates, investments in foreign financial instruments issued in foreign currencies subject sovereign funds to significant currency risks.

Thus, under the conditions of the crisis, both the volume of sovereign funds and their optimal structure (in terms of markets, instruments, currency, time structure, etc.) change. Nowadays sovereign wealth funds function in low-yield environment, that leads to lower profitability of the assets. Analysis shows that, in general, the majority of sovereign funds operate under uncertainty and lack of instruments that are suitable for placing funds on the basis of safety, liquidity and profitability criteria.

4   Application of Fuzzy Logic in Evaluation of Sovereign Wealth Funds' Management

There are difficulties with setting benchmarks and evaluation of sovereign funds. Under such conditions interesting possibilities are provided by the usage of models based on mathematical fuzzy modeling and fuzzy logic. The apparatus of the theory of fuzzy sets, proposed by L. Zadeh, is suitable for taking into account various kinds of uncertainties, the qualitative nature of requirements and estimates [13].

Fuzzy Logic (or Fuzzy Sets theory) is a relatively new approach applied since the 1960s years to the description of business processes in which there is uncertainty, making it difficult and even precluding the usage of exact quantitative methods and approaches. A distinctive feature of this method is the introduction of linguistic variables (subjective categories), that is, variables which values are words or sentences of natural language.

Speaking of fuzzy logic, most often researchers refer to systems of fuzzy inference. The task of fuzzy inference is to define a clear value for the output variable expressed in linguistic units.

The main stages of fuzzy inference are:

(1) formation of the base of rules for the system of fuzzy inference;
(2) fuzzification of input parameters;
(3) aggregation;
(4) activation of subconditions in fuzzy production rules;
(5) defuzzification.

Let's imagine the construction of a fuzzy model that allows us to evaluate the management of sovereign wealth fund. It is necessary to introduce the linguistic variable "effectiveness of sovereign fund management". A universal set for the introduced linguistic variable is the interval [0; 1]. As the set of values of the variable, we can consider the following linguistic variables (term sets) {low, medium, high}. Suppose that the expert specified the following values of the parameters: (0; 0; 0.25); (0.15; 0.525; 0.9); (0.85; 1; 1).

As input variables, it is advisable to consider: (1) the revenues derived from the investment of a sovereign fund; (2) costs of the sovereign fund (losses from investing funds, as well as costs incurred in administering the management of a sovereign fund).

As linguistic variables for "Revenues" and "Costs", you can use the words {low, medium, high}.

Thus, this model will consist of three parameters (two input variables and one output). This model has a MISO-structure (Multiple Input - Single Output).

The process of forming the base of rules of fuzzy inference is a formal representation of the empirical knowledge of an expert in a particular problem area.

For example, suppose that the expert formulated logical rules that are expressed in the form of pairs of parcels and conclusions such as "IF..., THEN...".

The system of fuzzy output will contain 9 rules:

1. If the Revenues are low and the Costs are low, then the management effectiveness of the sovereign fund is average;
2. If the Revenues are low and the Costs are average, then the management effectiveness of the sovereign fund is low;
3. If the Revenues are low and the Costs are high, then the management effectiveness of the sovereign fund is low;
4. If the Revenues are average and the Costs are low, then the management effectiveness of the sovereign fund is average;
5. If the Revenues are average and the Costs are average, then the management efficiency of the sovereign fund is average;
6. If the Revenues are average and the Costs are high, then the management efficiency of the sovereign fund is low;
7. If the Revenues are high and the Costs are low, then the management efficiency of the sovereign fund is high;
8. If the Revenues are high and the Costs are high, then the management effectiveness of the sovereign fund is average;
9. If the Revenues are high and the Costs are average, then the management efficiency of the sovereign fund is high.

Fuzzification, or the introduction of fuzziness, is the process of finding the membership function of fuzzy sets on the basis of the usual initial data. At this stage a correspondence is established between the numerical value of the input variable of the fuzzy inference system and the value of the membership function of the corresponding linguistic variable.

Using the membership function, the translation of linguistic variables into a mathematical language is carried out. For example, for the linguistic term sets {low, medium, high}, the parameter values for the input variable "Revenues" can be set (0; 0;

2); (1,75; 5; 8); (7; 10; 15). For the variable "Costs" (0; 0; 0,2); (0,1; 0,35; 0,6); (0,5; 1; 1). Units of measurement are percentages.

When defining the form of membership functions associated with each variable, you can choose from several types, and there are 2 main groups of methods for constructing the membership function: direct and indirect. With the direct method, the expert directly sets the rules for determining the values of the membership function. Examples of direct methods are the direct assignment of the membership function by a table, graph or formula. The disadvantage of this group of methods is a high level of subjectivity.

In indirect methods the values of the membership function are chosen in such a way as to satisfy the pre-formulated conditions. The expert's information is only the initial information for further processing. To this group of methods one can include such methods of constructing membership functions as methods based on paired comparisons, usage of statistical data, based on rank estimates, etc.

To assess the management of a sovereign fund, it is advisable to choose the triangular form of the membership function, which is one of the most common. The triangular form of the function is used with a small amount of information.

The goal of the aggregation stage is to determine the degree of truth of each of the sub-conclusions for each of the rules of fuzzy inference systems. Further, this results in one fuzzy set that will be assigned to an output variable for each rule.

At the stage of defuzzification the results obtained at the previous stages of fuzzy output are converted to the usual quantitative value of the output variable.

The process of developing a fuzzy-logical system can be implemented in the MatLab environment using the Fuzzy Logic Toolbox and the interactive fuzzy module, which will allow obtaining concrete results and visualizing them.

In addition to evaluating the management of the sovereign fund in general, the fuzzy logic method can be used to solve other tasks, for example, modeling the discount rate, risk assessment or analysis of a separate investment project in the sovereign fund portfolio.

Thus, the management of sovereign funds based on fuzzy logic allows the researchers to combine the subjective preferences of the decision-maker with quantitative estimates, including approximate ones, to ensure the efficiency of decisions.

5   Conclusion

The main contribution of this paper is the suggestion of approach to evaluate the activities of sovereign funds. In this paper we have revealed the essence of sovereign wealth funds, their main aim and tasks. It was proved that profitability of sovereign wealth fund is not the only important indicator of fund's management. It was shown that there are different types of sovereign wealth funds, they have different preferences in terms of safety, liquidity and other factors. The problem of managing sovereign funds refers to optimization tasks with several objective functions.

We have analyzed Russia's practice in sovereign wealth funds' management and came to the conclusion that there are difficulties with proper evaluation of their management taking into consideration different external and internal factors.

The results show that usage of fuzzy logic approach contributes to the improvement of sovereign wealth funds' management. This approach provides an opportunity to analyze information about investments under different conditions on the global financial market, make predictions and choose rational solutions. The approach expands the apparatus of management decisions in the sphere of sovereign wealth funds' management under different scenarios.

Acknowledgement. The paper is dedicated to the 100th anniversary of the Financial University under the Government of the Russian Federation.

References

1. Sovereign Wealth Fund Institute. https://www.swfinstitute.org/sovereign-wealth-fund-rankings/
2. Al-Hassan, A., Papaioannou, M., Skancke, M., Sung, C.: Sovereign Wealth Funds: Aspects of Governance Structures and Investment Management. IMF Working Paper (2013). https://www.imf.org/external/pubs/ft/wp/2013/wp13231.pdf
3. Bodie, Z., Briere, M.: Sovereign Wealth and Risk Management—A New Framework for Optimal Asset Allocation of Sovereign Wealth (2013). http://research-center.amundi.com/page/Publications/Working-Paper/2013/Sovereign-Wealth-and-Risk-Management-A-New-Framework-for-Optimal-Asset-Allocation-of-Sovereign-Wealth?search=true
4. Brown, A., Papaioannou, M., Petrova, I.: Macroeconomic Linkages of the Strategic Asset Allocation of Commodity-Based Sovereign Wealth Funds (2010). http://www.imf.org/external/pubs/ft/wp/2010/wp1009.pdf
5. Hentov, E., Petrov, A.: How Do Sovereign Funds Invest (2018). https://www.ssga.com/investment-topics/asset-allocation/2018/how-do-sovereign-wealth-funds-invest.pdf
6. Kunzel, P., Lu, Y., Petrova, I., Pihlman, J.: Investment Objectives of Sovereign Wealth Funds—A Shifting Paradigm (2011). https://www.imf.org/external/pubs/ft/wp/2011/wp1119.pdf
7. Scherer, B.: Portfolio Choice for Oil-Based Sovereign Wealth Funds (2009). http://www.edhec-risk.com/edhec_publications/all_publications/RISKReview.2009-10-27.1539/attachments/EDHEC%20working%20paper%20Portfolio%20choice%20for%20oil-based.pdf
8. The 2018 Preqin Sovereign Wealth Fund Review (2018). http://docs.preqin.com/reports/The-2018-Preqin-Sovereign-Wealth-Fund-Review-Sample-Pages.pdf
9. Kazakevich, P.A.: Stabilization funds as a special category of participants on the securities market. Finan. Credit. 29(269), 38–45 (2007)
10. Tsvirko, S.E.: Modeling of sovereign funds' management. Bull. Volgogr. State Tech. Univ. 1(180), 147–155 (2016)
11. Global Financial Stability Report. The Quest for Lasting Stability. International Monetary Fund. April 2012 (2012). http://www.imf.org/external/pubs/ft/gfsr/2012/01/pdf/text.pdf
12. Ministry of Finance of the Russian Federation. https://www.minfin.ru/ru/document/?id_4=122130
13. Zadeh, L.A.: Fuzzy sets. Inf. Control **8**, 338–353 (1965)

Value Formation of Innovative Product: From Idea to Commercialization

Lyudmila Popova, Irina Maslova, Irina Korostelkina$^{(\boxtimes)}$,
Elena Dedkova, and Boris Maslov

Orel State University, Naugorskoe Highway, 40, 302020 Orel,
Russian Federation
{LVP_134, cakyra_04}@mail.ru

Abstract. In the article the conceptual framework of «value» and «innovations» is explored and the theoretical basis of the value approach is revealed at the beginning of the article. The definition of an innovative product is given and the development process of the very innovative product and the mechanism of its value formation at each development phase are revealed. The expenses for the calculation items and the development phases of the innovative product are estimated. The value-added elements are specified, from the idea generation to the commercialization of the innovative product. The problematics for the further research of value formation of innovative products depending on their specific nature is put.

Keywords: Innovative product · Value · Innovation · Expenses
Innovative production

1 Introduction

Technological development and digital technologies play a significant role in the innovative economy against the background of a competitive environment, the actions of the sanctions policy and external challenges. Domestic business entities form an innovative pathway of their activities, modernize the production and invest in the development of innovative products to enhance an entrepreneurial activity, expand market positions, and conquer new markets. In modern market conditions of economic globalization it is insufficient to just develop an innovative product, it is necessary to provide a well-established mechanism for its commercialization, updating the product value to market environment.

The analysis of the current activity of innovative enterprises allows us to talk about the problems of the value formation of an innovative product, measurement and analysis of its elements. The methodological analysis toolkit and value estimation of an innovative product are separately put into practice, not related to current analytical activities, which often leads to an increase in the cost of the analysis process, excessive use of the potential resources, untimely data acquisition and so on.

The mechanism for the value formation of an innovative product is partially examined in the works of foreign and domestic authors but a comprehensive model of value formation and its accounting and analytical support is not presented in the

© Springer Nature Switzerland AG 2019
T. Antipova and A. Rocha (Eds.): DSIC 2018, AISC 850, pp. 331–338, 2019.
https://doi.org/10.1007/978-3-030-02351-5_38

studies. This has contributed to the choice of the subject of research. The main goal of this article is to develop theoretical foundations, elaborate conceptual provisions and create a mechanism for the value formation of an innovative product.

Traditional methods of scientific research and specific techniques were used in this article: structural-logical analysis, comparative-economic analysis, system-historical analysis, graphic scientific classification, modeling, and decomposition.

The article suggests a specific mechanism for creating an innovative product and a model for the value formation at each stage of its creation: from idea to commercialization.

2 Theory

The formation of the scientific economic concept of value began many centuries ago, but this issue still proves its relevance. Thus, according to the labour theory of Smith and Ricardo [9], the concept of «value» means that the human labour is the creator of value.

Marx [5] continued the research of A. Smith within the framework of the labour theory of value and suggested that surplus value can be created only by means of labour. Therewith, Marx identifies the cost of labour with the value created in the process of its consumption, speaking about the difference that exists between them due to the process of use.

Russian and foreign authors currently continue dealing with a categorical framework regarding value. Maslov considers the concept of «value» in the framework of such categories as cost price and price, while accounting includes such concepts as «manufacturing cost», «cost of sales» and «sales value» (selling price), etc. [7]. The category of «value» can be represented as a multi-component category, expressing the industrial relationships arising from internal and external continuous interaction of cost elements.

At enterprises of different economic sectors the movement of resources leads to the gradual value creation as a result of the transformation of raw materials and materials of one value into a finished product, work or service more valuable to the client. The newly created value will be increased unevenly. The process of value formation is accompanied by the process of worth generation, which can be understood as a product or a service. The main form of value distribution between economic sectors is the budget, which fulfills its key function redistributing the received value in the economic sectors. At the stage of redistribution the value is considered as the sum of goods (works, services) that were sold during the reporting period [10]. The distribution of value passes through two interrelated stages. At the first stage, the financial resources available to the enterprise are distributed and the tax part of the budget is formed.

Regardless of how the process of consuming value is carried out, new products are continuously created in the production process, which can act as analogous to previously produced goods (works, services), or they can be a new, more advanced product. In a modern high-tech economy, innovations play a huge role in ensuring the competitiveness of enterprises.

The concept of innovation is related to universal categories - extremely broad and structurally complex, with many approaches to disclosing its content. In the literature there are a significant number of definitions of «innovation». The concept of «innovation» first appeared in scientific studies of the XIX century. The Theory of Economic Development by the Austrian economist Schumpeter [15] put new life into the concept of "innovation" in the early twentieth century. J. Schumpeter considered innovation (clumps of *reality*, in which new *combinations changes* in development are carried out) as a change, conceiving that the main function of innovative activity is the function of change management [2].

Innovation as a product is considered, as a rule, in the narrower sense of the word, within the subject, segmented areas. Analysis of domestic and foreign literature shows that in practice the term «innovation» is often used in relation to any sphere of human activity. Drucker [1], Porter [12] and other scientists have proved the role of innovations in securing the economic growth and the progressive development of the economy. They have formed the conceptual and categorical framework of the theory of innovations and have described the mechanism for the formation of innovations.

In the modern Russian economy innovation (novelty) is seen as the final outcome of innovative activity, which has been implemented as a new or improved product on the market, or a new or improved technological process used in practice. Kleiner [4], Kharin [14], and Fatkhutdinov [13] have made a significant contribution to the study of innovation, innovation systems and innovative activity, as well as factors that directly affect the effectiveness of innovative development.

Maslova [6], Popova [11], Vasilyeva [3] and others have paid considerable attention to the issues of accounting, estimation and analysis of value, value elements of the innovative activity and the performance evaluation of production and implementation of an innovative product.

For the purposes of this study the value of an innovative product is viewed as a complex accounting and economic category that has an assessment characteristic and combines production processes occurring in the society between the subjects of value relationships in the context of the required labour costs for the production and sale of an innovative product, the components of which do not exist independently, but constantly interact with each other through the market.

3   Main Findings of the Study

The process of developing an innovative product is carried out through the implementation of successive stages from the idea generation to profit making. The generation of the idea of developing an innovative product presupposes the inception of a thought of innovation, its necessity and prospects for implementation. Then the process of development of an innovative product begins. A clear construction of the information system structure of an innovative project on the innovative product development greatly simplifies the estimation and planning of costs in the context of the growing complexity of the production, transport and distribution systems of the economic entity. Creation and implementation of an innovative product includes research, scientific and technical, organizational, financial, investment, production and marketing activities.

The most important component of the introduction of innovation into the real economy is to bridge the gap from a fundamental scientific idea to a commercially attractive product. An important role in solving this problem is played by the well-formed feasibility for the development and manufacturing application of an innovative product, the key part of which is the determination of the economic targets of the project and the calculation of the cost of its production. The innovative process accumulates expenses from the time of the generation of a new idea until the time of its commercialization. It is important to properly assess and systematize the expenses for innovative activity, which will enable to manage them and take into account the factors of their minimization [1].

The value formation of an innovative product is characterized by its specific features due to the fact that innovative products are new; they are often science-intensive, technically complex, involving significant material, time and intellectual expenses.

The formation of expenses for calculation items is carried out at different stages of the innovative product development. At the *stage of the pre-project analysis*, general business expenses and other manufacturing expenses are formed. *Stages of the innovative product development* include the following cost items: material resources; recyclable waste; purchased products, semi-finished products and productive services of third-party organizations; fuel and energy for technological purposes; wages of production workers; insurance premiums; production start-up and development costs; maintenance and equipment operation costs; general business expenses; general production costs; waste losses; other production costs. *The stage of an innovative product launch* includes expenses for material resources, general business expenses, other production costs and commercial expenses.

The results of innovation activity can be broadly divided into two parts. The first part includes the material results of innovation activity, which can be expressed in the form of movable and immovable property, improved products having a concrete material form, created, mastered, modernized or modified machines, equipment, devices, machinery, tools, instruments, etc. All material components of innovation activity have a concrete reflection in business accounting; standard methods of valuation and accounting, which do not cause much controversy and questions, have been developed for them [8].

The second, but not less significant part of the results of innovation activity is represented by intangible outcomes. They mean exclusively intangible assets of the enterprise.

When the cost of an innovative product is formed, a significant proportion of the total costs is spent on research and development, research expenses of a marketing nature, expenses for the technical re-equipment of production, and so on. The complexity of the market value formation of an innovative product is connected with the fact that it is associated with the task of consolidating in a certain market place and getting its circle of long-term customers, as well as the difficulties in determining the perspective dynamics of the market of the innovative product and accumulating objective analytical information on the new market.

Effective management of innovative activity requires that decision-makers have a clear understanding of the process of value formation of an innovative product unit, which can be obtained through analysis and evaluation of accounting data on the formation of expenses for its development. An efficient analysis of the value of an innovative product is based on reliable accounting information.

The production of an innovative product provides for a wide range of different processes and accounting operations, which requires the formation of an optimal accounting and analysis space within each innovation project. The development of an innovative product is recommended to be carried out within the framework of an innovative project that provides a list of activities to achieve the set goals through the implementation of the resource potential in the relevant activity, with the necessary information and analytical support (Fig. 1) for effective implementation of the investment process.

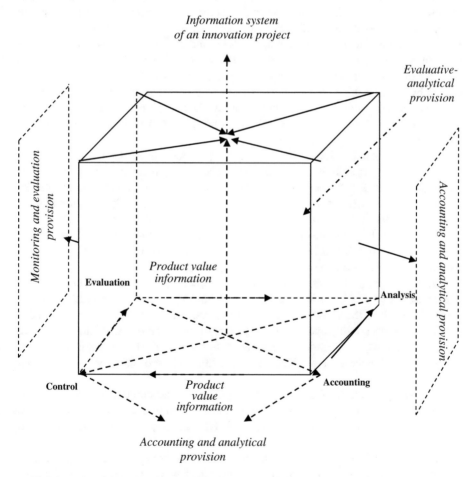

Fig. 1. Information-analytical provision of information system of an innovation project

The information system of the innovation project is based on information that is formed within the framework of such concepts as «evaluation», «accounting», «analysis», «control».

Herewith, analysis and accounting are organised under the action of accounting and analytical provision, and control over them is carried out within the framework of accounting and control provision. The advantage of this information system is its structuredness. Within the framework of accounting and analytical provision, information on each element of an innovative product is formed on the basis of information on the innovative product contained in primary documents, accounting registers, as well as on the basis of all types of reporting, which allows to evaluate the process of formation of each cost element. Thus, accounting is an information base for the analysis of an innovative product and allows to identify areas of greatest risk, bottlenecks in the activities of innovative enterprises.

The interaction between accounting and control forms accounting and control provision, in which information about the innovative product is carefully checked, compared with the planned indicators, etc. The evaluation of the information system of the innovation project includes such indicators as material expenses and income, on the basis of which the efficiency and profitability indicators are calculated. At the same time, the obtained values will be substantially adjusted taking into account the schemes chosen by the enterprise for the development and commercialization of the innovative product. Thus, depending on the terms of supply, the parameters of the storage systems and the selected channels for the distribution of innovative products, the cost, volume and time of operations will change.

Evaluation and control of the innovative product ensure the transparency and complexity of the operation of the enterprise. The management and control process of the innovative product formation is necessary in the conditions of ensuring the effective movement of the material resources of the economic entity. In these conditions, the process of planning expenditure items of financial resources for reimbursement of expenses, as well as organization of a process for attracting additional sources of funding and a control receipt of money compensation for the products sold to the participants of the information process, are particularly important.

When analyzing the value of innovative products, for the assessment of the effectiveness of spending it is necessary to preliminarily specify an indicator for each goal - the result of the work - which will determine whether the goal was achieved or not. Thus, the accounting and analytical system of the enterprise will allow to quickly track the intra-project dependence of the level of costs on the results achieved, as well as to track the process of product value formation at each stage of the innovation project.

The implementation of the target principle of cost accounting for the innovative product development is possible through the introduction of a separate production cost account, for example, account 22, «Innovation Production», which assumes all costs of innovation production.

In this case it is advisable to open sub-accounts separately for each type of innovative products, the development and production of which is carried out within the framework of a separate innovation project. Analytical accounting is proposed to be carried out at separate stages of the innovative product development, starting with a pre-project survey, which is necessary to analyze the feasibility of the innovative product development, then taking into account all the expenses at the stages of the innovative product development, and ending with the stage of launching the innovative product to the market. Within each of the stages, it is necessary to keep records on individual cost items.

Such model of accounting will allow to quickly and efficiently analyse the value formation of the product at each stage of its development and timely assess the efficiency of the production process and the rational use of resources.

In the conditions of the rapid development of the economy and the unceasing pace of innovative development, innovation-oriented enterprises are required to constantly maintain the efficiency of their activity through an accounting and analytical system. To monitor the value formation of an innovative product it is advisable to carry out cost accounting and analysis for each stage of the innovation project. In this regard, further work is required on creating an integrated mechanism for accounting, analysis and control of the value elements at each stage of the innovative product development that includes modern management technologies and analytical tools, allowing to evaluate the cost of the innovative product both in terms of producer's interests (cost analysis) and from the point of view of consumer interests (analysis of utility and functionality).

Thus, the development of an innovative product, the process of its value formation from the idea generation to commercialization in an accounting and analytical system will allow us to quickly evaluate the cost of each stage in the total cost of an innovative project and determine the correlation of the value and the significance of each stage. The analysis and control of the cost of the innovative product value will allow us to identify unnecessary expenses, find ways to minimize costs, improve production and management processes, adequately determine the resource potential, thereby reducing the value of the innovative product.

4 Conclusion

The article presents the author's vision of the categorical framework, in particular, the concepts of «value», «innovation», «innovative product value». Innovative product is an accounting unit for the purposes of accounting and analysis. The value of an innovative product is defined as an integral category for assessing the performance of an innovative enterprise in the framework of an accounting and analytical system. A mechanism for the value formation of an innovative product is presented, reflecting its structural elements and helping to determine the motion vector of value flows within the framework of innovation production activities. The necessity of the development of an integrated mechanism for accounting, analysis and control of the value elements of an innovative product, which is necessary for the development and adoption of tactical and strategic managerial decisions concerning the production and implementation of innovative products through the stages of the innovation life cycle, is grounded for future research.

Acknowledgement. The article was prepared in the course of carrying out research work within the framework of the project part of the state task in the field of scientific activity in accordance with the task No. 26.2758.2017/PCh (26.2758.2017/4.6) for 2017-2019 on the topic «System for the formation and distribution analysis of the value of innovative products based on the infrastructure concept».

References

1. Drucker, P.F.: Management Challenges for the 21st Century: (translated from English). Publishing House, Moscow «Williams» (2003)
2. Ilyenkov, S.D., Gokhberg, L.M., Yagudin, S.Yu., Kuznetsov, V.I., Bandurin, A.V., Ilyenkova, N.D., Pudich, V.S., Smirnov, S.A.: Innovative management: a textbook for universities. Edited by S.D. Ilyenkova. Banks and Stock Exchanges, Moscow «UNITI» (1997)
3. Korostelkina, I.A., Vasilyeva, M.V.: The system of internal control over the value of an innovative product. Management Accounting, no. 6, pp. 72–78 (2017)
4. Kleiner, G.B.: Production Functions, 239 p. Finance and Statistics, Moscow (1986)
5. Marx, K.: Capital, vol. 1–3. Essays, vol. 23–25
6. Maslova, I.A., Korostelkina, I.A., Dedkova, E.G.: Retrospective study of the category «value». Pridneprovsky Scientific Herald, no. 4, pp. 73–75 (2017)
7. Maslov, B.G.: Construction of the model of management accounting of value formation in the conditions of integrated data processing. Economic and Humanitarian Sciences, no. 6, pp. 28–34 (2009)
8. Nazarova, L.A., Elokhova, I.V.: A conceptual approach to the definition of the concept of «non-material result of the enterprise's innovative activity». Vestnik of the South Ural State University, no. 1, pp. 42–50 (2013)
9. Smith, A., Petty, W., Ricardo, D.: Anthology of economic classics. Econom-key, Moscow (1993)
10. Sorokina, M.S.: Correlation of the categories «price cost» and «cost» for the purposes of element-wise estimation of value added. Management Accounting, no. 10, pp. 71–76 (2013)
11. Popova, L.V., Korostelkina, I.A., Gudkov, A.A.: State regulation and budgetary support as a means of ensuring the effective functioning of the agricultural industry in Russia [Text]. Actual Problems of Economics, no. 10, pp. 111–120 (2014)
12. Porter, M.: Competitive strategy. The analysis of competitor industries, 454 p. Albina Publisher, Moscow (2011)
13. Fatkhutdinov, R.A.: Innovative management: a textbook for high schools, 6th edn., 448 p. Peter, St. Petersburg (2008)
14. Kharin, A.A., Rozhdestvensky, A.V., Kolensky, I.L.: Innovations. Part 1: Innovative Activity: Basic Concepts. RGUITP, Moscow (2009)
15. Schumpeter, J.A.: History of economic analysis. Origins, no. 1. Redkoll: V. Zhamin (Ed.), A. Baranov, J. Kuzminov and others. Economics, Moscow (1989)

Digital Media

Are Latin American YouTubers Influential?

Abel Suing[✉], Geovanna Salazar, and Carlos Ortiz

Department of Communication, Communication and Audiovisual Culture Group,
Universidad Técnica Particular de Loja, San Cayetano High s/n,
Champagnat Street, CP 11-01-608, Loja, Ecuador
{arsuing,gesalazar2,ccortiz}@utpl.edu.ec

Abstract. The rise and visibility of social networks has created a new scenario for the creation and mass diffusion of audiovisual products. In this environment, YouTube is positioned as the audiovisual network par excellence where thousands of Internet users express themselves through videos on various topics; this is how YouTubers arise, users that upload their own production videos and capture a representative number of subscribers and visualizations, even with important sponsorships. The aims of the research are: to determine the trends on YouTube of the five main YouTubers in Latin America and to carry out a comparative analysis with their accounts on Facebook and Twitter, to establish if there is similarity and approximation, or if their rise and popularity only have effect on YouTube. The methodology that was used is quantitative based on the data and indexes provided by the SocialBlade tool for YouTube and Fan Page Karma tool for Facebook and Twitter. It is concluded that YouTube is positioned as the ideal platform for the diffusion of audiovisual content capturing millions of visualizations in themes focused on entertainment, beauty, comedy, horror; the index of followers on Facebook and Twitter accounts of the Latin American YouTubers varies substantially compared to YouTube.

Keywords: Internet · Social networks · Video · Interactivity
Digital communication

1 Introduction

Digital technology and the phenomenon of technological convergence have led to the integration of all traditional media, previously separated, into a global multimedia market that produces deep transformations in the media outlook, where the media will become one of the most important sectors of the Information Society, of great cultural and social influence [1]. The phenomenon of social television is framed within the concept of connected television, both in its technological aspect as well as in contents. Social TV is closely linked to the growth in importance and visibility of technology in today's society: mobile applications, social networks, and second screens [2]. The interaction between the audience and their participation in the generation of content will characterize the television offer that will be developed in the future; this is how Internet TV is born with a component of social participation and interactivity that probably will not be able to, or want to get rid of, among other factors, because it constitutes one of the main advantages of this mean of transmission compared to others [3].

© Springer Nature Switzerland AG 2019
T. Antipova and A. Rocha (Eds.): DSIC 2018, AISC 850, pp. 341–348, 2019.
https://doi.org/10.1007/978-3-030-02351-5_39

Frameworks such as transmedia narration or cross-media production set up a context from which to work. When the metrics are each time more social, some sectors point out that their products are the emotions of the users and the television programs begin to be valued because of the quality of the experience in social networks [4]. The dominant circuits of information and communication are marked by technological use and by social practices that could be synthesized into four main causes [5]: media communication, transnational media communication, social discourses and individual perception of communication and information.

The new services are divided between those associated with the content (information associated with the channel or programming guide) and the pure interactive ones. The pure interactive services had not been initially thought for television, although later they have been adapted in view of their possibilities [6]. It will appear a new relationship of intermediation between content providers and search engines, which will involve sharing the advantages of network distribution [7]. At the moment there is so much desire and creativity to administer a channel of YouTube as there were in the old times in the television, a time in which the limits were nonexistent. YouTube, like television at some time, lives the same adventure of defining the target audience and what the audience wants. [8]. YouTube video portal is a large store or support of millions of videos broadcasted by users, companies or media, which offer very different values. YouTube is positioned as a brand, with much power, of a new format to inform, a new communicative support, the one of online videos [9].

YouTube is currently a machine that reflects the tastes of the public, for better or for worse, in real time. Few television executives could have foreseen the madness that would inspire the simple act of unpacking, videos in which all users do is to open the toy packaging and play with them [10]. Content creators who have appeared on YouTube are called YouTubers, they have diverse backgrounds, create content with some frequency. The YouTubers have important hearings because of the way to convey their messages, and have become celebrities with the power to influence, "it is not only in its scope but also the phenomenon of YouTubers as opinion leaders (…) the influencer phenomenon is an intermediate step between brands and consumers that is attracting interest from marketers" [13].

The YouTubers feed on the digital environments in which young people, but above all have developed an emotional connection with them [14], where it matters most the relationship and identification with the messages that the recording quality of the same. In addition the YouTubers begin to assume a role of tutors and trainers for young people, with a colloquial and interspersed with jokes the YouTubes influencers act as teachers and guides to guide the processes of appropriation practical knowledge, like in a video game [15], are therefore an emerging area in the academic field and with ample possibilities of development in communication and education.

The objectives of the research are: (1) Determine the trends on YouTube of the five major YouTubers in Latin America (Hola Soy Germán, Yuya, Enchufe TV, Werevertumorro, DrossRotzank) with a cut on June 20, 2018; and, (2) Carry out a comparative analysis with the social networks Facebook and Twitter, establishing whether Youtubers accounts in these networks denote the same trend as in YouTube. The hypotheses are: (1) The YouTubers of Latin America achieve a denoted positioning in YouTube social network, which is considerably reduced in networks such as Facebook and Twitter. (2) The contents issued by YouTubers are focused on comedy and entertainment themes.

2 Methodology

The methodology that was used is quantitative. Through the tool SocialBlade [11] (analytical tool of YouTube) indexes and values of the profile are obtained, in addition to the activity carried out by the YouTubers, like subscribers, views, type of channel, increase in the last 30 days (in subscriptions and views) and future projections (in subscriptions and views), with a cut to June 20, 2018. For the comparative analysis with Facebook and Twitter, the Fan Page Karma tool [12] is used, a monitoring tool that allows obtaining graphs and data statistics on the main variables of these social networks with regarding to their activity.

3 Results

The YouTuber Hola soy Germán is consolidated as the most representative in Latin America, with almost 34 million subscribers and surpassing in visualizations 3 billions, despite not registering activity in the last year, but with a positioning that comes from long ago. His channel focused on entertainment videos increases in the last 30 days (with a cut to June 20) more than 100 thousand followers and exceeds 26 million views. In the space of a year, the channel Hola Soy Germán will reach approximately 36 million subscribers and almost 4 billion views (Table 1).

Table 1. Profile and trends of the Youtuber Hola Soy Germán

Name of the channel	Subscribers	Views	Type of channel	Increasing in the last 30 days		Future repercussions (1 year)	
				Subscribers	Views	Subscribers	Views
Hola Soy Germán	33.966.637	3.526.965.805	Entertainment	141.719	26.137.915	35.755.113	3.927.765.273

The YouTuber Yuya ranks second in influence, surpassing 21 million followers and achieving more than 2 billion views. This channel with a theme of fashion and beauty increases in the last 30 days just over 170 thousand followers and 15 million views, making it one of the channels with greater reception since its creation. In a year, with a rate of frequent posts, the channel of Yuya will increase to more than 24 million followers, significantly exceeding 2 billion views of the content exposed (Table 2).

Table 2. Profile and trends of the YouTuber Yuya

Name of the channel	Subscribers	Views	Type of channel	Increasing in the last 30 days		Future repercussions (1 year)	
				Subscribers	Views	Subscribers	Views
Yuya	21.499.816	2.247.251.536	Fashion and beauty	170.802	15.073.107	24.622.589	2.513.571.197

The Ecuadorian channel Enchufe TV with a high technical standard is positioned in third place in terms of its level of dominance in the YouTube audience, with more than 17 million views and approaching to the 6 billion views in their videos. With a comic style, Enchufe TV presents an increase in the last 30 days that exceeds 200 thousand followers and more than 101 million views. This channel would get approximately more than 20 million subscribers and 7 billion views in a year. The channel offers its audience content with a daily frequency or at least more than two audiovisual products a week, reflecting a social reality typical of the Ecuadorian idiosyncrasy (Table 3).

Table 3. Profile and trends of the YouTuber EnchufeTV

Name of the channel	Subscribers	Views	Type of channel	Increasing in the last 30 days		Future repercussions (1 year)	
				Subscribers	Views	Subscribers	Views
Enchufe TV	17.376.236	5.810.798.972	Comedy	215.992	101.961.844	20.549.903	7.438.684.974

The channel Werevertumorro, although it has a lower rank than the rest of YouTubers, shows a significant reach, with almost 16 million subscribers and more than 2 billion views in the content exposed. Its comic style with sketches, short films, videoblogs, series and varied content reaches in the last 30 days to increase more than 116 thousand followers and 24 million views, predicting that in one year Werevertumorro will reach 18 million subscribers and will be closer to the 3 billion views of the channel. The channel maintains a constant update with two or three videos a week, productions with which seeks to innovate, including conjunctural issues, travel and others (Table 4).

Table 4. Profile and trends of the YouTuber Werevertumorro

Name of the channel	Subscribers	Views	Type of channel	Increasing in the last 30 days		Future repercussions (1 year)	
				Subscribers	Views	Subscribers	Views
Werevertumorro	15.812.426	2.467.276.019	Comedy	116.692	24.726.379	18.130.882	2.794.717.654

Finally, the fifth most popular channel in Latin America, DrossRotzank, passes the 13 million subscribers with an average of almost 3 billion views to its content. Its format allows the last 30 days to register an increase of more than 200 thousand subscribers and a considerable index of approximately 53 million views. At the end of 12 months the channel will have exceeded 18 million subscribers and 3 billion views. Unlike the other YouTubers that presented themes focused on fashion, beauty or comics, DorssRotzard presents terrifying audiovisual content, framing common situations in a genre capable of attracting millions of followers (Table 5).

Table 5. Profile and trends of the YouTuber DrossRotzark

Name of the channel	Subscribers	Views	Type of channel	Increasing in the last 30 days		Future repercussions (1 year)	
				Subscribers	Views	Subscribers	Views
DrossRotzsrk	13.651.169	2.879.584.636	Horror	244.360	53.113.100	18.223.117	3.745.477.633

Now, the positioning of YouTubers on Facebook is modified compared to the videos platform. Werevertumorro is the profile which number of fans exceeds by far 22 millions, ranking it as the most popular within the social network, unlike the fourth place it maintains on YouTube. Germán Garmendia's account has almost 18 millions of followers, but it drops to a second place compared to his YouTube channel. Yuya is approaching 13 millions of followers, however, descends one place compared to her popularity with videos. The profile of Enchufe TV and El Diario de Dross present in terms of number of fans, more than 9 millions and just over 4 for El Diario de Dross. The first three channels carry out an important management of their Facebook accounts, so that they accumulate millions of followers on their profiles and show that they perform a parallel job with the two social networks (Fig. 1).

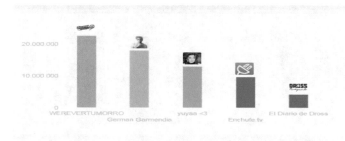

Fig. 1. Fans on Facebook of the Latin American Youtubers

The "engagement" shown by YouTubers on Facebook constitutes a reflection of the vinculation maintained with the audience. In this way, it is seen that el Diario de Dross, although it is positioned as the fifth most influential channel, is the first in terms of commitment, for its effective management, added value and fluidity within the profile that reaches 33%. Werevertumorro with 23% also shows a management and interaction with the users, in addition to a fluid dialogue. The profiles with the lowest percentage of engagement correspond to Enchufe TV, Yuya and Germán Garmendia, with 3.4%, 1% and 0.6% respectively. It is verifiable that the low engagement of these accounts does not affect their level of impact on the platform (Fig. 2).

Twitter presents a unique situation, the indexes in terms of followers are decreasing in reference to YouTube and Facebook. Although Germán Garmendia has more than 11 million, it does not compensate the more than 30 that has in the audiovisual

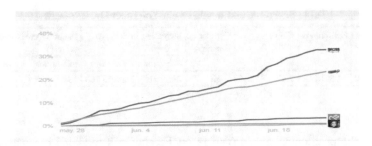

Fig. 2. Engagement on Facebook of the Latin American YouTubers

platform. Yuya presents half of followers that on YouTube with more than 10 millions, a pretty wide difference. Werevertumorro, like the previous channel, reaches basically to the half of the fans in the account with just over 8 millions. El diario de Dross and Enchufe TV present the greatest differences with more than 2 and 1 million followers each one, less than the half on YouTube. Although the five YouTubers have a presence on Twitter, it is far from achieving dimensions such as the ones of YouTube, perhaps due to the conditions of the network (Fig. 3).

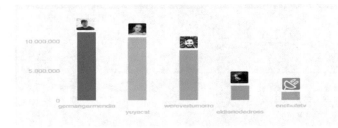

Fig. 3. Followers on Twitter of the Latin American YouTubers

The commitment index that YouTubers expose on Twitter is similar to the one of Facebook. Again, el Diario de Dross is presented as the channel with the best engagement, which represents a loyalty with the audience. Unlike this profile, the other YouTubers maintain a minimum percentage of commitment in a period of six months, without being greater changes in the course. Yuya gets 15%, Werevertumorro 12%, Germán Garmendia 7.5% and EnchufeTV with 1.7%. These values show again that, although the percentage of management, administration and other resources that encompasses the engagement is reduced both in Facebook as well as on Twitter, this does not affect the influence achieved by each YouTuber (Fig. 4).

Cristopher Medina, well-known YouTuber from Loja (Ecuador), highlights that the content uploaded to the platform responds to a script, although he also improvises. He points out that "the characteristic present in each YouTuber is the content that reflects the daily aspects of the people, a fluid and normal action in front of the camera that leads to an easy identification with them". The YouTuber also notes that now young

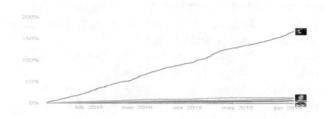

Fig. 4. Engagement on Twitter of the Latin American YouTubers

people no longer use Facebook as a first option, but instead they prefer to YouTube or Instagram. In his case, he recognizes that YouTube and Instagram have been the platforms that have benefited him the most.

On the challenges that a YouTuber can experience, Medina believes that "there would not be major inconveniences, it is enough to speak in front of a camera and be able to identify the themes that may have the greatest reception among the public. You need to have the desire, taste to produce and record", he says.

Ramón Salaverría, expert in digital media, states that what is happening with YouTube and other similar platforms is that they are becoming an alternative, a competition to traditional television; YouTube has managed to attract an audience profile that traditional TV channels are losing. He recognizes that on YouTube there is a tendency to amateurism towards the video that follows perfectly defined rules, generating an undergroud or alternative aesthetic that sometimes combined with a kind of graphic and sounding aesthetics quite strident, which is not common in television.

"There are a number of characteristics that can be identified as common denominators of the popular YouTubers, the self-confidence, the lack of traditional stiffness, the direct relationship with the public, the constant invitation to participate," he says. However, Salaverría sees that in networks there exist trends that are maintained for a time, begin to decline and are replaced by another. "I'm not saying that the current YouTubers will not remain in the future as referent people in the network, but they will surely evolve towards other ways of telling information, of sharing it".

4 Conclusions

- The Latin American YouTubers reflect an ascending growth on YouTube, with audiovisual content that captures millions of views by the audience, in themes focused on entertainment, beauty, horror and humor.
- From the analyzed channels, Hola soy Germán, is constituted as of greater acceptance, with a capturing of the audience, that goes beyond the frequency of his postings.
- All the channels will experience a significant growth in the next year, but it is DrossRotzark who will increase the greater number of subscribers in total, which marks an important evolution pattern, to be taken into account and to be followed up on. While some channels attract more subscribers than others, it is noteworthy

that YouTubers as EnchufeTV and DrossRotzark accumulate the largest number of followers in the last 30 days, which could result in that the current positions of YouTubers in Latin America could be modified.

- The YouTubers with the greatest reception (Hola Soy Germán, Yuya and EnchufeTV) show a low percentage of engagement on Facebook and Twitter. The number of followers on both networks responds solely to their positioning on YouTube.
- The hypotheses are confirmed. The Latin America YouTubers achieve a good positioning in the network, but this is reduced on Facebook and Twitter. The majority of contents focus on comedy and entertainment. Nowadays Latin America YouTubers consolidate as a social phenomenon, however, everything will depend on their ability to adapt to technological changes.

References

1. Hellín, P., Rojo, P., San Nicolás, C.: La televisión digital terrestre en Murcia. Informe técnico sobre la situación y portunidades de implantación de la televisión digital terrestre en la Comunidad Autónoma de la Región de Murcia. Comunicación Social Ediciones y Publicaciones (2009)
2. Gallego, F.: Twitter y televisión: nuevas audiencias, nuevos sistemas publicitarios, nuevos negocios. In: Francés, M., Galvalda, J., Llorca, G., Peris, A. (coord.) La Televisión de la crisis ante el abismo digital, 179, Gedisa Editorial (2014)
3. Huget, M.: La televisión 2.0: un marco favorable para el género documental. In: Francés, I. M., Galvalda, J., Llorca, G., Peris, A. (coord.) El documental en el entorno digital, p. 16, UOC Ediciones (2013)
4. Aguado, J.: La comunicación móvil. Hacia un nuevo ecosistema digital. Gedisa Editorial (2013)
5. Benítez, L.: Juventud y democracia. TELOS: Cuadernos de Comunicación e Innovación N° 89, 99 (2011)
6. Arrojo, M.: La configuración de la televisión interactiva. De las plataformas digitales a la TDT. Netbiblo Editorial (2008)
7. Álvarez, J.: La televisión etiquetada: nuevas audiencias, nuevos negocios. Fundación telefónica: Editorial Ariel, S.A. (2011)
8. Ciampa, R., Moore, T., Carucci, J.: Youtubers para dummies. Editorial Grupo Planeta (2016)
9. Marcé, B.: YouTube: Las claves para aprovechar todas sus potencialidades. Profit Editorial (2012)
10. E & N. El crecimiento explosivo de YouTube: Vale el doble que Netflix. http://www.estrategiaynegocios.net/lasclavesdeldia/916559-330/el-crecimiento-explosivo-de-youtube-vale-el-doble-que-netflix
11. SocialBlade (2018). https://socialblade.com/
12. Fan Page Karma (2018). www.fanpagekarma.com
13. Elorriaga, A., Monge, S.: La profesionalización de los YouTubers: el caso de Verdeliss y las marcas. Revista Latina de Comunicación Social 73, 37–54 (2018)
14. Meseguer, J.: El increíble éxito de los youtubers. IEEM Revista De Negocios 19(3), 88–91 (2016)
15. Gewerc, A., Fraga, F., Rodés, V.: Niños y adolescentes frente a la Competencia. Digital Entre el teléfono móvil, youtubers y videojuegos. Revista Interuniversitaria De Formación Del Profesorado 31(2), 171–186 (2017)

Digital Medicine, Pharma and Public Health

Dynamic Artificial Neural Networks as Basis for Medicine Revolution

Leonid N. Yasnitsky$^{(\boxtimes)}$, Andrew A. Dumler,
and Feodor M. Cherepanov

Department of Applied Mathematics and Informatics, Perm State University,
Bukirev Str. 15, 614600 Perm, Russia
yasn@psu.ru

Abstract. This article proposes a method of constructing dynamic neural network mathematical models that allow not only to diagnose the disease at the current time, but also to simulate the appearance and development of diseases in future periods of time, as well as to control their appearance and development by selecting the optimal lifestyle and optimal intake of drugs. It is assumed that the use of dynamic neural network medical systems, instead of static, allow doctors, before prescribing courses of treatment to patients, to test the effect of drugs not on patients, but on their virtual mathematical models. The action of the system is demonstrated by examples.

Keywords: Neural network · Diagnosis · Prediction · Heart attack
Recommendation · Lifestyle

1 Introduction

At the XII Russian National Congress of physicians the report of Yasnitsky [1] it was stated that modern medical science for its methodology behind the technical disciplines more than 100 years. Doctors practically do not use the method of mathematical modeling to the extent that engineers, physicists, mathematicians do. Instead of creating dynamic mathematical models of their patients and, through scenario prediction, using these models to select the optimal courses of treatment and prevention of diseases, doctors prescribe medicines to patients, and then observe the patient: "will help – will not help". If "does not help", doctors prescribe other drugs.

From the point of view of modern physics, such experimentation on living objects is an anachronism. Representatives of technical disciplines used experimentation on real objects as the only method of research, until the end of the XIX century. In the XX century there was a method of physical modeling, and then it was replaced by a more progressive method of mathematical modeling.

Now, in the XXI century, engineers, before making any decisions, as a rule, first create mathematical models of real objects and perform virtual computer experiments on mathematical models. In this way, engineers investigate the possible options for the operation of the simulated objects and select the most optimal conditions to ensure their long-term trouble-free operation.

It would seem that doctors should do the same.

T. Antipova and A. Rocha (Eds.): DSIC 2018, AISC 850, pp. 351–358, 2019.
https://doi.org/10.1007/978-3-030-02351-5_40

Indeed, in medicine there is currently an avalanche-like growth of publications devoted to the use of neural network modeling. Solid reviews are published, devoted to the list of successes, the analysis of opportunities and prospects of application of neural networks. For example, in [2] on the basis of 56 publications the experience of neural networks application for diagnostics of various diseases is analyzed. In the review works [3, 4] about 40 publications are analyzed, in [5] 23 publications are reviewed.

In some publications, for example, in [5–8], in addition to the term "diagnosis", the term "forecast" is used. However, a closer study of such publications usually reveals that the term "prognosis" is understood only in the narrow sense of the word-as the outcome of the disease: "the patient will survive or will not survive", "there will be complications, or there will be no complications", "heart attack will happen, or not", etc. In these models there is no time factor and therefore they are called static. Static models diagnose only at the time of application of the diagnostic system. Human diseases are not considered to be processes that develop over time. Therefore, using static models it is impossible to estimate how long and what will result in the future use of a drug. With the help of such models it is impossible to choose the best courses of prevention and treatment of diseases. As a result, doctors have to put not virtual, but real experiments – to assign their patients treatment courses and observe what the prescribed treatment will lead to, and if it does not help – try to use another drug.

Apparently, for the first time the possibilities of neural networks for modeling the development of diseases in time were shown in the early publications of the authors of this article [9, 10]. The results presented below are a continuation of these studies.

2 Dynamic Neural Network Modeling

2.1 Methodology

In the above-mentioned works [9, 10], the authors of this article proposed a method of ideological combining of the capabilities of two technologies of artificial intelligence: neural networks and expert systems. An original algorithm was developed that allows to correct the results of neural network scenario prognoses of coronary heart disease (CHD) development with the help of knowledge incorporated in the international SCORE scale.

The disadvantage of the proposed algorithm [9, 10] for creating dynamic neural network models was that this algorithm could not be applied to other diseases, because the SCORE scale was developed and verified only for CHD.

Another solution to this problem is seen in the application of the method of sliding Windows, but the use of this method is associated with high labor costs and also has a number of disadvantages.

This report presents the experience of another option for creating dynamic neural network models for scenario prediction of a wider range of diseases. The essence of this option is that we strive to use, if possible, a small number of uncorrelated input parameters, leaving many other input parameters "behind the scenes". For example, in the development of neural network system for diagnosis and prediction of cardiovascular diseases, we fundamentally refuse to use such important data for diagnosis, as the results of biochemical analysis, electrocardiography, coronary angiography, as well as many specific methods of disease verification. Instead, we introduce the maximum possible number of parameters characterizing the patient's body: gender, age, eye color, hair color, blood group, the presence of a transverse fold on the earlobe, place and time of birth, genetic parameters, heredity, etc. We inform the neural network about the environment in which the patient lives, how he or she eats, what lifestyle he or she leads, his or her brothers and sisters, his or her profession, bad habits, physical education and sports. We inform the neural network of information about previous diseases, information about the presence of diseases of blood relatives, and we report the minimum number of complaints, the patient. Sometimes, with the skillful selection of input parameters, this data is enough to train the neural network to diagnose some diseases very accurately. However, it should be borne in mind that this method can be successfully used only by researchers who have sufficient experience in the application of neural networks and are very well versed in the simulated subject area. Even in this case, the success of the method is not always guaranteed and depends on the characteristics of the simulated domain.

Taking into account the above, we have created a neural network diagnostic and prognostic system consisting of a complex of neural networks, each of which corresponded to different nosological forms of cardiovascular diseases and contained 27 input neurons to enter parameters characterizing the General information about the patient, history and lifestyle, as well as the minimum number of patient complaints. For the training of neural networks a lot of data about 2000 patients of the Department of emergency cardiology of the city clinical hospital № 4 of Perm were used.

Generation of neural networks, their training, optimization and testing were carried out in the author's neuropackage for each nosological unit according to the method of Perm scientific school of artificial intelligence www.PermAi.ru. It turned out that the best neural networks are perceptrons with one hidden layer containing from one to four sigmoid neurons.

After training, neural networks were combined with the help of the user interface into a single neural network diagnostic and prognostic system and posted on the website www.PermAi.ru in section "Projects". The user interface is designed so that the results of the system are displayed graphically in the form of eight columns, the height of each of which reflects the degree of development of the corresponding nosological form of cardiovascular disease, as shown in Figs. 1, 2, 3 and 4.

2.2 Estimation of Model Error

The diagnostic and prognostic properties of the system were tested on a test sample of 400 data on patients not included in the training set. The relative mean square error of testing neural networks was 30% for myocardial infarction (sensitivity 81.4% and specificity 90.0%), for stable angina 22%, for unstable angina 18%, for coronary heart disease 20%, for hypertension 12%, for arrhythmia and heart blockages 22%, for chronic heart failure 12%, for acute heart failure 35%.

It should be noted that the accuracy of the diagnosis could be improved by entering additional input parameters, for example, add data of cardio -, echo - coronary-graphic studies, current General and biochemical blood analysis, etc., as it is done in our early

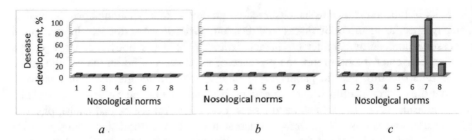

Fig. 1. Degrees of development of diseases diagnosed by the system: *a* – to patient No. 1, *b* – to patient No. 2, *c* – to patient No. 3. Risks of diseases: 1 - myocardial infarction, 2 - stable angina, 3 - unstable angina, 4 - ischemic heart disease, 5 - hypertension, 6 - arrhythmia and heart block, 7 - chronic heart failure, 8 - acute heart failure.

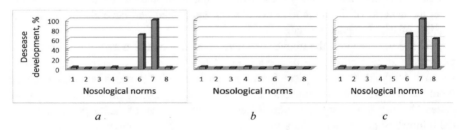

Fig. 2. The same provided that the age of the patients has increased by 30 years, and the weight has increased by 20 kg.

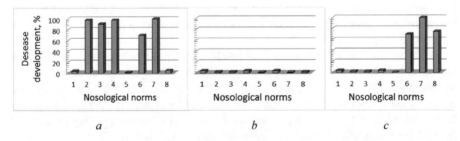

Fig. 3. The same provided that in addition to increasing age and weight, patients acquired diabetes mellitus.

publications [9, 10]. But then neural networks would become unsuitable for scenario forecasting, because these additional input parameters would have to be "frozen" during forecasting, which is contrary to reality. For example, increasing age is necessarily going to change the biochemical analysis of blood, data of the echo-cardiographic studies, etc.

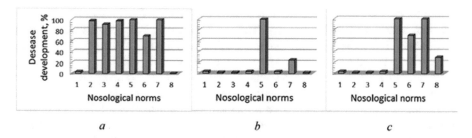

Fig. 4. The same provided that instead of diabetes, patients acquired hypertensive disease.

2.3 Virtual Experiments

Virtual computer experiments were performed on mathematical models of three patients:

Patient 1. The man aged 50 (born on 09.05.1967), height - 177 cm, weight - 80 kg, blood type - two, Rh factor positive, smoking, not doing physical exercises, there are heart diseases in blood relatives, no hypertension, no diabetes, no cerebral blood flow violations, no diagnosis of heart disease, no cardiac surgery, thrombophlebitis, no chest pains, complaining of shortness of breath with physical exertion, no asthma attacks at night, heartbeat, no sensations of interruptions in the work of the heart, no swelling of the limbs and face, not complaining of dizziness and headaches.

Patient 2. The woman aged 39 (born on 20.11.1977), height - 160 cm, weight - 60 kg, blood type - first, Rh factor positive, smoking, not doing physical exercises, no heart diseases in blood relatives, no hypertension, no diabetes, no cerebral blood flow violations, no diagnosis of heart disease, no cardiac surgery, no varicose disease or thrombophlebitis, no pains in a thorax, no shortness of breath, no asthma attacks at night, no palpitations, no sensations of faults in work of heart, having edemas of the face, complaining of frequent dizziness and headache.

Patient 3. The man aged 50 years (born on 15.08.1966), height - 180 cm, weight - 75 kg, blood type - fourth, Rh factor negative, smoking, not doing physical exercises, with heart diseases in blood relatives, no hypertension, no diabetes, no cerebral blood flow violations, earlier diagnosis of heart disease, no cardiac surgery, thrombophlebitis, no chest pains, complains of dyspnoea at rest, there are no attacks of choking at night, no palpitations, having sensations of interruptions in the work of the heart, having edemas of the extremities, complaining of frequent dizziness and headache.

After entering the parameters of patients, the neural network system has made the diagnoses presented in graphical form in Fig. 1, *a* - to patient No. 1, in Fig. 1, *b* - to patient No. 2, in Fig. 1, *c* - to patient No. 3.

As can be seen from Fig. 1, the system did not reveal the risks of cardiovascular diseases for patients No. 1 and No. 2, whereas it was diagnosed for patient No. 3: cardiac arrhythmia and blockade - 70%, chronic heart failure (CHF) - 100%, acute heart failure (AHF) - 20%.

In Fig. 2 in a similar form, the results of scenario forecasting of the development of diseases are given, if the age of patients has increased by 30 years, and the weight - by 20 kg. All other parameters of patients were kept unchanged. As can be seen from the picture, in patient No. 1 the system forecasted the appearance of a risk of arrhythmia and heart block - 70% and the risk of CHF - 100%. Patient No. 2 had no cardiovascular disease risks, and in patient No. 3 the system predicted the risk of arrhythmia and heart block - 70%, the risk of CHF - 100%, and the risk of AHF - 60%.

Figure 3 shows the results of scenario predictions of the development of cardio-vascular disease provided that in addition to increasing age and weight, patients were ill with diabetes mellitus. As can be seen from the figure, in patient No. 1 diabetes mellitus stimulated the appearance of a 98% risk of stable angina, 91% risk of unstable angina, and a 98% risk of coronary heart disease. Risks of arrhythmia and cardiac blockade, as well as CHF remained at the same high level - 70% and 100%, respectively. In patient No. 2, diabetes did not cause any risk of cardiovascular disease, while in patient No. 3 diabetes led to an increase in the risk of AHF to 75%.

Figure 4 results are different for diagnosis forecasting of diabetes mellitus, and instead of it hypertensive disease has been added. This forecast corresponds to the fact that during the diagnosis (30 years ago) it was recommended to the patients to abstain from sweets, and they did not follow the arterial pressure. As can be seen from Fig. 4, the addition of hypertension instead of diabetes mellitus in patient No. 1 did not change the pattern of scenario predictions, in patient No. 2 a 26% risk of CHF appeared, and in patient No. 3 the risk of AHF decreased from 75% to 30%.

In Fig. 5 the results of scenario forecasting differ in that, among other things, patients have been regularly engaged in physical exercises or light sports all this time (for 30 years). As can be seen from the figure, in this case, in patient No. 1, the risk of stable angina would decrease from 98% to 2%, the risk of unstable angina would remain at the same level, the risk of coronary artery disease would decrease from 98% to 91%, the risk of arrhythmia and blockade of the heart would decrease from 70% to 4%. In patient No. 2, CHF would decrease from 26% to zero, and in patient No. 3 the risk of AHF would decrease from 30% to 8%.

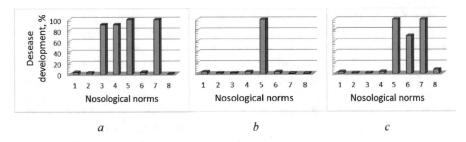

a   b   c

Fig. 5. The same provided that the patients would have regularly engaged in exercise or light sports during the whole period of 30 years.

Figure 6 provides the results of scenario forecasting on the assumption that patients are still not engaged in physical exercises, but stop smoking. In this case, patient No. 1 would have zero risk of both angina and IHD, but the risk of arrhythmia and heart block would remain at 70%, as in Fig. 4. In Patient No. 2, the pattern of scenario forecasting would be preserved, as in Fig. 5, i.e. for him to stop smoking or to do

physical exercises instead equally positively affects the state of the cardiovascular system, namely, the risk of CHF in both options of the patient's behavior would be avoided.
For patient No. 3, the stopping of smoking will lead to an increase in AHF from 8% to
63%. It is not recommended to stop smoking for such patient. Apparently, the habit of
smoking allows him to level out the resulting stresses, which has a positive effect on the
state of his cardiovascular system, in particular, on the predisposition to AHF.

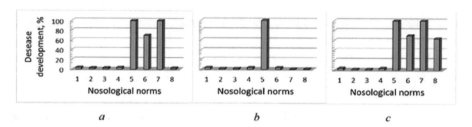

Fig. 6. The same is true provided that patients are still not engaged in physical activity, but have
stopped smoking.

Figure 7 presents the results of scenario forecasting on the assumption that patients
have reduced their weight by 20 kg, that is, returned to their original weight, which
they had 30 years ago. In this case, the risk of cardiac arrhythmia and heart blocking of
the first patient would decrease to 4% from 70%, the risk of cardiovascular disease of
the second patient would remain the same, and the risk of the third patient's AHF
would decrease from 22% to 10%.

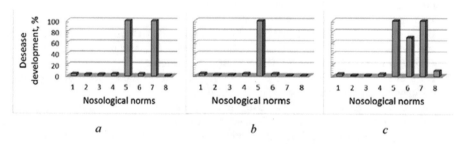

Fig. 7. The same is true provided that patients would reduce their weight by 20 kg, i.e. would
return to their original weight.

3   Conclusion

As can be seen from the above examples, the developed dynamic neural network model
allows not only to diagnose diseases at the current time, but also to predict the
appearance and development of diseases in the future periods of life of patients
depending on their lifestyle and medication. The use of such models allows doctors

before prescribing courses of treatment, first conduct virtual computer experiments, checking in this way the effect of drugs, select the most effective courses of treatment and prevention of diseases, and only then touch the health of patients.

Thus, in our opinion, the prerequisites for the transition of medical science and practice to a qualitatively new level – from archaic and immoral experimentation on living patients to experimentation on mathematical models of these patients will be created.

Acknowledgments. The publication was prepared with the financial support of RFBR: Grant № 16-01-00164.

References

1. Yasnitsky, L.N.: New possibilities of application of artificial intelligence methods in medicine: diagnosis, forecasting and modeling of development of cardiovascular diseases [Новые возможности применения методов искусственного интеллекта в медицине: диагностика, прогнозирование и моделирование развития заболеваний сердечно-сосудистой системы]. In: XII National Congress of therapists, Moscow, pp. 153–154 (2017)
2. Amato, F., López, F., Peña-Méndez, E.M., Vaňhara, P., Hampl, A., Havel, J.: Artificial neural networks in medical diagnosis. J. Appl. Biomed. **11**, 47–58 (2013). https://doi.org/10.2478/v10136-012-0031-x
3. Sandhu, I.K., Nair, M., Shukla, H., Sandhu, S.S.: Artificial neural network: as emerging diagnostic tool for breast cancer. Int. J. Pharm. Biol. Sci. **5**(3), 29–41 (2015)
4. Narang, S., Verma, H.K., Sachdev, U.: A review of breast cancer detection using ART model of neural networks. Int. J. Adv. Res. Comput. Sci. Softw. Eng. **2**(10), 311–319 (2012)
5. Awwalu, J., Garba, A.G., Ghazvini, A., Atuah, R.: Artificial intelligence in personalized medicine application of AI algorithms in solving personalized medicine problems. Int. J. Comput. Theory Eng. **7**(6), 439–443 (2015)
6. Kojuri, J., Boostani, R., Dehghani, P., Nowroozipour, F., Saki, N.: Prediction of acute myocardial infarction with artificial neural networks in patients with nondiagnostic electrocardiogram. J. Cardiovasc. Dis. Res. **6**(2), 51–60 (2015)
7. Basit, A., Sarim, M., Raffat, K., et al.: Artificial neural network: a tool for diagnosing osteoporosis. Res. J. Recent Sci. **3**(2), 87–91 (2014)
8. Raji, C.G., Vinod Chandra, S.S.: Graft survival prediction in liver transplantation using artificial neural network models. J. Comput. Sci. **16**, 72–78 (2016)
9. Yasnitsky, L.N., Dumler, A.A., Bogdanov, K.V., Poleschuk, A.N., Cherepanov, F.M., Makurina, T.V., Chugaynov, S.V.: Diagnosis and prognosis of cardiovascular diseases on the basis of neural networks. Biomed. Eng. **47**(3), 160–163 (2013). https://doi.org/10.1007/s10527-013-9359-0
10. Yasnitsky, L.N., Dumler, A.A., Poleshchuk, A.N., Bogdanov, C.V., Cherepanov, F.M.: Artificial neural networks for obtaining new medical knowledge: diagnostics and prediction of cardiovascular disease progression. Biol. Med. **7**(2), BM-095-15, 1–8 (2015). http://www.biolmedonline.com/Articles/Vol7_2_2015/BM-095-15_Artificial-Neural-Networks-for-Obtaining-New-Medical-Knowledge-Diagnostics-and-Prediction-of-Cardiovascular-Disease-Progr.pdf

Neural Network System for Medical Diagnostic of Gastrointestinal Diseases

Olga V. Khlynova[1]([✉]), Leonid N. Yasnitsky[2],
and Irina V. Skachkova[2]

[1] Department of Hospital Therapy, E.A. Vagner Perm State Medical University,
Petropavlovskaya Str. 26, 614000 Perm, Russia
`yasn1949@mail.ru`
[2] Department of Applied Mathematics and Informatics, Perm State University,
Bukirev Str. 15, 614600 Perm, Russia

Abstract. This article describes development experience of the neural network system for medical diagnostic of gastrointestinal diseases. There was used patient's practical medical information for its creation. As input parameters were taken into consideration different factor groups, include demographic, patient's complaints, life history, medical history and additional methods of research. Neural network model allowed making a significance assessment of factors, which have disease's development influence. As a result, was designed neural network system of differential diagnosis, allowing diagnoses "gastritis", "peptic ulcer". In the future, developed diagnostic system can be used as a "provisional diagnosis of gastrointestinal diseases".

Keywords: Gastrointestinal diseases · Neural network system
Gastritis · Gastric ulcer · Provisional diagnosis

1 Introduction

For the last few years interest in neural networks has significantly increased: they are applied in finance, business, medicine, the industry, and other areas. Neural networks can be used everywhere where it is required to solve problems of prediction, classification or control because they are applicable practically in any situation when there is a communication between the input and predicted variables even if this communication has the difficult nature. The medical statistics marks that in the last decades pathology of digestive tract occupies one of the leading places in the list of diseases [1]. Questions of diagnostics in gastroenterology one of the most debatable moments in the modern medicine, first of all because of uncertainty of data; wide personal spacing of parameters; low specificity of the majority of diagnostic signs. Absence of accurate diagnostic criteria and abundance. Therefore, in the circumstances, enhancement of system of diagnosis of diseases of digestive tract using highly sensitive, but at the same time not invasive tests is rather demanded. The classical diagram of creation of a network assumes presence of group of input parameters (neurons) which are information on a research object, and several output neurons which amount is set equal to quantity of various diseases [2]. This diagram is used in many neural network diagnostic system,

© Springer Nature Switzerland AG 2019
T. Antipova and A. Rocha (Eds.): DSIC 2018, AISC 850, pp. 359–365, 2019.
https://doi.org/10.1007/978-3-030-02351-5_41

but during the experiments was clarified that, changeover of one network N with outputs on N networks with one output allows to lower an error of setting of diagnoses therefore the research of each diagnosis separately is possible.

In the articles devoted to medical diagnostics, there is information about the application and development of artificial intelligence systems, in particular, the tool of neural networks. There are works using neural network technologies for diagnosing diseases of the lungs, blood diseases, cardiovascular system, cancer [3–6].

In articles published on this topic, the quality of the created neural networks is estimated, both with the help of traditional neural network estimation of the testing error, and with the help of medical indicators of the informative value of diagnostic methods: sensitivity, specificity, accuracy. Sensitivity - the proportion of really sick people in the survey sample, which according to the test results are identified as patients (a measure of the likelihood that any case of the disease will be detected). Specificity is the proportion of really healthy people who have not been diagnosed with a disease (a measure of the probability of correct identification of people without a disease). Accuracy - the proportion of correct results among all examined patients [9, 10].

2   Research Objective

Design of the neural network models allowing to make preliminary diagnoses of diseases of digestive tract with a fine precision, based on the parameters received by means of questionnaires, containing data of the inspected patients.

3   Methodology

To create a system of employees of the E.A. Wagner Perm State Medical University was given a data set consisting of 588 examples. Each example contained information in the form of signs about patients who had gastrointestinal disease of various degrees, coded as follows: 0 - no disease,…, 4 - the heaviest form. In the set presented, the maximum possible number of parameters characterizing the degree of the disease is displayed: demographic data, patient complaints, anamnesis of life, anamnesis of the disease, objective status and additional research methods. The total number of parameters is 208. The whole set was broken up into training and testing, in a ratio of 90%:10%. Thus, as a structure for each neural network (a separate neural network was designed for each of the diseases), a multilayer perceptron with 208 input parameters and one output was used. Under the error of testing is understood the root-mean-square error calculated on the test sample according to the formula:

$$\varepsilon_T = \frac{\sqrt{\frac{\sum_{n=1}^{N}(d_n - y_n)^2}{N}}}{|\max(y_n) - \min(y_n)|} 100\%$$

where N - the number of examples of the testing sample, d_n - the degree of the disease calculated by the neural network, and y_n - the actual degree of the disease.

The logarithmic function and the hyperbolic tangent function were used as the activation functions of the neurons of the hidden and output layers.

Designing, training and testing of neural networks was carried out with the Neuroimulator neuroseset [7] and methods for data analysis were used with the help of machine learning libraries in the Python language.

As the difference between the received mean square errors for a disease "Gastritis" in Neyrosimulyator and STATISTICA is insignificant, and for a disease "Ulcer" was succeeded to receive the best error in Neyrosimulyatore [8], the models created by means of Neyrosimulyator were selected as the best.

The diagnostic properties of the neural networks were tested on the test set using various estimates: calculation of the mean square error of the test and calculation of the accuracy of the diagnosis (the percentage of correctly diagnosed patients among the patients of the test sample). A correctly diagnosed diagnosis is one in which the difference between an actual diagnosis and a diagnosis of a neural network does not exceed one in modulus.

4   Results and Its Discussion

Results of check of operation of neural network diagnostic system on the testing set are presented in Figs. 1 and 2 in the form of comparison of the diagnosis made by doctors and the diagnosis received as a result of computation of a neural network.

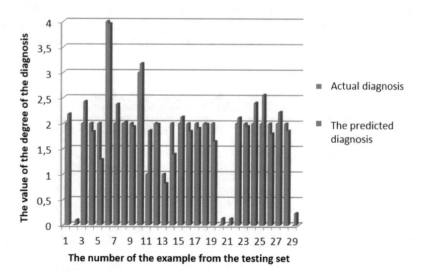

Fig. 1. Comparing of the actual and predicted setting of a disease for a disease "Gastritis"

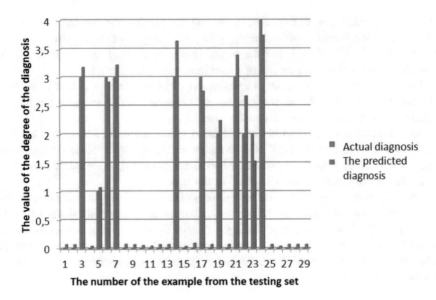

Fig. 2. Comparing of the actual and predicted setting of a disease for a disease "Ulcer"

By means of a certain technique, namely by serial switching on (activation) of input neurons and observation over result of a network significances of input parameters of models [9] (Figs. 3 and 4) were defined. Assessment of the significance of input parameters showed that for diagnosis the greatest influence the data connected to a status mucous a stomach, a supply and complaints of the patient to belly-aches had "Gastritis". For diagnosis "Ulcer" the most significant were parameters connected to a status mucous a stomach and a duodenum, complaints of the patient to pain in epigastralny area and a symptom of "niche".

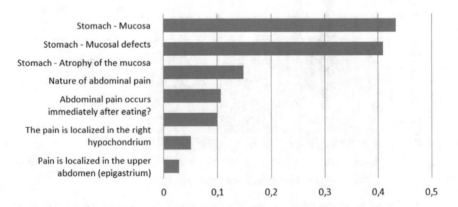

Fig. 3. The most significant parameters for the diagnosis "Gastritis"

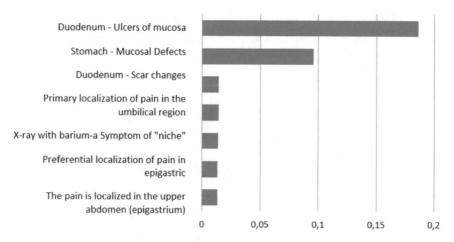

Fig. 4. The most significant parameters for the diagnosis "Ulcer"

The most significant parameters revealed by a neuronet confirm the medical facts that speaks about adequacy of the constructed models.

At the initial stage, testing errors were: 24% for gastritis, 24% for cholecystitis, 23% for GERD, and 18% for gastric or duodenal ulcer. The accuracy of the diagnosis was 52% for gastritis, 68% for cholecystitis, 62% for GERD, and 88% for gastric or duodenal ulcer. Thus, initially neural networks showed insufficient quality of diagnosis. Therefore, it was necessary to analyze the baseline data for emissions and inconsistent examples. Detection of such examples in the training set occurred using statistical methods, and using the method, the definition of suspicious examples, based on the neural network model [10].

Its essence lies in the analysis of examples for which the trained neural network finds it difficult to put the right degree of development of the disease, that is, the module of the difference between the diagnosis of the doctor and the diagnosis of the neural network is maximal. It was suggested that such examples fall out of the general pattern, but does not mean that they should be removed. Since there was a possibility that initially in the compilation of the original set, an error could be made in stating the degree of development of the diagnosis. Suspicious samples were re-examined by medical specialists and, if necessary, corrected. After optimization of neural network models and detection of emissions, it was possible to reduce the error in testing and improve the accuracy of diagnosis. Table 1 presents the calculated root-mean-square errors in testing and the accuracy of diagnosis.

Based on the calculated Spearman correlation coefficients between the input signs and diseases, as well as the coefficients of significance obtained during the training of neural networks, 30 parameters were determined that do not affect any disease from the list. These parameters carry information about the patient's personal life (education, number of children, etc.), some external symptoms (skin pallor, pigmentation, etc.). These parameters were excluded from the sample, and thus it was possible to reduce the number of parameters from 208 to 178 (Table 2).

Table 1. Learning outcomes of neural network models for each disease

Disease	Mean square error on the test set (%)	Accuracy (%)	Sensitivity (%)	Specificity (%)
Gastritis	12.87	83.6	91.1738	100
Ulcer	13.19	88.52	76.92	100
GERD	15.15	88.52	83.33	97.56
Cholecystitis	16.34	86.88	92.59	95.83

Table 2. Learning outcomes of neural network models for each disease after removing insignificant parameters

Disease	Mean square error on the test set (%)	Accuracy (%)	Sensitivity (%)	Specificity (%)
Gastritis	10.38	90.16	98.2138	100
Ulcer	11.61	88.68	90.3	100
GERD	9.71	90.16	86.36	95.12
Cholecystitis	12.92	88.36	96.12	94.22

5   Conclusion

Trained neuronet models show high accuracy of diagnosis, besides, after removing insignificant parameters, the quality of diagnosis has improved by 2–6% in the mean square error of testing. In the prospects of this study, the expansion of the class of predictable diseases of the gastrointestinal tract and the conduct of research to identify new medical knowledge [4]. Also, a preliminary version of the interface was designed, which allows diagnoses based on parameters from the original set for four GI diseases. In addition to diagnosing, it is possible to save completed questionnaires, which in the future will automate the collection of data and pre-training models.

To start using the interface, you must register in the system, this will enable you to save the completed questionnaires. Then it is necessary to fill in 6 groups of symptoms: personal data, complaints, anamnesis of life, medical history, objective status, additional research methods. The diagnosis is made only on the basis of completely filled forms. This system is mainly provided for students of medical institutions for training, for testing and assessing the quality of diagnosis. Link to the interface: http://gastro.permai.ru/.

Acknowledgments. The publication was prepared with the financial support of RFBR: Grant No. 16-01-00164.

References

1. Khlynova, O.V., Kokarovceva, L.V., Beresneva, L.N., Kachina, A.A.: What are the dangerous diseases of civilization for the cardiovascular system? (Чем опасны болезни цивилизации для сердечно-сосудистой системы?). Bull. Perm Sci. Cent. UB RAS **3–4**, 4–10 (2012)
2. Karmazanovskij, G.G.: Evaluation of diagnostic significance of the method (Оценка диагностической значимости метода). Ann. Surg. Hepatol. **2**, 139–142 (1997)
3. Utomo, C.P., Kardiana, A., Yuliwulandari, R.: Breast cancer diagnosis using artificial neural networks with extreme learning techniques. Int. J. Adv. Res. Artif. Intell. **3**, 7 (2014)
4. Pournik, O., Dorri, S., Zabolinezhad, H., Alavian, S.M., Eslami, S.: A diagnostic model for cirrhosis in patients with non-alcoholic fatty liver disease: an artificial neural network approach. Med. J. Islam. Repub. Iran **28**, 1–6 (2014)
5. Yasnitsky, L.N., Dumler, A.A., Bogdanov, K.V., Poleschuk, A.N., Cherepanov, F.M., Makurina, T.V., Chugaynov, S.V.: Diagnosis and prognosis of cardiovascular diseases on the basis of neural networks. Biomed. Eng. **47**(3), 160–163 (2013). https://doi.org/10.1007/s10527-013-9359-0
6. Yasnitsky, L.N., Dumler, A.A., Poleshchuk, A.N., Bogdanov, C.V., Cherepanov, F.M.: Artificial neural networks for obtaining new medical knowledge: diagnostics and prediction of cardiovascular disease progression. Biol. Med. **7**(2), BM-095-15 (2015)
7. Cherepanov, F.M., Yasnicky, L.N.: The neurostimulator 5.0 (Nejrosimulyator 5.0). Certificate of state registration of computer programs No. 2014618208. Application Rospatent No. 2014614649. Registered in the Register of computer programs on August 12 (2014)
8. Cherepanov, F.M., Yasnicky, L.N.: Neural network filter to exclude outliers in statistical information (Нейросетевой фильтр для исключения выбросов в статистической информации). Bull. Perm Univ. Ser. Math. Mech. Inf. **4**, 151–155 (2008)
9. Yasnicky, L.N.: Neural networks is a tool for obtaining new knowledge: successes, problems, prospects (Нейронные сети – инструмент для получения новых знаний: успехи, проблемы, перспективы). Neurocomput. Dev. Appl. **5**, 48–56 (2015)
10. Yasnicky, L.N.: Intelligent systems (Интеллектуальные системы). Knowledge Laboratory, Moscow (2016)

Digital Public Administration

The Relationship Between Smart Cities and the Internet of Things in Low Density Regions

Isabel Maria Lopes[1,2(✉)] and Teresa Guarda[2,3(✉)]

[1] UNIAG (Applied Management Research Unit),
Polytechnic Institute of Bragança, Bragança, Portugal
isalopes@ipb.pt
[2] Algoritmi Centre, Minho University, Braga, Portugal
tguarda@gmail.com
[3] Universidad Estatal Peninsula de Santa Elena – UPSE, La Libertad, Ecuador

Abstract. In these times of digital transformation, cities have overcome the challenges of the past and are building the future. The use of technological resources as a means of efficiently delivering various services and improving citizens' quality of life has transformed regions and cities into smart regions and cities. There have been a remarkable amount of projects implemented by the Municipalities in the last years, taking the technologies to the cities. However, for a project to be interesting, it must have a positive impact on society, that is, citizens. This evidence gave rise to the present study whose goal was to find out if citizens living in inner cities, labeled as smart cities, actually consider them that way, and whether their city uses innovative solutions that optimize their daily lives. The results are discussed in the light of the literature and future work is identified with the aim of shedding some light on a field as emerging, promising and current as this of Intelligent Cities and the Internet of Things.

Keywords: Internet of Things · Smart cities · Smart regions
Information technologies

1 Introduction

At present, companies with world-renowned prestige that operate in the area of Information and Communication Technologies are betting on innovative solutions that optimize the daily life of citizens, providing a better quality of life for the population. Cisco Systems and IBM, Microsoft already develop new solutions and initiatives for smart cities. CISCO launched the "Global Intelligent Urbanization initiative" to help cities around the world using the network as the fourth utility for integrated city management, enabling a better quality of life for citizens and development economic. Microsoft is working with Coventry University and Birmingham City Council on the Intelligent City Proof of Concept Project, an interoperable technology platform focused on transportation. IBM has announced its Smarter Cities to stimulate economic growth and quality of life in metropolitan cities and cities by activating new approaches to thinking and acting in the urban ecosystem [1].

© Springer Nature Switzerland AG 2019
T. Antipova and A. Rocha (Eds.): DSIC 2018, AISC 850, pp. 369–378, 2019.
https://doi.org/10.1007/978-3-030-02351-5_42

The territory can create an environment conducive to innovation, provided its sustained by technology and technological innovation. The concept of smart region appears as an alternative to assist the territorial organization of space, in order to generate an innovative environment capable. Regions function as collectors and repositories of knowledge and ideas, facilitating the flow of ideas, knowledge and learning. In the smart region becomes urgent, the capacity for innovation and adoption of new knowledge, techniques, and technologies. Being these the main vectors of the regional development process [2].

The grouping of people in specific places is inevitable, this generates the need to prevent or reduce the creation of waste in all aspects, infrastructure, resources, management, pollution, health, traffic, among others. The definition of a smart city is linked to an efficient city in which its concept encompasses sustainable urban development that is capable of being able to respond to any political, economic, operational and social environment that may arise.

Until now, cities have been changing, on a greater or lesser scale, and the discussion about their role in the economy and social welfare has been intensified. An intelligent city is aimed at people, whose main objective is the well-being of the population.

The present paper is structured as follows. After this introduction, a brief outline is given of the Portuguese Smart Cities Network, followed by a definition a smart city and Internet of Things. Subsequent to this is the presentation of the research method used in this study, followed by the main results obtained. The paper ends with the conclusions and with suggestions for future works.

2 The Portuguese Smart Cities Network

The Portuguese Smart Cities Network (RENER) was started in 2009 with 25 municipalities as a pilot network for electric mobility launched by the Portuguese government. The cities acted as test sites for electric charging points and intelligent mobility systems [3]. Among these 25 municipalities, 18 are the Portuguese district capitals, and the others are the cities of Vila Nova de Gaia, Almada, Loures, Sintra, Cascais, Torres Vedras and Guimarães.

In 2013, RENER invested in the thematic extension of its action, incorporating other areas such as energy efficiency, renewable energies, water and waste management, governance and citizenship, culture and tourism, all in the sense of a holistic model of smart cities. Over the last year, it has also promoted the geographic extension of the network with the integration of 18 more national municipalities, thus gaining scale, critical mass and cooperation capability. The Portuguese Smart Cities Network was thus formalized in November 2013 as a natural evolution of the Renewable Energy Living Lab (RENER), which had been created in 2009 by INTELI – Association Intelligence in Innovation. In 2015, the network incorporated three more municipalities, namely Macedo de Cavaleiros, Miranda do Douro and Lagoa. The Portuguese Smart Cities Network is currently composed of 46 municipalities throughout the national territory (see Fig. 1).

Fig. 1. Map of the Portuguese municipalities integrating RENER (Source [3]).

With the creation of the RENER the Network of Intelligent Cities of Portugal (www.smartcitiesportugal.net), all the cities integrating the RENER network became as well associated with this new platform. INTELI is the managing entity of the RENER Network, made up of 46 municipalities and a member of the European Network of Living Labs. In the Portuguese Smart Cities Network each municipality works as a site of development and experimentation of urban solutions.

This idea of experience sharing among municipalities has already reached Spain with the creation of the Iberian Smart Cities Network, which is currently composed of 111 cities.

3 Smart Cities

The term Smart Cities has been widely used over the last years. The main goal of the smart cities initiative is to enable cities to manage their assets efficiently, investing in innovation and creativity as a way to promote sustainable and inclusive urban development. Initially, the model of a smart city applied to information technologies that could be used to plan city development. The first publication on the subject matter is considered to be the book by Ishida and Isbister [4] on methods the information society applied to create the virtual space of the city [5], using the Internet and IT infrastructure. Subsequent papers evolved towards the city management method [6], the ability to attract top class specialists [7] or the ability to develop and absorb innovation [8]. The concept of "smart cities" currently has come to dominate both the academic literature and the public policies agenda. Several worldwide projects are being conceived and implemented, with different characteristics, motivations, maturity levels, government models and funding sources. However, the motto is always the use of information and communication technologies to make urban life easier [9].

International Data Corporation defines a Smart City as a city which has stated its intention to use information and communication technologies to transform its modus operandi in one or more of the following areas: energy; environment; governance; mobility; buildings and services. The main goal of a Smart City is to improve the quality of life of its citizens ensuring sustainable economic growth [10]. An smart city can be defined as a multidisciplinary domain gathering several fields of action and skills in order to achieve development. These fields are at their core supported by information technologies, hence the designation of intelligent, but they must also be strongly targeted at a governance model anchored in civic participation and they must be a source of economic development [11].

Nowadays, cities are facing several challenges related to climate change, demography, energy dependency and social exclusion, which calls for new urban development paradigms. In this context, concrete smart city projects are being implemented around the world and an exponential growth is expected in the smart city market [12].

Sassen [13] gives primacy to people, claiming that if a smart city does not mobilize its citizens' intelligence, then it is not that smart and it is no more than the plain implementation of technical services. When people are added to this equation, everything gets more complicated. People do not all have one same shape and we cannot control their opinions, wishes or concerns.

Smart cities are complex systems, often called "systems of systems," including people, infrastructure, and process components (see Fig. 2).

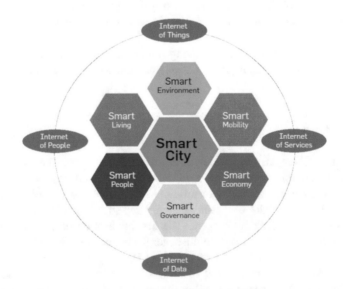

Fig. 2. A smart city model (Source [14]).

Most smart cities models consist of six components: government, economy, mobility, environment, living, and people [14].

4 Internet of Things

Since the beginning of the Internet in the late 1960s, when the number of linked sites was reduced, through the 1990s when 1 billion people connected to the Internet with their desktops and laptops, and by the first decade of the 21st century, where more than 2 billion people connected to the Internet through their phones cell phones, and Cisco Systems has predicted that 25 billion things will be connected to the Internet and to each other in 2015, being this number in 2020 doubled (50 billion) [15, 16].

Innovation at the information and communication technologies level is happening at fast speed. Since the rapid development of wireless technologies, sensor networks, smart networking, and wearable's are creating new sources of business value [17].

IoT represents the first real evolution of the Internet, with a major advance in the ability to collect, analyze and distribute data. It represents a breakthrough that will lead to the use of revolutionary applications. The number of devices like tablets, smart phones, personal computers, laptops, PDAs and even other portable embedded devices connected to the Internet is increasing and a large part of these mobile devices incorporates different sensors and actuators that can detect, make intelligent decisions, perform calculations and transmitting collected information via the Internet. A network of such devices with different sensors can give rise to numerous amazing applications and services that can bring significant personal, professional and economic benefits [18].

The Internet of Things (IoT) is a global infrastructure for the information society, enabling advanced services by interconnecting (physical and virtual) things based on existing and evolving interoperable information and communication technologies [19].

5 Research Approach

The unprecedented urban growth rate gives rise to an urgent need to find more intelligent ways of managing challenges [20]. However, little has been said about what residents/citizens understand by smart city or about internet of things and what technologies should be implemented for a city to be identified under that denomination.

With a view to empirically characterize citizens' opinion on the relationship between smart cities and internet of things, the application of the survey technique seemed appropriate, since it enhances a clear, straightforward and objective answer to the questions presented to the respondents. Moreover, since the aim was to characterise a high number of people, such number made the use of alternative research techniques impossible or not recommended.

5.1 Population

Since the inland cities integrating RENER are in lower number than the coastal ones, our interest in the inland cities grew. Nonetheless, the major question was: How do we get to the citizens of those cities? In the first place, among the 10 possible cities, the choice fell on those which might eventually be considered the most remote due to the fact that they are further from big urban centers and smaller.

Secondly, which citizens should be inquired? Since many selection criteria were found possible, a relatively deep analysis was made and one criterion was chosen. The survey would be conducted in higher education institutions.

Therefore, two inland cities in Portugal were selected and the study sample was composed of the students and teaching and non-teaching staff of two Polytechnic Institutes. Since the email addresses are available in the portal of each institution, the selection of the sample was relatively easy. In order to carry out the survey, 450 online questionnaires were sent to the 450 citizens constituting the sample. Among those, 273 people answered the questionnaire, which corresponds to a response rate of 61%.

5.2 Structure

The structure of the survey resulted from the review of literature regarding smart cities. The survey questions, to be answered individually and confidentially, were organized into three groups. The first group corresponds to the characterization of the respondents. The second group is related to the main questions leading this study, which are what they understand by smart city and what technologies they consider important for a city to be considered a smart one. The third group aims to assess what areas the respondents consider as priorities, what technological resources they find important in the transformation of a city and which they find to be the most imperative. The last question asked was whether or not they consider the city where they live a smart city.

6 Results

The data analyzed for the presentation of these results was produced based on a representative sample of students, teachers and other staff from polytechnic higher education institutions. Among the sample, 59% of the respondents are male and 41% female.

The ages of the respondents range from 19 to 63 years old. Among the 273 people surveyed, 198 are students, 58 are teachers and 17 are other staff. As was mentioned in the section on the structure of the questionnaire, a number of questions were put to the respondents; however, given the incidence of this work, it is the issues related to the second and fourth survey groups that will merit our attention.

The second group of questions in the survey inquired the respondents over what they understand by smart city. From the 273 respondents, 62 stated that they did not know what a smart city was and therefore could not define it.

Among the remaining answers (211), a number of definitions were given, some of them being transcribed as follows: "They are environmentally friendly cities, where the use of public transport and other means which do not pollute so much is promoted"; "They are cities with projects which enable to spend less electrical energy, investing in other energy sources"; "They are cities considered to be intelligent, which invest in the use of information technology to make their residents' life easier"; "… it is a city where there are easy, accessible and sustainable ways to provide quality of life at the disposal of the citizen". It is clear that it is difficult to reach one single concept to define a smart city. However, from the answers obtained, it can be said that some aspects are mentioned more often, which enables us to state that for the respondents of this survey, a smart city is one where a number of projects are implemented using technology as a means to improve the life of the people who live in those cities.

Another approach was if they could define what the Internet of things is. Of the 273 respondents, 152 replied that they did not know what the Internet of Things was and therefore could not define it.

The remaining 121 developed some definitions: "…sensors installed on the treadmills"; "It is the Internet directed to" things ", considering" things "such as buildings, public transportation, parking lots etc"; "They are technologies that, for example, installed in a city lead to smarter city management"; "They are digital network connections, interactions and controls of everyday objects, appliances and vehicles"; "They are technological devices and tools that make life easier for people, for example streamlining processes in public administration"; "… allows the connection of everyday objects between us and the internet".

When talking about Smart City and Internet of Things there is a term that stands out, "connectivity", as this is what will allow us to interconnect various devices to the Internet. Asked how many devices are connected to the Internet, the responses can be observed in Fig. 3. Most respondents have between 1 and 3 devices connected to the Internet.

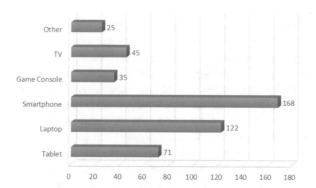

Fig. 3. Connectivity of devices.

Asked about the integration of different aspects and the benefits provided by the intelligent city, the answers were dissolved by the various options (see Fig. 4). It should be noted that the answers in this question were multiple choice.

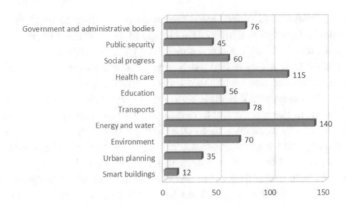

Fig. 4. Aspects to consider.

As can be seen, the highlights are Energy and Water, followed by Medical Assistance, Government and Administrative Bodies and the Environment.

Asked if they considered that there is a relationship between the use of the Internet of Things in cities and their classification as smart cities, the answer was almost unanimous, since 92% of respondents consider that yes, against 8% who think that a city can be considered a smart city without using any kind of information and communication technology.

Was observed that 235 (96%) of the respondents considered that there is a relationship between the applicability of the Internet of Things and a city considered to be intelligent, 22 (8%) do not consider there is any relation and 16 (6%).

7   Conclusions

One of the main IoT goals is to make the Internet more immersive and pervasive. As a network of highly connected devices, IoT technology works for a range of heterogeneous devices (such as sensors, RFID tags, and smartphones). Multiple forms of communications are possible among such "things" and devices. IoTs must be designed to support a smart city's vision in terms of size, capability, and functionality, including noise monitoring, traffic congestion, city energy consumption, smart parking meters and regulations, smart lighting, automation, and the salubrity of public buildings [21]. They must exploit the most advanced communication technologies, thus supporting added-value services for a city's administration and citizens [14].

With the possibility of everything connected, developments aimed at intelligent cities are gaining more and more expression, as they generate benefits and consequently a better quality of life for the citizens who inhabit it. Without doubt, the Internet of Things is the new technological period that will eloquently change the way people interact, cities and "things". One of the limitations of this research work is the delimitation of the study to two regions. While it is believed that sufficient data has been generated for the purposes of this paper, it will readily be accepted that a greater number and scope could result in a richer and more sustained data set.

Among the future work to be done, the scope of the study is highlighted to other regions of the country or even to other regions of the globe.

Acknowledgment. UNIAG, R&D unit funded by the FCT – Portuguese Foundation for the Development of Science and Technology, Ministry of Science, Technology and Higher Education. Project No. UID/GES/4752/2016.

This work has been supported by COMPETE: POCI-01-0145-FEDER-007043 and FCT – Fundação para a Ciência e Tecnologia within the Project Scope: UID/CEC/00319/2013.

References

1. Schaffers, H., Komninos, N., Pallot, M., Aguas, M., Almirall, E., Bakici, T., Hielkema, H.: Smart cities as innovation ecosystems sustained by the future internet. Int. J. Commun. Syst. **25**(9), 1101–1102 (2012)
2. Guarda, T., Díaz-Nafría, J., Augusto, M.F., et al.: Territorial intelligence in the impulse of economic development initiatives for artisanal fishing cooperatives. In: Springer, C. (ed.) Smart Innovation, Systems and Technologies book series (SIST) in International Conference of Research Applied to Defense and Security, vol. 94, pp. 105–115 (2018)
3. RENER: Rener Living Lab—Rede Portuguesa de Cidades Inteligentes
4. Ishida, T., Isbister, K.: Digital Cities: Technologies, Experiences, and Future Perpectives. Springer, Berlin (2000)
5. Komninos, N.: The Age of Intelligent Cities. Smart Environments and Innovation-for-all Strategies. Routledge, New York (2015)
6. Meer, A., Winden, W.: E-governance in cities: a comparison or urban information and communication technology policies. Reg. Stud. **37**(4), 407–419 (2003)
7. Murray, A., Minevich, M., Abdoullaev, A.: The future of the future: being smart about smart cities. http://www.kmworld.com/Articles/Column/The-Future-of-the-Future/The-Future-of-the-Future-Being-smart-about-smart-cities-77848.aspx
8. Florida, R.: Cities and the Creative Class. Routledge, London (2005)
9. Silva, C., Selada, C., Guerreiro, D., Afonso, P., Melo, R.: Índice de Cidades Inteligentes—Portugal. INTELI—Inteligência em Inovação, Centro de Inovação (2012)
10. Coimbra, G.: Smart cities benchamark 2015. International Data Corporation (IDC) (2015)
11. Torrinha, P.: Cidades Inteligentes: Sobre evolução – Uma perspetiva. Centro de Computação Gráfica (2016)
12. Selada, C., Almeida, A.: Smart Cities Portugal Roadmap, Technical Datasheet. Edition and Coordination: INTELI (2014)
13. Sassen, S.: É fantástico olhar para a longa vida de uma cidade, Entrevista dada a F. Cardoso, Smart Cities, n. 12, Jul/Ago/Set. Smart Cities, n. 12, Jul/Ago/Set, 22–25 (2016)
14. Khatoun, R., Zeadally, S.: Smart cities: concepts, architectures, research opportunities. Commun. ACM **59**(8), 46–57 (2016)
15. McAuley, D., Mortier, R., Goulding, J.: The Dataware manifesto. In: Proceedings of the Third International Conference on Communication Systems and Networks (COMSNETS), Bangalore, pp. 1–6. IEEE (2011)
16. Nordrum, A.: Popular internet of things forecast of 50 billion devices by 2020 is outdated, vol. 18 (2016)
17. Guarda, T., Leon, M., Augusto, M., Haz, L., Orozco, W., Alvarez, J.: Internet of Things challenges, Lisbon, Portugal, pp. 14–17 (2017)

18. Zheng, L., Wang, J.: Technologies, applications, and governance in the internet of things. Internet of things-Global technological and societal trends. From smart environments and spaces to green ICT (2011)
19. Leloglu, E.: A review of security concerns in Internet of Things. J. Comput. Commun. **5**, 121–132 (2016)
20. Talwoo, N., Pardo, T.: Conceptual smart city with dimension of technology, people, and institutions. In: Proceedings of the 12th Annual International Conference on Digital Government Research, pp. 282–291 (2011)
21. Khorov, E., Lyakhov, A., Krotov, A., Guschin, A.: A survey on IEEE 802.11ah: an enabling networking technology for smart cities. Comput. Commun. **58**(1), 53–69 (2015)

Implementation of IT Security and Risk Management Process for an Academic Platform

Lidice Haz$^{(\boxtimes)}$, Manuel Eduardo Flores Morán, Ximena Acaro,
Carlos Julio Guzman, and Luis Espin

Facultad de Ciencias Matemáticas y Física, Ciudadela Universitaria
"Salvador Allende", Universidad de Guayaquil, Av. Delta y Av. Kennedy,
Guayaquil, Ecuador
victoria.haz@hotmail.com, {manuel.floresmo,
ximena.acaroc, carlos.guzmanr, luis.espinp}@ug.edu.ec

Abstract. Nowadays, technological risks and information security are important processes of management and administration for public and private institutions. This is due to the obliquity of technology in the development of business processes, and its level of impact on organizational goals and objectives. This work implements a process for identifying, measuring, controlling and monitoring the IT risks that would allow the prevention and reduction of the losses due to the materialization of these types of risks in a Higher Education Institution. The risks analysis was applied to the information assets associated to two institutional critical processes that could cause material, financial, operational and image damages. In general, the results allowed to identify the high, medium and low level risk. Also, it was presented an action plan that included mitigation control to counteract the effects of identified risks, as well as its probability of occurrence, an estimated budget, and feasibility the analysis of implementing these countermeasures.

Keywords: Technology risk management · Information security
IT platform · Academic management

1 Introduction

The deployment of technology is one of the most common strategies used for any type of organization, and it has increased the dependency in the use of electronic channels and information within the institutions [1]. This fact has allowed technology to move up in category from a passive element in order to become an important asset for the operations of the organizational process [2]. Indeed, Technology is key part of the institutional process, and its performance cannot be isolated from those of the remaining elements of the process.

For a higher education institution, it is necessary to continually improve its quality in order to outstand among the institutions of its kind; an important factor for the accomplishment of this objective is the implantation of new strategies, techniques, and

T. Antipova and A. Rocha (Eds.): DSIC 2018, AISC 850, pp. 379–386, 2019.
https://doi.org/10.1007/978-3-030-02351-5_43

tools that improves aspects as: planning, administration, management and operations, technology, among others [3, 4].

One of the most important processes that has to be defined and controlled in any type of institution is the process of Risk Management, because the organizations could be exposed to significant losses when an adequate risk management of the critical activities and processes is not performed; therefore affecting the normal flow of operations and not achieving its objectives [5].

The present project was applied in a higher education institution in Ecuador. The implementation of the IT Security and Risk Management Process allowed the university to control –to a reasonable level– IT risks, assuring the highest degree of mitigation possible. In other words, the university was able to identify and control the vulnerabilities and threats in order to lower the impact of the risk in case of an occurrence. Institutions must control risks without significantly modifying its structural characteristics and functionality, and allow returning to its original structure once the risk event has finished [6].

2 Justification

Currently, the complex, dynamic and competitive institutional environment has taken the institutions to a point of technifying and automating their processes, implementing strategies that would allow the improvement in the quality of the products and services they offer the community [7]. These changes are produced due to the demands of the business environment, the new organizational models, the requirements established by consumers, the regulatory changes established by the control entities which aim is to generate value added to the products and services improving their quality; nevertheless, every change –as little it may seem– always carries a risk, and there is the need of guaranteeing the availability and operations of the services, avoiding interruptions and scenarios that could cause losses and affect the image of the institution [7, 8].

The "Model of Institutional Evaluation of Universities and Polytechnic Schools" – issued in September 2015 by the CEAACES– defines criteria and sub-criteria of quality in the higher education system through attributes that are measured by standards. This model defines the criteria of "Resources and Infrastructure" that evaluating the characteristics of the information technologies in order to determine adequacy for guaranteeing the activity flow of the academic community [9].

In the Number 300 about the Risk Assessment of the Internal Control Norm issued for Office of the General Comptroller of the State [10]. The public institutions use as a reference this framework with the purpose of setting their own internal control, giving special importance in risk management, and mechanisms to identify, analyze, and deal with potential events that could affect the execution of the processes and the achievement of the objectives.

This situation makes mandatory for the universities to implement a process to manage the IT security and risks. These risks could affect the availability, confidentiality, trustfulness, accuracy, and transparency of the information of their academic management platforms –reasonably ensuring the compliance with the criteria, sub-criteria, and standards mentioned beforehand, and evaluated by CEAACES, and the Internal Control Norms of the General Comptroller of the State.

Currently, the institution has not formally defined the technological and security processes that are under the responsibility of the Direction of IT and Communication. This makes it difficult for identifying the technological risks that could stop these processes, and the prevention of these events.

In previous years, there were technological risk events that got materialized and – last time– produced a loss of information from the academic management platform; this event pushed the university to take reactive measures to restore the information within several days, affecting the productivity of academic staff. Additionally, there are certain managerial reports with errors from the academic management platform that are corrected manually in the database, increasing the IT security risk.

The present work was made taking into account this background; in which the IT Security and Risk Management Processes were developed and implemented, with the purpose of identifying, measuring, controlling, and monitoring the risk, mitigating the probabilities and impact of possible events that could affect the activities of the processes that are supported by the academic management platform [11].

3   Methodology of Development

The present work belongs to the work model of Research of a Feasible Project, because it has as an objective the execution of a proposal. A Feasible Project is defined as "a study that consists in the investigation, elaboration, and development of a proposal of feasible operative model for the requirements or needs of organizations or social groups" [12].

During the development of this work, qualitative techniques were used for the comprehension and description of the critical processes of the business, placing them as part of the knowledge of the real context of the organization [13]. Other techniques were additionally used, such as observation, interviews, and documental gathering [14].

Interviews were used for the collection of information; for this purpose, there were questionnaires designed and directed to different staff members of the institution depending of their functions.

The documental gathering was developed through the selection of information based in descriptive and expository field bibliographic investigation about the topic of study. This allowed expanding the knowledge of the topic using primary and secondary sources of information. National and International standards and norms were consulted such as ISO 31000 Risk Management, ISO/IEC 27005 IT Security Risk Management; ISO 22301 Business Continuity Management; "Model of Institutional Evaluation of Universities and Polytechnic Schools" from CEAACES; The Internal Control Norm issued by the Office of the General Comptroller of the State, among others.

The IT Security and Risk Management Process –also known as PGRTI in Spanish– defines 4 generic phases as it is shown in Fig. 1. Simultaneously, these phases contain sub-processes that could be applied depending of the environment and the activity in which the organization develops.

Fig. 1. Phases of the operative risk management process

Phase one defines and understands the institution itself, and its context. Phase two establishes the macro-processes for the administration of the IT security and risk management. Phase three specifies the action plans in order to mitigate and manage the identified risks; and finally, phase four deals with the feedback and improvement process of the results from previous phases.

4   Results

After applying the Operative Risk Management Process factor IT [15], the results obtained in each of the phases were the following:

Phase 1: Organization and its context.- The process map of the university was evaluated –process map oriented to achieve the highest satisfaction of internal and external customers, by offering academic, and research services, and attachment with the society.– The institution classifies its processes in:

- Strategic Processes: Executed by top administration of the university; they offer direction.
- Mission Processes: Productive Processes oriented to the compliance of policies and strategies related to the quality of academic services offered by the institution –in other words, these enable the achievement of the institutional objectives.–
- Support and Advisement Processes: Processes that support the mission and strategic process. They recruit competent staff, reduce labor risks, keep the best operative and functional conditions, coordinate and keep the optimization of resources and the administrative efficiency.
- Evaluation and Follow-up Processes: Processes oriented to the internal control and evaluation of the processes mentioned previously.

Phase 2: Definition of the IT and information security risk management.- In order to define the technological risks, it was necessary to identify all the processes that take place in the academic management macro-process. These are detailed in Table 1.

These processes are related mainly with the academic management of the university, and they are supported by several information systems the institution has put in place; these systems are administered by the process owners and by the Direction of Information Technology and Communication. The processes identified as critical are:

- MSO-DC-PA Academic Planning.
- MSO-DC-MT Registration.

Table 1. Processes of the academic management macro-process

Academic management macro-process	
Macro-process: Strategic	Process: Quality assurement. • Selection and recruitment of faculty staff
Macro-process: Mission	Process: Teaching. Academic planning: • Opening and closure of the academic period • Registration of teachers in platform • Definition of class schedule • Organizing student class groups • Evaluating and updating the micro-curriculum programs • Administering the policies and strategies of the academic evaluation process • Planning and executing teacher evaluations
	Registration: • Registering student admissions
	Process: Titulat (awarding a degree). Titulation: • Following-up development of thesis and graduation • Controlling development of intership practice plan • Keeping updated alumni information

For this purpose there were several considerations: type of process, maximum tolerated period of interruption, objective recovery time, and objective recovery point, with the purpose of establishing the critical level and the importance of the institution. Then, a list of possible threats was made. These threats could affect the information assets associated with the critical processes, which were analyzed to evidence the vulnerabilities and possibilities of exploiting them [16]. In Table 2, a list of high risk threats is shown. These could affect one or several of the information assets for critical processes.

In Fig. 2, the complete map of the forty-seven evaluated risks is shown, in terms of their occurrence probability and impact over the information assets, and associated critical processes. Twenty risks are labeled high risk (red color), twenty-five medium risks (orange color), and two low risks (green color).

Phase 3: Risk Treatment.- After identifying high risk threats, an action plan was made to mitigate those risks, and to reverse them to acceptable tolerance levels. The mitigation strategies of the risks –and the reaction to them– depended of the risk appetite of the institution. The proposed controls in the action plan are associated with 40 controls of the ISO/IEC 27001:2013 norm [17].

During the elaboration of the action plan, tasks were determined, as well as the necessary resources, and the accountability. The implementation period of all selected controls is 17 months; and the initial budget for implementation is USD$58,400.00. It is important to highlight that the initial budget did not take into consideration the dollar amount already spent by the institution in some controls up to date.

Phase 4: Monitoring and Revision.- The last phase of the risk management process is the monitoring of the execution of the controls, and the establishing of indicators marks

Table 2. High risk threats

Threat description	Threat agent
Malicious code	Hacker
Robbery	Organized crime
	Discontented staff member
Unauthorized modification of information	Inexperienced or discontented staff member
	Supplier/contractor
	Former staff member
Network or computer viruses	Hacker
	Inexperienced or discontented staff member
Attack/terrorism	Subversive group
Sabotage	Discontented staff member
Structural failure of the building	Inexperienced staff member
Defected back-up	Inexperienced staff member
Earthquake	Natural
Broken or unmaintained UPS	Discontented staff member
Electric power interruption	Material (failure)
Fraud	Discontented staff member
Uneven power supply	Material (failure)
Rejected service	Inexperienced staff member

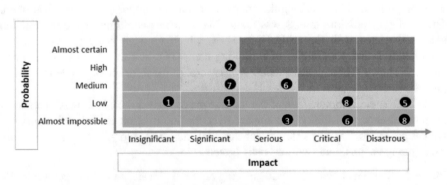

Fig. 2. Institutional risk map

that would allow verifying the application and effectiveness. Monitoring activities are mainly the following:

- Periodic revision and updates of the internal policies.
- Periodic follow-up of the budget compliance.
- Evaluations made by the Internal Auditor.
- Independent evaluations performed by the External Auditor.
- Contracts' auditing performed by the legal attorney.
- Revisions performed by the owners of the information assets.
- Verification of function segregation.

- Review of the compliance of the levels of service agreements made with technology providers.
- Reports of level of compliance of the action plan.
- Revision of mitigation plans of newly detected risks and threats.

Institutional Risk Management and the continuity of regular operations of the institution are under the responsibility of the top executive authority of the university and of the University Counsel; therefore, they must offer their formal commitment and support in order that the Information Technology and Security Risk Management Process –which has been named PGRTI– could be executed with the established periodicity defined in the present work.

5 Conclusions

Information Technology and Security Risk Management enables the creation of risk awareness within the top authorities and participating administrative units; it is imperative the formal commitment and support of the top administration in order to start a project related to the information technology and security risk management.

It is necessary to create and to formally approve a process, as a key factor for the execution and compliance of it. Its construction must be based in worldwide accepted standards and norms in order to ease the implementation and acceptance by the parties involved.

The IT Security and Risk Management Process can be replicated in any type of institution and industry. The process is generic and its execution periodical, at least once per year. This process allows for the comparison of risk levels in the information assets between each period. The PGRTI implementation requires: updated process map and process inventory with its respective levels of urgency, person responsible, and the information assets related with these processes.

After the implementation of PGRTI, it was necessary to interview the owners of each of the critical processes, and it was then determined that: IT management is perceived afterwards as an investment; the IT plans are focused in the most important aspects and greater risks; and it was possible to change the IT activities from "extinguishing fires" to the prevention and anticipation of the risks.

In future projects associated with risk management, the use of commercial risk management software is feasible. The purpose is to generate a centralized information repository of the risks associated with critical processes and the business objectives; also to observe metrics and incidents that could generate any losses to the organization.

References

1. Isaca: Risk IT, Marco de Riesgos de TI, basado en COBIT (2009)
2. Solís, G.: Cobit y la Administración de Riesgos (2008)
3. Asamblea Nacional del Ecuador: Constitución del Ecuador (2008)
4. Asamblea Nacional del Ecuador: Ley Orgánica de Educación Superior (LOES) (2010)

5. Alhawari, S., Jarrah, M.A.A., Hadi, W.E.: Implementing risk management processes into a cloud computing environment. In: Global Business Expansion: Concepts, Methodologies, Tools, and Applications, pp. 951–963. IGI Global (2018)
6. Instituto Nacional Electoral de Mexico: Metodología de Administración de Riesgos – Procesos. Sistema de control interno institucional INE (2014)
7. Schou, C., Hernandez, S.: Information Assurance Handbook: Effective Computer Security and Risk Management Strategies. McGraw-Hill Education Group, New York City (2014)
8. Fuenzalida, R., Ambrosio, E.: Riesgo Tecnológico. Su medición como prioridad para el aseguramiento tecnológico (2011)
9. Consejo de Evaluación: Acreditación y Aseguramiento de la Calidad de la Educación Superior CEAACES. Modelo de Evaluación Institucional de Universidades y Escuelas Politécnicas (2015)
10. Contraloría General del Estado de Ecuador: Normas de control interno para las entidades, organismos del sector público y personas jurídicas de derecho privado que dispongan de recursos públicos (2009)
11. Chaitanya, B., Kodukula, K., Tai-hoon, K.: A dependency analysis for information security and risk management. Int. J. Secur. Appl. 9(8), 205–210 (2015)
12. UPEL: Venezuela: Universidad Pedagógica Experimental Libertador, p. 7 (1998)
13. Ferrer, J.: Conceptos básicos de Metodología de la Investigación (2010)
14. Hernández, C.R.: Metodología de la Investigación. McGraw-Hill, New York City (2011)
15. Comité de Supervisión Bancaria de Basilea: Buenas prácticas para la gestión y supervisión del riesgo operativo (2003)
16. Chanchala, J., Umesh Kumar, S.: Information security risks management framework–a step towards mitigating security risks in university network. J. Inf. Secur. Appl. 35, 128–137 (2017)
17. ISO/IEC 27001:2013 Information Technology: Security Techniques – Information Security Management Systems – Requirements. International Organization for Standardization (2013)

Digital Technology and Applied Sciences

Severe Software Defects Trigger Factors: A Case Study of a School Management System

Nico Hillah[✉]

University of Lausanne, Lausanne, Switzerland
nico.hillah@unil.ch

Abstract. In this paper, we identify the groups of triggers that are responsible for severe software failures. These failures prevent any essential operation or activity to be conducted through the concerned system or other systems connected to it. In fact, the occurrence of these failures causes a double financial cost to organizations: one in fixing them and the other one because of the unavailability of the system or systems. We targeted three types of software defects as sources of these failures. We conducted this study by classifying 665 software defects of a school management system and we found that the top two trigger groups are the technology and the IS architecture groups.

Keywords: Software defect severity · Software defect triggers
Software defect classification

1 Introduction

In this age of information, every organization uses software systems to perform all types of activities in different domains. Unfortunately, most of the time, they are subject to failures [1, 2]. Software failures are defined as *"Termination of the ability of a product to perform a required function or its inability to perform within previously specified limits"* [p. 5, 3].

In fact, depending on their severity, these failures induce not only financial loss to organizations, but also time and resource loss in correcting them. Software defects (SDs) are the sources of these failures. IEEE standard 1044-2009 [p. 5, 3] defines a defect as: *"An imperfection or deficiency in a work product where that work product does not meet its requirements or specifications and needs to be either repaired or replaced"*. Different studies have investigated the sources and factors triggering SDs [4, 5]. Nevertheless, there is no existing literature on the types of triggers associated with the level of SDs severity that they generate. Knowing which types of triggers generate which level of SDs severity will help systems administrators in particular and organizations in general to better allocate their resources in order to address software failures. In this regard, the question we address in this paper is *which types of trigger factors generate the most severe SDs?*

To answer this question, we conducted a case study on a software system. In fact, we performed two main classifications of its SDs. The first classification was to identify the severity of SDs; then we classified the same SDs based on the trigger factors using EVOLIS framework [4].

© Springer Nature Switzerland AG 2019
T. Antipova and A. Rocha (Eds.): DSIC 2018, AISC 850, pp. 389–396, 2019.
https://doi.org/10.1007/978-3-030-02351-5_44

The paper is structured as follows: first, we will introduce existing techniques software defects classification; secondly, we will present the methodology that we used in conducting this study; then, we will show the results that we obtained; and finally we will present our contribution.

2 Related Works

In the software life cycle, the classification of defects presents many advantages [6]. The classification of defects helps the software development teams to reduce the cost of correcting them, to detect defective modules and to have efficient resource planning. Various studies have proposed and evaluated different approaches to collect and to analyze these SDs. The main approaches are (1) taxonomies [7, 8], root cause analysis [9], schemes [3] and the classification of these SDs [10].

There are different existing schemes in classifying SDs [10]. (1) The Orthogonal Defect Classification (ODC) of IBM [11] was developed in 1992 by Chillarege et al. [11] and it classifies defects across *"the dimensions (1) defect type, (2) source, (3) impact, (4) trigger, (5) phase found, and (6) severity"* [12]. (2) The HP Defect Origins, Types and Modes, the approach of Hewlett Packard, was developed by the HP software metrics in 1986 [13] and this scheme classifies the defects according to their types, their origins and their mode [12]. (3) The IEEE standard 1044-2009 is the scheme we retain for our first classification project. We selected this approach because it proposes the most complete definition of the SDs severity types among the three schemes. Moreover, this severity attribute is one of the most used attributes in SDs classification in practice [12]. The main advantage of choosing the severity attribute is the possibility for managers to identify which defect to correct first [12]. We retain the severity attribute for our first classification.

3 Methodology

To be able to understand the relation existing between triggering factors for SDs and the severity of SDs, we conducted a case study of a system that we will name system A. This system is developed by an educational organization and it is a school management system. Its purpose is to help schools in managing the grades of their students. It is used for managing more than 90000 students' grades. The first version of the system A had been released middle 2012. We classified 665 SDs of system A. The collection of SDs covers a period of one year and four months from January 2015 to April 2016. System A has nine released versions over this period. The defects repository tool used by this organization is Jira [14].

Our objective is to classify SDs according to their severity and then classify these same SDs according to the factors that trigger them. In fact, we analyzed the SDs of this system A by classifying them according to the defect severity attribute of IEEE 1044-2009 standards [3] and then by classifying them with EVOLIS framework [4]. The software team in charge of maintenance of the system A and a member of our research team had conducted both classification.

3.1   The Classification of SDs Based on Severity

Our first classification is done based on the severity attribute of the IEEE standard (see Table 2). The IEEE's standard defines this attribute as *"The highest failure impact that the defect could (or did) cause, as determined by (from the perspective of) the organization responsible for software engineering."* [3]. The five values of severity are classified from the most significant to the least significant ones (see Table 1). For the purpose of this study, we define any software defect (SD) as severe as long as it belongs to one of these severity levels: (1) Blocking (B), (2) Critical (C) and (3) Major (Maj). The Minor (Min) and the Inconsequential (Inc) are not considered as severe SDs.

Table 1. Severity values [3]

Attribute	Value	Definition
Severity	Blocking (B)	Testing is inhibited or suspended pending, correction or identification of suitable workaround
	Critical (C)	Essential operations are unavoidably disrupted, safety is jeopardized, and security is compromised
	Major (Maj)	Essential operations are affected but can proceed
	Minor (Min)	Nonessential operations are disrupted
	Inconsequential (Inc)	No significant impact on operations

Table 2. Classification of system A's SDs based on their severity

	B	C	Maj	Min	Inc	Total
Jan	10	12	46	22	0	90
Feb	5	8	24	17	0	54
March	12	17	49	31	0	109
April	9	8	50	16	2	85
May	3	1	20	5	0	29
June	0	2	25	14	0	41
July	0	1	5	4	0	10
Aug	2	7	8	7	3	27
Sept	5	11	18	15	4	53
Oct	7	4	22	6	0	39
Nov	15	8	37	20	3	83
Dec	7	5	18	12	3	45
Total	75	84	322	169	15	665

3.2   The Classification of SDs Based on the EVOLIS Framework

For our second classification project, we chose the EVOLIS framework [4] (see Table 3). This framework proposes a technique to classify SDs according to the factors that trigger them. *"EVOLIS can be caused by a large variety of factors: bugs that need to be fixed, users that wish to have new functionalities, new market opportunities that require new software features, performance standards that the system must reach, technical changes in the environment with which the system must interact, obsolescence of applications and so on"* [3]. EVOLIS identifies four main groups of factors that triggers SDs: (1) IS/users fit (U.F) triggers that are defined as any failure related to the user interface, the user documentation and aptitude to use the system. (2) The technology (TCH) triggers are related to defects that concern the software as well as the hardware platforms as information system components. (3) According to the authors, the IS architecture (ACH) triggers concern *"different types of integration evolution, namely an evolution of integration among components of the system, among business functionalities, or an integration with systems outside of the company."* [3]; and finally (4) the Business-IS (Bs-IS) alignment triggers that *"address the co-alignment between business and information systems"* [3].

Table 3. Classification of system A's SDs based on their trigger factors

	ACH	Bs-IS	TCH	U.F	Total
Jan	13	23	21	33	90
Feb	13	9	18	14	54
March	15	26	25	43	109
April	27	26	13	19	85
May	4	8	10	7	29
June	9	2	19	11	41
July	1	1	2	6	10
Aug	1	4	9	13	27
Sept	15	8	8	22	53
Oct	16	4	9	10	39
Nov	23	19	18	23	83
Dec	5	10	12	18	45
Total	142	140	164	219	665

Subsequently, we grouped these results into a two-dimensional table (see Table 4). Each dimension represents the results obtained for each classification project.

Table 4. Two-dimensional classification of SDs of system A

	B	C	Maj	Min	Inc	Total
ACH	17	12	99	14	0	142
Bs-IS	24	21	66	28	1	140
TCH	19	30	87	27	1	164
U.F	15	21	70	100	13	219
Total	75	84	322	169	15	665

4 Discussion and Contribution

We analyzed the results threefold: the results of severity classification, followed by the results of EVOLIS and then we combined and analyzed both results together. First, the severity classification showed us that the top three high types of SDs are respectively the major type of SDs with 322 SDs, followed by the minor type with 169 and the critical type with 84 SDs (see Table 2). Second, the EVOLIS classification showed us that the top three groups of factors that trigger SDs are respectively the IS/users factors with 219, followed by the technology factors with 164 and then the factors related to the IS architecture and business-IS alignment (see Table 3). These last trigger groups have almost the same number of SDs: 142 SDs for the IS architecture and 140 SDs for the business-IS alignment SDs. Further, the analysis of both combined results showed us that technology triggers represent respectively 12%, 18% and 53% for blocking SDs, critical SDs and major SDs (see Fig. 1). In total, the technology triggers are responsible for 83% of the severe SDs. Similarly, the architecture triggers represent respectively 12%, 8% and 70% for blocking SDs, critical SDs and major SDs (see Fig. 1). The business-IS alignment represents respectively 17%, 15% and 47% for blocking SDs, critical SDs and major SDs. Finally, the IS/users triggers represent 20% of the total of severe SDs (see Fig. 1).

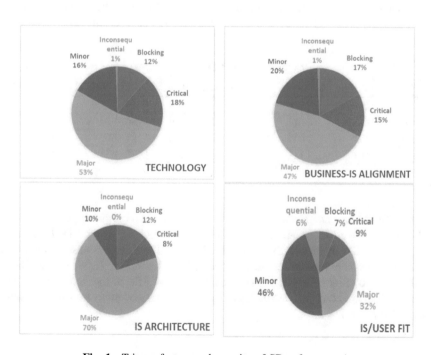

Fig. 1. Trigger factors and severity of SDs of system A

Analyzing these results separately did not give so much information to organize the SDs management. However, when we put them together, we found that some type of SDs trigger factors are the source of some specific severe groups of SDs. In fact, we observed that the majority of the inconsequential SDs are triggered by the IS/users fit factors.

This implies that the probability of an inconsequential SDs to be triggered by either the IS architecture, the business-IS alignment or the technology factors are very low or barely existent. Furthermore, we assigned weighting factors to each severity level according to their importance as followed (see Table 5):

Table 5. The weighting factors for the severity level.

Severity level	Weighting factor
Blocking	40%
Critical	30%
Major	20%
Minor	8%
Inconsequential	2%

We then apply this weighted scoring model to our two-dimension table to calculate the weighted scores (W) for severe SDs per trigger factor (see Table 6).

Table 6. Severe SDs weighted score.

Severe SD	Weight (W)	Bs-IS	W-Bs-IS	ACH	W-ARC	TCH	W-TCH	U.F	W-U.F
Blocking	0.4	24	9.6	17	6.8	19	7.6	15	6
Critical	0.3	21	6.3	12	3.6	30	9	21	6.3
Major	0.2	66	13.2	99	19.8	87	17.4	70	14
Total	0.9	111	**29.1**	128	**30.2**	136	**34**	106	**26.3**

Looking at these results, we can conclude that the technology trigger factors, with the highest weighted score 34, are responsible for most of the severe SDs followed by the IS architecture factors, with 30 weighted score. Then the business-IS alignment, with 29.1, and finally the IS/users fit triggers, with 26.3 (see Fig. 2). We can also notice that there is a considerable gap between the number of SDs of the first two groups of triggers (IS architecture and technology and the last two of them (business-IS alignment and IS/User fit).

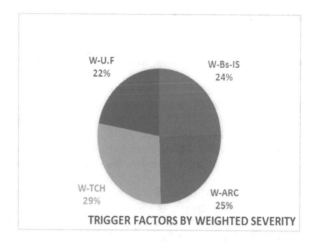

Fig. 2. Trigger factors by weighted severity of system A

5 Conclusion

To the question of which groups of SDs triggers generate the most severe SDs, we answered that the technology triggers are at the head position with a total of 34 weighted score. In the second position is the IS architecture triggers which comes with 30.2 weighted score, and then followed by the business-IS alignment triggers with 29.1. The last position is occupied by the IS/users fit triggers with 26.3 weighted score of the total severe SDs analyzed. We also found that the majority of the defects triggered by IS/user factors are either minor or inconsequent types of SDs.

The results obtained from this study will help software managers to improve the management of SDs by allocating the SDs correction resources more accurately thus reduce the cost of managing SDs. In our future work, we will analyze other software systems and compare their results to the ones we obtained in this study. We will also investigate in depth this close relation between the first couple of trigger groups (IS architecture and technology) and the last couple of trigger groups (business-IS alignment and IS/users fit).

References

1. Charette, R.N.: Why software fails [software failure]. IEEE Spectr. **42**(9), 42–49 (2005)
2. Kaur, R., Sengupta, D.J.: Software process models and analysis on failure of software development projects. Int. J. Sci. Eng. Res. **2**(2), 4 (2013)
3. 1044-2009 IEEE Standard Classification for Software Anomalies (2009)
4. Métrailler, A., Estier, T.: EVOLIS framework: a method to study information systems evolution records. In: 2014 47th Hawaii International Conference on System Sciences (HICSS), pp. 3798–3807 (2014)
5. Alshazly, A.A., Elfatatry, A.M., Abougabal, M.S.: Detecting defects in software requirements specification. Alex. Eng. J. **53**(3), 513–527 (2014)

6. Rajbahadur, G.K., Wang, S., Kamei, Y., Hassan, A.E.: The impact of using regression models to build defect classifiers. In: Proceedings of the 14th International Conference on Mining Software Repositories, pp. 135–145 (2017)
7. Binder, R.: Testing Object-Oriented Systems: Models, Patterns, and Tools. Addison-Wesley, Reading (2000)
8. Vallespir, D., Grazioli, F., Herbert, J.: A framework to evaluate defect taxonomies. In: XV Congreso Argentino de Ciencias de La Computación (2009)
9. Leszak, M., Dewayne, P.E., Stoll, D.: Classification and evaluation of defects in a project retrospective. J. Syst. Softw. **61**, 173–187 (2002)
10. Mellegård, N.: Improving Defect Management in Automotive Software Development, LiDeC—A Light-weight Defect Classification Scheme. Chalmers University of Technology, Gothenburg (2013)
11. Chillarege, R., et al.: Orthogonal defect classification-a concept for in-process measurements. IEEE Trans. Softw. Eng. **18**(11), 943–956 (1992)
12. Wagner, S.: Defect classification and defect types revisited. In: Proceedings of the 2008 workshop on Defects in large software systems, pp. 39–40 (2008)
13. Huber, J.T.: A comparison of IBM's orthogonal defect classification to Hewlett Packard's defect origins, types, and modes. Hewlett Packard Company (1999)
14. Atlassian: Jira—Logiciel de suivi des tickets et des projets. Atlassian. https://fr.atlassian.com/software/jira. Accessed 06 Apr 2018

A Design for Blockchain-Based Digital Voting System

Mahmoud Al-Rawy[✉] and Atilla Elci

Electrical Electronics and Computer Engineering, Aksaray University,
Aksaray, Turkey
mahmoud.rawie@gmail.com, atilla.elci@gmail.com

Abstract. In recent years, blockchain technology has affected to a large extent all aspects of life, however, until now paper-based elections have been practiced. It is time to upgrade the election scenario using modern technology such as blockchain and advanced cryptography methods. Actually, both of Estonia and New South Wales have been using i-voting systems, but after developing an example software of their systems for the purpose of analysis, it is discovered that it has weaknesses against many kinds of attacks, such as malwares, network attacks, and servers attacks. The fact that the blockchain technology has demonstrated infinite immutability and resistance against hacking, so it is possible to use it to secure election results from fraud by saving every single piece of data, record or transaction with unchangeable history. By abandoning the traditional database and compensating it with two blockchains instead of one ensures voter/vote privacy, as well as, safeguards the results from manipulation. Also, using blockchain's distributed network reduces the load on the network. Finally, solutions to problems of impersonation and vote selling are suggested. The technology behind the digital voting system design is explained in terms of the processes involved, such as ID creation, authentication, voting, and vote tallying.

Keywords: i-Voting · Vote/voter privacy · Blockchain
Public-private key algorithm anonymization

1 Introduction

Unsurprisingly, history is littered with examples of elections being manipulated in order to influence their outcome; scammers and rulers have developed means of manipulating elections to achieve purposes other than conducting democratic elections. Despite these concerns, in 2005, Estonia introduced an online election system to be the first country in the world to implement one. In the most recent elections, over 30% of participators cast their ballots online [1, 5]; over 280,000 votes were submitted through iVote in New South Wales; 70,090 Norwegian votes submitted online in 2013. However, security vulnerabilities, such as in voters' client devices, servers and in voter authentication process, have been demonstrated by security analysis on these systems [12, 13].

© Springer Nature Switzerland AG 2019
T. Antipova and A. Rocha (Eds.): DSIC 2018, AISC 850, pp. 397–407, 2019.
https://doi.org/10.1007/978-3-030-02351-5_45

Previously, it has been difficult to achieve fair election without running the risk of fraud and manipulation of the results [11]. With the emergence of Bitcoin [6], many researchers on the blockchain technology and smart contracts suggested that it is a suitable basis for e-voting, furthermore, it could have the prospect to make e-voting more acceptable and reliable in the society [4]. Some papers have explored blockchain-based e-voting [20, 21] including now this paper.

Blockchain technology in recent years has attracted much attention, too, due to its extreme resistance against hacking. Fundamentally, blockchain uses the distribution principle (distributed replication, decentralization), therefore, there are many copies of the same identical data spread over many nodes, but none is the golden copy. Each of these nodes has a copy which gives us fast, secure and transparent peer-to-peer transfer of transactions [3]. Also, blockchain is not an isolated technology, but a mixture of cryptographies, mathematics, and algorithms that have been beautifully combined to complete one another. Eventually, most important advantages of using blockchains-based e-voting are: (i) anonymity, (ii) accuracy, and security particularly against DoS Attacks (Denial of Service Attacks) and (iii) strong integrity (immutability).

It is fundamental that a proposed e-voting scheme contributes to preventing any violation of voter-ballot anonymity, and absence of any possibility to manipulate the results. Some of the main vulnerabilities usually e-voting systems have faced are highlighted below as follows:

Vote Privacy. Voter's choices must be anonymous (secret ballot). Non-observance of the secrecy of the ballot leads to attempts to influence the voter either by intimidation or potential vote buying.

Identity Theft. Impersonation and multiple vote casting were real issues in the past with traditional-based voting systems. Deceptive voters used to register themselves multiple times or manipulate their eligibility of voting. In Australia, 217 ineligible voters cast votes in 1996 elections, and in 2010 elections a family cast more than 150 votes by impersonating others. The security analysis of the Estonian Internet Voting System demonstrated that attackers can plant malware in the voter's client and read the National-ID card's PIN, then impersonate an eligible voter to cast an unqualified vote as they desire. In fact, due to the lack of biometric devices (online) electronic voting machines can allow identity theft. A biometric device would not be available in every home on the polling day, and, supplying such equipment for each house is very expensive and unpractical. Therefore, there must be a solution to overcome this issue.

Immutability. The biggest election concerns lie in the immunity of the electoral system from manipulation and counterfeiting [17].

1.1 Problem and Solution Approach

Privacy must be preserved in an electronic voting system to dispel the threats highlighted above. This paper proposes a novel e-voting architecture to properly facilitate digitalized elections maintaining vote and voter privacy while preventing fraud through employing firstly the blockchain technology to insure result immutability, secondly, through the RSA public-private key algorithm [2] to prevent identity theft.

The proposed voting scheme uses the blockchain technology to store the cast eIDs and votes, thus it can act as an immutable database. As usual, the current Web-based system uses Secure Hyper Text Transfer Protocol (HTTPS) yet there is need for a further configuration affected by the system administrators to secure the connection between the server and the clients' computers, and thus preventing Logjam and Freak attacks. By disabling the support for TLS (Transport Layer Security) export cipher suites and using a 2048-bit Diffie-Hellman group, also disables other cipher suites that are known to be insecure and enable forward secrecy to obtain the required server immunity [14, 15].

Moreover, it is necessary to relinquish use of third-party server to read and verify votes by voters. Studies proved that it opens the door to different opportunities for privacy violations as in the cases of elections in Australia and some other countries. So, any kind of reading the vote or overriding it is prevented just as in traditional paper-based voting [16].

A major challenge in this work is allowing people to cast their votes right from their home, without the need of going to the polling stations. Usually, digital voting systems rely on the use of standalone electronic voting machines (EVM) which also perform user (/voter) verification as well as the entire election process. In this proposal, we give up on such machines and allow the voters to use just an internet browser to cast ballots and votes.

Our design of the blockchain-based digital voting system is explained in terms of the processes involved, such as, voter logging in, digital ID creation, re-logging, Electoral Authority preparation, vote casting, and vote tallying.

2 Voter Authentication Process

Ballot casting process of a digital voting system requires the creation and signing of an electronic identity after a secure logging into the system, as described below.

2.1 The Access Authority

Assuming the country where a digital voting based election will be held has its own e-government where every single citizen is recorded and has his/her own individual private e-number. Therefore, it is possible to make the very first voter's authentication stage via (TLS client authentication [19]) starting with asking the voter full name, National ID number, and e-number. Then, encrypting the requested information all together thus computing the resultant hash by the following equation.

$$\textbf{Hash} = (\text{Enc}(\text{eNumber} + \text{FullName} + \text{NIDNumber}))$$

The computed hash is used in the browser cookie to identify a voter. Web applications that use cookies are known to be vulnerable to XSS (Cross-site Scripting)

Attacks [18], so voter's privacy might be violated by unauthorized accesses. Using HTTPs protocol is insufficient, attacker can request a page with HTTP, so, the same cookie will be sent without any protection. So to overcome this issue, system pages must only be sent to HTTPS calls before starting the session; furthermore, browser should be configured not to allow access JavaScript into the session cookie.

2.2 Digital ID Card Creation

As the voters would identify themselves in the login step, they must have first created an electronic identification (eID), consisting of an ID and digital signature. "eID" is an electronic identification solution of citizen [7]. The idea behind creating digital ID card is to prove voter identity and to be verified by Electoral Authority's Representative (EAR, more on this below) while using it as an identity in the voting processes.

In public-private key algorithm users need a set of two complementary keys; one of them is published for use by the public, and the other is kept secret by the individual for personal use. If information encryption is done by using the public key, so only the person who has the second prime of the key (private key) will be able to decode that information. On the other hand, if the person uses his private key to encrypt, anyone who has the public key can decrypt; this process also makes sure the signer (encryptor) is the owner of that key pair. Using public-private key algorithm provides full protection of the data from being modified (immutability), as well as authenticates the private key owner.

Creating a digital ID card is essential to generate the public-private key and use them to maintain vote privacy. This process composed of three steps is introduced below.

2.2.1 eID Generation

After voter logs into the system a public-private key set will be automatically established. The public key will be stored in the system, open to the public. The second key (private one) will be known and used only by the voter. Indeed, it must not fall into the hands of others, because someone might use it to impersonate the voter or use it for signature purposes. After that, some information about the person will be requested to be matched with the stored information in the system's database, in order to verify the voter's credibility. The requested information consists of the following fields:

1. Voter's National Identification Number and its expiration date.
2. Three images of the voter, captured right from the system using a computer's webcam. The first one of the voter himself. The second image of the first side of the voter's personal identification card and the last image of the other side of the voter's personal identification card.
3. Voter's phone number.

However, voter's personal identification card number is requested for the second time for confirmation purpose. In order not to depend on the availability of usable biometric devices (fingerprint, eyeprint), it is necessary to invoke another solution, such as these required images can be photographed, encrypted and stored in the database. Clearly, these will prove that the voter has created own electronic identity.

Voter's phone number will be used in the process of the time-based one-time password. It will be used while logging into the system, also in every process after that including casting a vote, to verify voter's identity. Finally, all the requested fields are combined and encrypted with the voter's public key to obtain verifiable eID hash.

$$\mathbf{eID} = \begin{pmatrix} \text{RSA.Enc(Voter's Public Key, \{FullName + IDNumber} \\ \text{+ IDExpirationDate + Images + PhoneNumber\})} \end{pmatrix}$$

Voter will create a unique signature by using his private key, allowing others to check the validity of the signature using his public key for any purpose of verification [8]. The signature will be used to sign the ballot by the voter to prove ownership of the private key without requiring to reveal it (Fig. 1).

$$\mathbf{Signature} = (\text{Sign (Voter's Private Key, eID)})$$

Fig. 1. Voter authentication.

2.3 Re-login

It will be possible to keep the election system open to the public some number of days before the election in order to allow voters to enter and establish their eIDs. In this way, the Electoral Authority representatives will have enough time to correct voter data if there will be any mistakes. Meanwhile, voters will get familiar with the election system and check whether someone else used their data to create eID. If so they can inform the Electoral Authority representatives to receive necessary assistance. Even after establishing the eID, attackers who can hack the eID of an eligible voter and change the phone number to redirect the privileges to his favor will be exposed. The moment that voters try to log into the system for the second time, their eID will be corrupted, due to the fact that his data has been changed, thus they can easily report to EAR immediately. As the voter successfully completes establishing his/her eID, he can log out of the system, and re-login during the polling day in order to vote.

Protecting voter data from being used by others must be given serious consideration; time-based one-time password algorithm [9] may be used to preclude this issue. In the time-based passwords, time synchronization is very important, a secret key and a timestamp will be defined, and everything will be synchronized via a standard protocol such as network time protocol. Once the password token will be created, it can be used in a limited duration just to ensure that the voter is using his own data in the login process, or the ongoing operation is within his knowledge, because he is the one who receives the verification code on his phone.

Once voter logs into the system for the second time in order to vote, the system will check whether his electronic identity has been successfully established or not. In case that everything has been done in a proper order, the voter will receive a time-based verification code on his phone. The code could last for two minutes. Also, the voter must confirm his identity using his own private key.

2.4 Summary

So far we highlighted the use of the eID and electronic signature instead of relying on the database management system's self-created IDs to increase the e-voting system's security level. Now, let us summarize accomplishments of this approach highlighted in this section as follows:

- Obligating the voters to capture photos of themselves and their ID cards is part of the system's processes. These images allow the EAR to verify whether the voters have established their own eID or there are impersonations, especially the impersonating a deceased people. These will be converted to bits format, encrypted and stored in the database. In addition, it will be part of the eID hash, so it will be impossible to decrypt, change or manipulate images.
- Nowadays almost every individual has his own cell phone, which gives us the possibility to use it to secure the voters' data from getting exposed. Otherwise, supposing that we did not use the time-based verification code method, and someone's private key got exposed anyone who has the private key and the individual's information can re-login into the system and cast a vote as he desires. That is why the verification code is used to add another level of protection. Even though an individual's information and the private key could fall in other's hand, they will not be able to open the system unless the verification code can be received in time.

3 Voting, Verification and Tallying Process

Basically, in any election employing secret ballot-open counting, the following points must be considered: (1) Vote by secret ballot: a vote's originator must be unknown. More clearly, the voter's ballot shall not be violated by anyone anyhow, which means it must be completely concealed whether it is directed to a candidate or it is null. (2) The voters must be marked in the electoral registry that they voted to prevent duplicate voting or cheating and to allow only the legitimate voters to cast a ballot. (3) Open counting: the outcome of the election should be determined by an open verifiable counting of the votes.

Unlike the Bitcoin [19], which uses a single public blockchain, the proposed digital voting system will use two private/permissioned blockchains: Let's call the first blockchain as BL-v, the voters' identities and the second one as BL-b, the encrypted ballots. The use of permissioned blockchain limits the parties who can read the information, also, restricting the nodes and separating it in different locations around the country. Address generation, transaction creation, block creation and vote tallying processes are described below.

- **Addresses.** Every transaction requires candidates' new addresses in order to keep the vote-voter anonymity every single time voter cast a vote [10]. The generated address will be specified by the recipient to receive the vote credit.

Sending a checksum with an address permits to verify that address was not manipulated. In addition, preventing any possibility of a Men-In-The-Middle Attack by making the communications between voter and candidate going through a secure channel using a handshake protocol, as illustrated in Fig. 2.

Fig. 2. Voter-candidate authenticated communication.

- **Transactions and Blocks.** At this stage, voters' identity must be registered once and their votes should remain anonymous. To achieve this, the identity will be stored in the BL-b blockchain, and his vote in the BL-v blockchain. As mentioned previously, a voter can vote as many times as long as his first vote is registered and counted only. So the voter will cast a transaction containing his eID and the recipient's address encrypted and signed by his own public key. Then post it to the BL-b. Once the BL-b block is successfully created the recipient's address will be decrypted by the voter and encrypted again with the recipient's public key to create a BL-v transaction. Eventually, the transaction will be posted with the one value credit to the BL-v, as illustrated in Fig. 3.
- **Vote Tallying.** Vote tallying will be one of the responsibilities of the Electoral Authority Representative for that constituency. EARs are (virtual agent) persons or institutions permissioned by the network and directed by the state to perform audit and certification of the votes/voters. An EAR could be assigned for every polling station, or even one per district.

Candidates/parties will get a summary ballot tallying just the votes received; furthermore, it will be fully controlled by the EARs. That is how the vote tallying will

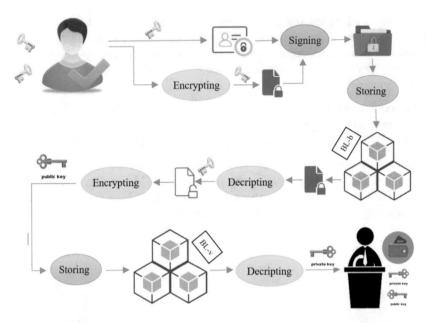

Fig. 3. Voting architecture.

happen automatically. EAR will verify both of the blockchains, also counting or listing all voters thus verifying who did cast his vote and who did not.

4 Threat Model

Maintaining sufficient security of the e-voting paradigm requires overcoming many difficult problems in cybersecurity, especially, with existing technology, a smallest lapse can affect the election result's integrity. Undesirable experiences in different places around the world (New South Wales, Estonia, India, etc.) in real-world have come up in online voting examples which demonstrated many security issues. There are many kinds of attackers (foreign countries, deceptive voters, funded criminals, etc.) who demonstrated realistic threats; many of these threats must be taken as an example while designing or setting online voting. Let's identify some potential attacks and any exposures that might exist to highlight the way to avoid and overcome it.

4.1 Cross-Site Scripting (XSS)

XSS vulnerabilities are a type of injection, simply attacker who can exploit an XSS attack could gain the ability to act as the victim. Moreover, both the voter and the vulnerable system, often will not be aware of the attack. Using the well-known practices below all together considered as a great way to defeat the majority of XSS vulnerabilities [23]. Let's outline these methods below:

Escaping. Escaping data means insuring that the data is secure before rendering it to the end user. So escaping user input and key characters prevent received data from being interpreted in any malicious way.

Validating Input. Input validation prevents voters from inserting any special characters, instead of rejecting the request.

Sanitizing. Assuring the input data will not do any harm to users and database by cleaning the data from any potentially harmful markup, moreover, changing untrusted user inputs to an acceptable format.

4.2   FREAK Attack

The SSL/TLS vulnerability (Factoring Attack on RSA-Export Keys) may allow attackers to decrypt secure communications between vulnerable clients and servers (intercept HTTPS connections) and force them to use older and weaker encryption, also known as the export-grade key or 512-bit RSA keys [24]. Let's highlight the necessary precautions against the Freak attack by:

- Disable the support for TLS export cipher suites,
- Disable the support for insecure known ciphers (not only RSA export ciphers),
- Disable support for ciphers with 40- and 56-bit encryption,
- Enable forward secrecy.

4.3   Malwares

Malicious attackers who tend to manipulate systems to have more access by promoting their privileges, can install Trojans or backdoors by using pre-existing botnets and target a specific country or region [22]. In this case, it would be easy for them to fully control data of the infected voters' computers.

Voters must be very careful and familiar with the computer protection instructions to be able to protect their own computer. Some of the common critical steps to protect voters' computers are being highlighted below:

- Installing firewall.
- Installing security software from a reliable company.
- Setting the operating system and the web browser to update automatically.
- Making sure that the web browser's security setting is high enough to detect unauthorized downloads.

5   Conclusion

This paper proposes a novel e-voting architecture to properly facilitate digitalized elections maintaining vote and voter privacy while preventing fraud. The key processes of the architecture, namely, voter ID creation, voting, vote verification and vote tallying are introduced in introducing the logical design of the digital voting system. The key technologies used in affecting such results are the distributed ledger of Bitcoin, namely the blockchain technology, and RSA public-private key system. Moreover, finding some solutions to overcoming many server- and client-side attacks faced by earlier elections was indicated.

Fundamentally, the blockchain technology's ledger decentralization and blocks serialization provide great tools against results manipulation. So, using the proposed blockchain based architecture leads to achieve sound and fair election. Let's not forget the issue of impersonation. By deliberately opening the system some time before the election, voters will be allowed to establish their identity. If there is any impersonation, it will be discovered. Also, eID's legitimate owner only will be able to use it due to the time-critical one-time password algorithm.

Furthermore, an effective approach proposed to address the issue of vote-selling by making vote vendors unreliable is one of the most important solutions to this structure. Several ways and possibilities allow a candidate/party to buy a voter's ballot. Firstly, by taking all login information of the voter and login into the system in his place during the polling day and casting vote. Secondly, asking a voter to attend a specific place and make him cast his vote for them remotely. Thirdly, a voter can be asked to video himself during the voting process. Our proposed approach involves allowing voters to cast a vote unlimited times, with respect the first vote cast will be accepted due to the fact that the structure of the blockchain in itself prevents double voting.

It should be noted that the proposed scheme permits absentee ballot due to its web-based nature allowing remote vote casting; however mail-in ballot (postal vote) and proxy voting are not permitted. The identified approach involves tradeoffs and may not be suitable for all. Some citizens, especially older ones, may not be able to cope with the complexity of digital voting: eID creation, PKI use, or even using a browser. It can be solved by posting guidance videos or preparing help centers to guide voters who have difficulty following the voting procedures. As a stop-gap measure, postal and proxy vote may be authorized in advance in certain exceptional cases, such as inability and incapacity to reach the polling station.

References

1. Brightwell, I., Cucurull, J., Galindo, D., Guasch, S.: An overview of the iVote 2015 voting system, New South Wales Electoral Commission, Australia, Scytl Secure Electronic Voting, Spain (2015)
2. Rivest, R., Shamir, A., Adleman, L.: A method for obtaining digital signatures and public-key cryptosystems. Commun. ACM **21**, 120–126 (1978)
3. Nakamoto, S.: Bitcoin: a peer-to-peer electronic cash system (2008). http://bitcoin.org/bitcoin.pdf
4. Glaser, F.: Pervasive decentralisation of digital infrastructures: a framework for blockchain enabled system and use case analysis. In: Hawaii International Conference on System Sciences, Goethe University Frankfurt, Hawaii (2017)
5. Kizhakkedathil, N.: A Study into the Prospects of Implementing End-To-End Verifiability in Estonia Voting. Tallinn University Of Technology, Faculty of Information Technology, Department of Computer Science, Tallinn (2016)

6. Zyskind, G., Nathan, O., Pantland, A.: Decentralizing privacy: using blockchain to protect personal data. In: IEEE CS Security and Privacy Workshops (2015)
7. Lyon, D.: National IDs in a global world: surveillance, security, and citizenship. Case West. Reserve J. Int. Law **44**, 607–623 (2010)
8. Johnson, D., Menezes, A.: The elliptic curve digital signature algorithm (ECDSA). Technical report CORR 99–34, Department of C&O, University of Waterloo, Canada (1999)
9. M'Raihi, D., Machani, S., Pei, M., Rydell, J.: TOTP: time-based one-time password algorithm. Internet Engineering Task Force (IETF) (2011)
10. Dunphy, P., Adleman, L.: A First look at identity management schemes on the blockchain. In: IEEE, VASCO Data Security (2018)
11. Hastings, N., Peralta, R., Popoveniuc, S., Regenscheid, A.: Security considerations for remote electronic UOCAVA voting. National Institute of Standards and Technology, NISTIR 7770 (2011)
12. Springall, D., Finkenauer, T., Durumeric, Z., Kitcat, J., Hursti, H., MacAlpine, M., Halderman, J.J.: Security Analysis of the Estonian Internet Voting System, University of Michigan, Open Rights Group. ACM, New York (2014)
13. Halderman, J.A., Teague, V.: The New South Wales iVote System: Security Failures and Verification Flaws in a Live Online Election. University of Michigan, University of Melbourne [cs.CR]. arXiv:1504.05646v2 (2015)
14. Adrian, D., Bhargavan, K., Durumeric, Z., Gaudry, P., Green, M., Halderman, J.A., Heninger, N., Springall, D., Thome, E., Valenta, L., VanderSloot, B., Wustrow, E., Zanella-Beeguelin, S., Zimmermann, P.: Imperfect forward secrecy: how Diffie-Hellman fails in practice (2015)
15. Durumeric, Z., Adrian, D., Mirian, A., Bailey, M., Halderman, J.A.: Tracking the FREAK attack. https://freakattack.com/
16. Mckay, R.: Flaws in iVote's re-vote process which attempts to defeat coercers. http://www.bigpulse.com/governmentelections#changevoteaw
17. Jones, D.W., Simons, B.: Broken ballots: will your vote count? Stanford University Center for the Study of Language and Information, California (2012)
18. Cross-Site Scripting. http://shiflett.org/articles/cross-site-scripting
19. Parsovs, A.: Practical issues with TLS client certificate authentication, University of Tartu, Software Technology and Applications Competence Center, Estonia (2014)
20. Moura, T., Gomes, A.: Blockchain voting and its effects on election transparency and voter confidence. In: Proceedings of the 18th Annual International Conference on Digital Government Research, USA, pp. 574–575. ACM (2017)
21. McCorry, P., Shahandashti, S.F., Hao, F.: A smart contract for boardroom voting with maximum voter privacy. In: International Conference on Financial Cryptography and Data Security. Springer, pp. 357–375 (2017)
22. Danchev, D.: Study finds the average price for renting a botnet, ZDNet, May 2010. http://www.zdnet.com/blog/security/study-finds-theaverage-price-for-renting-a-otnet/6528
23. Vonnegut, S.: Preventing XSS: 3 Ways to Keep Cross-Site Scripting Out of Your Apps, October 2017. http://www.zdnet.com/blog/security/study-finds-theaverage-price-for-renting-a-otnet/6528
24. Vonnegut, M.: FREAK Attack: What You Need to Know, March 2015. http://www.zdnet.com/blog/security/study-finds-theaverage-price-for-renting-a-otnet/6528

Botnets the Cat-Mouse Hunting

Teresa Guarda[1,2(✉)], Samuel Bustos[1], Washington Torres[1],
and Freddy Villao[1]

[1] Universidad Estatal Península de Santa Elena – UPSE, La Libertad, Ecuador
tguarda@gmail.com, ing.samuelbustos@yahoo.com,
wtorresguin@gmail.com, fvillao@upse.edu.ec
[2] Algoritmi Centre, Minho University, Guimarães, Portugal

Abstract. The evolution of information and communication technology and the exponential growth of the Internet are been accompanied by the spread of malware, being currently one of the most serious security threats on the Internet. Today, given the technological power of a device connected to the Internet, where a large amount of confidential information is stored and, in many cases, without a security standard in relation to this information; it should be keep in mind that as new assistive technologies emerge, there are individuals or organizations that attempt to infiltrate each of these devices to obtain unethical information or confidential data to carry out a particular attack, one of the main threats is the botnets. This article aims to analyze the botnets forms of propagation within a network, violating the security measures implemented and advanced protection solutions.

Keywords: Cyber security · Malware · Botnets
Advanced Malware Protection

1 Introduction

Since the beginning of the Internet in the late 1960s, when the number of linked sites was reduced, through the 1990s when 1 billion people connected to the Internet with their desktops and laptops, and by the first decade of the 21st century, where more than 2 billion people connected to the Internet through their phones cell phones, and Cisco Systems has predicted that 25 billion things will be connected to the Internet and to each other in 2015, being this number in 2020 50 billion [1, 2].

The information and communication technologies innovation is happening at fast speed. Wireless technologies, sensor networks, smart networking, wearable's, accompanied by the growth of the Internet and the diffusion of its uses, are creating new sources of business value [3] and have brought numerous challenges for the protection of privacy, data integrity and security. The Internet has become an integral part of our lives, supporting a huge range of services, such as commerce, administration, health care, and banking, administration, from being just an information access tool to becoming a platform for disseminating opinions, sharing personal facts, socializing and working [4]. Among the various security challenges, one of the biggest

© Springer Nature Switzerland AG 2019
T. Antipova and A. Rocha (Eds.): DSIC 2018, AISC 850, pp. 408–416, 2019.
https://doi.org/10.1007/978-3-030-02351-5_46

threats is malware, an application with a variety of malicious behavior that endangers user data and can compromise infected computers.

Today the Internet attacks have undergone a remarkable transformation, while some time ago they focused on affecting the availability of infrastructure and services, today they also target people and organizations. Behind these new attacks there are compromised hosts, located in homes, schools, private and governmental organizations, that are infected with a bot that communicates with a controller and other bots that form what is commonly known as botnet [5]. There are several ways of propagation and control of botnets, which continue to increase their characteristics in order to avoid being detected by those affected. Malware nowadays seeks to be smarter and more careful in collecting information by camouflaging itself in the form of legitimate software in order to avoid the defense mechanisms that exist in today's systems. Based on this problem, several techniques are also used to detect botnets based on their behavior and traffic with the help of genetic algorithms.

This article contemplates background information on botnets, the means of propagation and control that are carried out on them and continuing with some ways to mitigate the latent danger to which the organizational networks are exposed on a daily basis and culminating with the conclusions of the work exposed.

2 Malicious Software

Based on the results of recent studies, that point to malware as being responsible for a significant portion of current cyber-security incidents. Malware is considered to be any program capable of causing damage to a user, computer or computer network, which can take various forms [6]. Malware is a malicious program used to violate system security policy with respect to the CIA triad (confidentiality, integrity, and data availability) [7, 8]. Malware is categorized into different types: Virus, Trojan, Warm, Spyware, Flooders, Drive by Download, Adware, Spammers, Keylogger, Backdoor, Rootkit and Bot/Botnet, according to how to impose threats to the system. We turn to a brief description of each type of malware [8, 9]:

- **Virus**: This variant is replicated whenever it is run looking to infect other machines or executable programs. Its propagation is made from machine to machine, leaving new infections as it spreads through executable files infected.
- **Trojan**: Program that acts discreetly, masquerading as a legitimate program in order to bypass the security mechanisms of the target machine. It can be used to obtain potentially relevant information contained in infected system files.
- **Warm**: Worms are able to operate independently, spreading to other computers or computer networks by exploiting vulnerabilities in those systems. The active demand for new machines and systems that can be compromised is the characteristic that best defines this variant, serving the already compromised systems as platforms for disseminating new infections.
- **Spyware**: Software used for the collection of information, from keystrokes, screen data, network traffic, webcams, among others. The information collected is then sent

to the malicious agent. This program is normally installed without the knowledge of the user, allowing the discreet collection of the information.

- **Flooders**: Used to generate considerable traffic volumes, aiming client denial of service (DoS) of the target system.
- **Drive by Download**: Attack used by code inserted in vulnerable web pages, aiming to attack the client of the system that accesses these pages.
- **Adware**: Malware type less evasive, but perhaps more profitable. The main function is the dissemination of various types of publicity, through its integration into programs. Advertising is presented in the form of banners, pop-up windows, web pages, email messages among other services available in the Internet.
- **Spammers**: Used to send abusive volumes of unwanted email. They can be installed on previously compromised machines in order to use them in sending unwanted email. Similar to Adware is a type of malware that can be very lucrative considering that the attackers are, in many cases paid by entities interested in disclose their products or services.
- **Keylogger**: It registers all the keystrokes of a infected machine, and the collected information can be stored locally or sent remotely to the malicious agent.
- **Backdoor**: Any mechanism that allows unauthorized access to features of that same program or system, and is not subject to any security procedures. Although commonly used by programmers for testing, this feature can be exploited so that a malicious agent guarantees unauthorized remote access to a machine. In other cases, it assumes the malicious code function installed on a victim's computer, in order for the attacker to remotely access the compromised system.
- **Rootkit**: Set of tools used by the attacker after having secured privileged access to a system. Privileged access gives the attacker complete control of the system's features, allowing it to change files, monitor processes, network traffic. These tools have the capacity to make changes in the infected system in order to circumvent the security mechanisms of the same system. They may contain other types of malware such as backdoors, in order to grant remote attacker access.
- **Bot/Botnet**: Program present in an infected machine that can be used to execute attacks on other machines. Usually these bots are part of a network, sometimes on a global scale of infected computers called botnet. Infected computers (zombies) are controlled by cybercriminals (botmasters). After being infected, the machines are often used to spread spam, distributed DoS attacks, among others.

Innumerous computer systems dispersed around the world are infected with several classes of malware, being one of the biggest threats the botnets. Despite the existence of antivirus software, intrusion detection and prevention systems, firewalls, etc., threats resulting from malicious software persist and continue to grow, finding new ways to evade defense mechanisms.

3   Botnets

Many computer systems scattered around the world are infected with several classes of malware. Despite the existence of various security techniques operated through firewalls, antivirus software, intrusion detection and prevention systems, and others; the threats resulting from malicious software persists and continues to grow, finding new ways to bypass the mechanisms of defense.

Is known a large variety of malwares, and among them are bot malwares that are used to raise botnets. These refer to a distributed set of computers and/or mobile devices controlled by an attacker to perform coordinated attacks [5].

One of the modern types of threats to the security of computer systems is the botnet. Botnet is a collection of robot software that works on host computers, autonomously and automatically, controlled remotely by one or more intruders.

Botnets are a complex and constantly evolving challenge that affects user confidence and Internet security. The fight against botnets requires cross-border and multidisciplinary collaboration, innovative technical approaches and the widespread deployment of mitigation measures that respect the fundamental principles of the Internet. The name "bot" stands for robot and "net" is the abbreviation for network, it is a robot that will perform its actions on the network through commands sent remotely by the network controller [10].

Among the components of the botnet we have: Bot (software installed on the victim's machine capable of performing a set of actions, usually malicious); Victim machine (machine infected by bot); Botmaster (user who has control of the network; and Command and Control (C & C) center (the means by which the botmaster sends the commands to the bots). Thus, the botnet is a bot managed by the botmaster through the command and control center, as shown in Fig. 1.

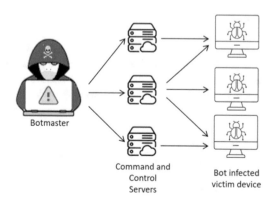

Botmaster

Command and
Control
Servers

Bot infected
victim device

Fig. 1. Botnet network components.

Botnets once inside the target network will facilitate several types of attacks, such as [11–13]: gathering information; distributed denial of service attack (DDoS);

spamming; sniffing traffic; malware repository; anonymity; illegal content; installation of Add-ons; etc. (see Table 1).

Table 1. Botnets different types of attacks.

Types of attacks	Action
Gathering information	Some bots have the ability to capture keys typed on the keyboard, points on the screen when mouse clicks, files, network traffic, and stored data. This ability can be used to capture information about bank cards, commercial strategies of large companies, documents of high level of secrecy, among others
DDoS	Botnets are used to launch denial-of-service attacks that can stop a target service or server. As the name indicates causes inaccessibility to a service by the original users. This attack commonly causes loss of connectivity due to excessive consumption of bandwidth or overload of internal resources of the sabotaged network
Spamming	Action that allows send spam to a set of electronic addresses, with the purpose of disseminates fraudulent or veracious (advertising) information. In this way, bots that have the ability to be introduced by a generic proxy over TCP/IP based networks to an infected machine arise, to an infected machine, and then proceed to use it to relay spam
Sniffing traffic	Use a sniffer of packages to see the data in clear and seize sensitive information such as users and passwords
Malware repository	Botnet controllers need resources to keep tools available. To this end, some machines are used as repositories (client bots). Placing these tools on multiple machines ensures botnet availability
Anonymity	When multiple machines around the world are used as bridges to access a compromised host, it is very difficult to perform a trace and identify the actual botmaster or intruder
Illegal content	Botnets can be used for illegal content storage such as files, documents, credit card numbers, and pornography
Add-ons installation	Creating a fake website with advertising and signing up for companies that pay for clicks on advertising is usually a great source of income for attackers. With the help of a botnet, these clicks can be automated so that a few thousand bots click on the advertisements

The life cycle of a Botnet is presented in five phases (Fig. 2). It begins with the Initial infection phase (1) in which the attacker seeks vulnerabilities in the technological equipment of a subnet to infect it, after this act the secondary injection phase (2) is executed in which a Shell (Script) is executed, and includes a bot executable in binary and installs itself. Continues the connection phase (3) in which a direct connection is established to be manipulated by the BotMaster, using the C & C so that the bots receive orders to be executed and culminating with the maintenance and update phase (5) in which it orders the bots to download updates or expand to other C & C equipment to avoid being detected by the security methods of said subnet [14, 15]. It should be noted that in phase (4) the malicious activities can be as wide-ranging as the types of attacks presented in Table 1.

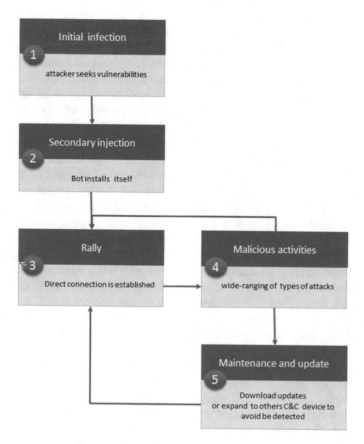

Fig. 2. Botnet life cycle (adapted from [15]).

Phases (3) and (5) are the most vulnerable and therefore the most important for detection schemes. If the botnet reaches stage (4), the detection scheme has failed and the cyber defense system must act at the next stage, which is the treatment (detection, mitigation, protection) of malicious actions [14].

4   Advanced Malware Protection

Currently, malware seeks to be more intelligent and judicious in collecting information, which allows a form of readable software to be used in the current systems. We are facing a hunting game with the usual two players, attackers and defenders. Attackers have increasingly irreverent and intelligent forms of attack that allow them to control systems; and defenders in turn have to innovate and upgrade systems to protect resources in the event of an attack. In this context, systems are hostage to the detection and mitigation of botnets strategies to prevent cybercrime, avoiding reaching a stage in which it is impossible to control [14].

The most recent cases of ransomware prove that even the most protected infrastructures are not safe from this type of threats increasingly capable of deceiving the defence mechanisms implemented. Then, the detection of new-generation malware and its mitigation, in the current context of threats increasingly sophisticated and intelligent, imposes new forms of defence in a species of cat-mouse hunting; being imperative the implementation of new defence's mechanisms, that enable Advanced Malware Protection (AMP) [16–18]. Nowadays there are several solutions of AMP developed by different manufacturers but with similar objectives and functionalities, as is the case: Palo Alto WildFire Networks; Cisco AMP; and FireEye Network Security.

The focus of this paper is the Cisco Advanced Malware Protection (AMP) platform. Cisco AMP is a security solution that analyzes the entire lifecycle of an advanced malware threat that not only prevents against intruders but also ensures the visibility needed to detect and mitigate them, and ensures protection across the spectrum of infection before, during and after a malware infection [19]. Prior to an attack, AMP uses the intelligence platform of the Cisco Talos Security Intelligence and Research Group and Threat Grid team, strengthening the system's defences against known or emerging threats. During an attack, the AMP uses known file signatures and the Cisco Threat Grid malware dynamic scanning technology. Identifies and blocks files that violate previously defined policies, attempts to exploit vulnerabilities, and malicious files that attempt to infiltrate the network. After an attack, or after the initial analysis of a file, AMP goes beyond real-time detection by continually monitoring and analysing all traffic and file activity, looking for possible indicators of malicious behaviour, regardless of the classification initially given to files. The AMP alerts the incident response team whenever it detects that a file, initially classified as clean or unknown, begins to display suspicious or malicious behaviour [19]. Cisco argues that most anti-malware solutions in network infrastructures and systems only inspect the files at your point of entry. The sophistication of malware allows them to bypass defence mechanisms. Techniques such as polymorphism, ciphering, or the use of unknown protocols are taken as forms adopted by malware to go unnoticed upon entry into a network. Cisco's AMP solution ensures continuous file and network traffic monitoring, even after initial inspection [20].

5   Conclusions

Currently, computer networks have been affected by various types of computer attacks, which leave their machines infected acting in an automated and unconscious way on the part of the users, forming what is called by Botnet.

Given the variety of network communication platforms, their potential vulnerability and the degree of sophistication of new malware threats, it is imperative that every organization have a more effective security strategy. This strategy should take into account the acquisition and use of recent and sophisticated technologies in order to combat threats and ensure the effectiveness of the defense system.

Nowadays there are several solutions like Cisco AMP developed by different manufacturers but with similar objectives and functionalities. It can be concluded that a platform such as Cisco AMP should not be considered as a substitute for all other existing technologies, but rather as a security complement to the threat of specific malware. These types of solutions are the best option for detecting and mitigating advanced evasion techniques.

References

1. McAuley, D., Mortier, R., Goulding, J.: The Dataware manifesto. In: Proceedings of the Third International Conference on Communication Systems and Networks (COMSNETS), pp. 1–6. IEEE, Bengalore (2011)
2. Nordrum, A.: The internet of fewer things [news]. Spectrum **18**, 12–13 (2016)
3. Guarda, T., Leon, M., Augusto, M., Haz, L., Orozco, W., Alvarez, J.: Internet of Things Challenges, Lisbon, Portugal, pp. 14–17 (2017)
4. Beigi, E., Jazi, H., Stakhanova, N., Ghorbani, A.: Towards effective feature selection in machine learning-based botnet detection approaches. In: International Conference on Communications and Network Security (CNS), pp. 247–255 (2014)
5. Feily, M., Shahrestani, A., Ramadass, S.: A survey of botnet and botnet detection. In: IEEE Third International Conference on Emerging Security Information, Systems and Technologies SECURWARE 2009, pp. 268–273 (2009)
6. Leloglu, E.: A review of security concerns in Internet of Things. J. Comput. Commun. **5**, 121–132 (2016)
7. Chakraborty, S., Stokes, J., Xiao, L., Zhou, D., Marinescu, M., Thomas, A.: Hierarchical learning for automated malware classification. In: IEEE Military Communications Conference (MILCOM), pp. 23–28 (2017)
8. Stallings, W.: Cryptography and network security: principles and practice. Pearson, Upper Saddle River (2017)
9. Alsulami, B., Srinivasan, A., Dong, H., Mancoridis, S.: Lightweight behavioral malware detection for windows platforms. In: IEEE 2017 12th International Conference on Malicious and Unwanted Software (MALWARE), pp. 75–81 (2017)
10. Singh, K., Guntuku, S., Thakur, A., Hota, C.: Big data analytics framework for peer-to-peer botnet detection using random forests. Inf. Sci. **278**, 488–497 (2014)
11. Al-Hakbani, M., Dahshan, M.: Avoiding honeypot detection in peer-to-peer botnets. In: IEEE International Conference on Engineering and Technology (ICETECH), pp. 1–7 (2015)
12. Kok, J., Kurz, B.: Analysis of the botnet ecosystem. In: 10th Conference of Telecommunication, Media and Internet Techno-Economics (CTTE), pp. 1–10. VDE (2011)
13. Iqbal, S., Kiah, M., Dhaghighi, B., Hussain, M., Khan, S., Khan, M., Choo, K.: On cloud security attacks: a taxonomy and intrusion detection and prevention as a service. J. Netw. Comput. Appl. **74**, 98–120 (2016)
14. Acarali, D., Herwono, I.: Survey of approaches and features for the identification of HTTP-based botnet traffic. J. Netw. Comput. Appl. **76**, 1–15 (2016)
15. Silva, S.S.C., Silva, R.M.P., Pinto, R.C.G., Salles, R.M.: Botnets: a survey. Comput. Netw. **57**(2), 378–403 (2013)
16. Acarali, D., Rajarajan, M., Komninos, N., Herwono, I.: Survey of approaches and features for the identification of HTTP-based botnet traffic. J. Netw. Comput. Appl. **75**, 1–15 (2016)

17. Guarda, T., Orozco, W., Augusto, M., Morillo, G., Navarrete, S., Pinto, F.: Penetration testing on virtual environments. In: 4th International Conference on Information and Network Security, pp. 9–12. ACM (2016)
18. Adamy, D.: EW 104: Electronic Warfare Against a New Generation of Threats. Artech House, London (2015)
19. Santos, O., Kampanakis, P., Woland, A.: Cisco Next-Generation Security Solutions: All-in-one Cisco ASA Firepower Services, NGIPS, and AMP. Cisco Press, Indianapolis (2016)
20. Cisco Advanced Malware Protection for Endpoints Data Sheet. In: Cisco Advanced Malware Protection for Endpoints Data Sheet. https://www.cisco.com/c/en/us/products/collateral/security/fireamp-endpoints/datasheet-c78-733181.html

Digital Platforms as Means of Social Interaction: Threats and Opportunities in Online Affective Relationships

Lidice Haz[(✉)], Ximena Acaro, Carlos Julio Guzman, Luis Espin, and Maria Fernanda Molina

Facultad de Ciencias Matemáticas y Física, Ciudadela Universitaria "Salvador Allende", Universidad de Guayaquil, Av. Delta y Av. Kennedy, Guayaquil, Ecuador
victoria.haz@hotmail.com, {ximena.acaroc, carlos.guzmanr, luis.espinp, maria.molinam}@ug.edu.ec

Abstract. Nowadays, the way in which human interactions are developed has been altered due to the diverse technologies available in the society. The Internet and social networks constitute tools that foster the development of human dimensions, the social and technological. Digital platforms promote social relationships by creating virtual spaces where the human interaction overcomes barriers such as time, distance, language and other variable that could limit communication. There are websites that ease the search for a person who matches criteria or parameters determined by the user; otherwise, finding an ideal match in the real-life situation would not be that simple. This work aims to describe the risks, threats and opportunities involved in the use of digital platforms, as well as the security mechanism that minimize the risks which the users are exposed to. Moreover, it analyzes the success or failure of the affective relationships established through digital means. Overall, the social interaction factors are described from the technological and cultural perspectives which boost that interaction in virtual spaces.

Keywords: Risk · Online dating · Relationships · Digital platforms
Social interaction · Virtual environments

1 Introduction

The current society encourages the use of information and communication technology as fundamental tools to interact through the web [1]. From that, it could be said that the use of technology has let to deal with the boundaries of time and space by creating complex environments, where people can interact in the virtual world with others regardless of their geographical location. The omnipresence, the functionality, the transcendence and the potentiality of the Information and communication technologies (ICTs) make them powerful instruments, which if utilized with a bad intention, could highly affect the life of people, especially, the ones' who are frequent users in the denominated online dating sites [2]. The great number of users in social networking sites, particularly, in websites that offer the service to find an ideal match to date or to

T. Antipova and A. Rocha (Eds.): DSIC 2018, AISC 850, pp. 417–425, 2019.
https://doi.org/10.1007/978-3-030-02351-5_47

set a sort of affective relationship lets this situation become a social phenomena that generates interest in the context of research due to the numerous threats and opportunities for the users. Among the opportunities that could exist, there is the emergence of successful affective relationships from friendship and professional relations, to even, romantic ones. This questions if it is the social network which exerts influence or if there are other factors implied, such as social, cultural, economic, or user's personality traits and gender [3]. Developing interpersonal relationships online exposes the users to threats and risks existing in the Internet, especially in these websites. The most palpable threat is related to the access to unreliable information, especifically, fake user profiles, fraud, lose of personal data, and even the possibility of being victims of human trafficking [4].

For the reasons mentioned above, this study presents an analysis of websites from technological and cultural perspectives in order to attempt to understand the factors that motivate their usage. For that, statistical data about online affective relationships index, the quantity of female and male users registered in these sites, and interviews addressed to users who have positive and negative experiences in their use.

2   Interaction in Social Virtual Spaces

In the virtual world, the users look for keeping in touch with a group of people through the interaction by messages and distribution of personal information. In this way, the social networking produces a debate regarding to the relation of the data privacy, the risks and the opportunities that are immersed in this type of relations, the addiction of keeping interconnected to known people and strangers, the proper use of free time, among others [3]. Consequently, the virtual spaces are means established in different contexts which aim the social interaction through the use of a digital platform. Because of that, it is set up a dynamic exchange among people, groups, and institutions that are related due to affective ties such as friendship, family, ideas, hobbies, religions, work relations, and others [5]. Additionally, they comprise an open system, in continuing construction, which involve people who feel identified with the same needs and problems, and their principles are to create, share, and collaborate in the virtual world [6]. It is important to determine the aspects that promoes the use of virtual spaces between a group of people who preten to establish affective relationships through the registration of a profile in a website. This information eases that different users have access or see the profile of one or various people according to the criteria of search set up [7].

2.1   Virtual Spaces and the Factors of Social Interaction

This paper describes two factors which nurture the social interaction in virtual environments, particularly in the social network sites. Also, it analyzes the technological component that determines the innovation and the ease of use of websites; and the cultural component that entails the level of social acceptance in the use of online dating websites without resulting in prejudice.

2.1.1   Technological Factor

The advances in technology and the evolution of the Internet have created new trends from the web 1.0, web 2.0 and web 3.0 that together let the creation of new websites which incorporate diverse elements to facilitate users' interaction, the center of information and content generator [8]. The ICTs become the transversal axis for socio-technological development, which, in turn, promotes the creation of virtual spaces where participation and interaction with other people is carried out from anywhere in the real world [9].

2.1.2   Cultural Factor

A virtual space simplifies access to the information available on the web. These virtual spaces allow to meet up and get to know individuals thanks to the information that is collected about them, which is stored in the user's profile [10]. In today's society, it is common that people interact with other human groups through virtual environments. In this sense, multiple studies have been conducted with the intention of analyzing the links between people, family groups, organizations and even social groups in different cities or countries [11]. Currently, online dating is accepted and increasingly more frequent. There are multiple websites that provide this service to users around the world, making online dating sites a very lucrative business and their industry growth exponential in recent years.

The question is: why do people prefer to interact in virtual spaces in order to establish emotional relationships? To try to answer this question, it is required to mention that it is known that today's society has undergone great changes by inte-grating ICTs in the development of their daily activities. This has generated different patterns of behavior and lifestyles. Among the variables that define participation in these virtual spaces, there is the time available for men and women to interact with people in their real environment. People spend a large part of their time working and academic activities, so their free time is diminished and also, the possibility of finding partners. In this context, various companies have taken advantage of this situation and with the help of ICT's, multiple digital platforms have been implemented to satisfy this need that is increasingly common in society.

2.2   Risks and Threats Existing in the Use of Virtual Spaces

The use of technological platforms and mainly social networks that foster social interaction in Internet-based spaces and the distribution of personal information of the user becomes a latent threat in relation to data privacy [12]. The criminals take advantage of the technologies and technify their methods of search and selection of potential victims. In most of cases, they utilize social networks, notably, the online platforms for seeking a spouse. These virtual spaces are often used as the first means of contact with their victims. The highest-level risks are human trafficking, kidnapping, rape, blackmail, identity theft, sexual harassment, pornography, fraudulent appropria-tion of personal data, among others. In this context, the users of these virtual media are exposed to threatening and dangerous situations for their physical and mental integrity [13].

In order to promote social interaction, the technological platforms, defined in this work as virtual spaces, make easier that the user communicate and share information based on the profile that is exposed in the cybernetic community [14]. In a first approximation, it is shown that the main reason to employ these online platforms correspond to the lack of time that people in the real world have to communicate or establish relationships with other people. In this group, there are users whose intentions to create affective bonds are real [15].

Other variables are related to crimes that can be explained from a social, economic and cultural context that the offender lives in his real life. For example, armed conflicts in different parts of the world, food shortages, poverty, insecurity among other aspects, promote migration and for many people these virtual spaces are ideal means to convince users to get out of their hometown.

There are also cases related to organizations whose aim is to attract mainly women to be recruited voluntarily and involuntarily for the purpose of sexual exploitation in forms of pornography, sex tourism, among others. Finally, there are users who simply find themselves in these media to satisfy their curiosities that can range from a simple conversation to an interaction of a sexual nature that is often related to psychological problems framed in hobbies or game addictions, morbid curiosity and sexual fantasies.

According to the scenario described above, it is necessary to ask: Why do people prefer to establish affective relationships through virtual spaces with respect to other forms of interaction and social communication? How is the information published in these virtual spaces used? What effects are evidenced in the behavior of people who use these means to date a person? Those are some of the questions that must be addressed to establish patterns of behavior and criteria about the different mechanisms of computer security that could be applied in these digital environments that are increasingly used by society.

2.3 Opportunities Existing in the Use of Virtual Spaces

The development of social networks and specifically the virtual platforms for seeking love took place around the year 2004. At present, it is a worldwide phenomenon within social networks due to their large number of active members. It is considered one of the virtual spaces whose community of users generates a large amount of Internet traffic. These platforms allow people to create profiles, share photos and videos, chat, send and receive messages and update information. Their interactive nature made these virtual spaces become a mechanism to establish affective relationships [16]. In other research, it is mentioned that social networks facilitate the creation of affective and emotional ties through visual, auditory and textual resources, such as: photos, videos and audios that allow a user's profile to be described in more detail. Therefore, it is imperative to use technology to articulate multiple communicative elements that in turn allow the construction of codes of emotional and affective exchange [17].

In the interviews made to members of the cybernetic community, they indicate that the messages that are sent and received through these platforms are from a simple "hello" to messages of sexual, religious and economic connotation that aims to attract the person that has matched the search criteria of a particular user. In this sense, virtual spaces promote dialogue and expression of ideas and thoughts freely and voluntarily among its members. That is, the development of social skills is encouraged through communication and group interaction.

Another group of users mentioned that after generating the first contact and maintaining it for some time it is possible to create ties of empathy and friendship, which although it does not end in an affective relationship, it was likely that another type of communication, including professional communication, can be established.

2.4   Security Mechanisms to Reduce the Risks Existing in the Use of Virtual Spaces

As mentioned, in relation to the use of virtual spaces, there are various computer and non-computer risks to which users are exposed during their interaction, mainly in social networks. This constitutes a latent threat that can affect users' physical and emotional integrity. To try to reduce this exposure, computer security controls should be implemented to filter the contents and websites to which users will have access, as well as controls on the privacy of data and the authenticity of the website [18]. Here are some of the mechanisms that can be used as controls to ensure user information:

- Avoid downloading multimedia information such as photos, videos or music. As well as, clicking on links that redirect to emails, social networks, or instant messaging programs that do not come from a legitimate user if that type of information has not been requested. This action is required for preventing the installation and execution of hidden programs that can take control of the electronic device;
- Evaluate periodically the security settings of the computer. The objective is to detect possible flaws in the configuration of the firewall, verify the updates of the antivirus and the operating system, among other aspects;
- Verify the security certificate of the website in order to verify the legitimacy of the site and the use of data encryption;
- Verify user data within the website. Contact website administrators through contact emails to report suspicious behavior, or to validate information of registered users at those sites;
- Do not deliver confidential personal information such as address, telephone numbers, financial information, family information, etc. To contact people whose real origin and intentions are not yet known, it is advisable to use electronic addresses created exclusively for this purpose.

The mechanisms described above can be applied as security controls that, although they do not eliminate the threat, try to reduce the security risk [19]. The main objective is to ensure user data to avoid exposing the own real information. In this sense, it is essential that the user be aware of the type of information provided in social networks during their interaction in the virtual space [20].

3   Digital Platforms for Social Interaction

Nowadays, one of the technological trends that has modified the social behavior of people are the so-called online dating sites. In this study, they are defined as virtual spaces for social interaction whose purpose is to help establish affective relationships between men and women. The factors that help explain this type of social interaction are mainly related to the personality of the users, which is observed through the profiles registered on the websites. Additionally, this industry is in constant growth generating great economic gains from the use and registration in these websites.

In the process of virtual interaction, there are multiple variables that are correlated through the personality characteristics, likes and preferences described in the profiles of each user and organized according to the objective to be achieved on the website [21], like searching for penpals, face-to-face friendship, romance-dating and even marriage. This information is necessary to increase the effectiveness of the search algorithms implemented by the websites. The results are obtained from the likes and preferences defined in the user's profile, which is taken as a basis to present potential partners, which is reflected in a liberal logic of choice [22].

In the virtual space, there are around 150 websites that provide the online dating service. Some are free and others accept the payment of memberships that allow you to upgrade to better search results and access more services, mainly online chat and message delivery. In Table 1, some of the websites that have a greater presence in virtual environments are shown.

Table 1. Online dating websites

Website	Year of creation	Active users	Origin of user	Average age
eDarling	2008	More than 14 millions	USA Spain	35 and 55 years
Meetic	2001	More than 30 millions	Europe	25 and 65 years
Latin american cupid	2000	More than 3 millions	North America South America Europe Asia Pacific Middle East	25 and 60 years
Badoo	2006	More than 40 millions	South America Spain Italy France	18 and 55 years
Loventine	2010	More than 2 millions	America Europe	25 and 45 years
eHarmony	2000	More than 30 millions	USA	25 and 45 years
Bumble	2014	More than 22 millions	USA	25 and 45 years
Edarling	2009		Spain	35 and 55 years
Okcupid	2004	More than 5 millions	USA	25 and 45 years

A first observation helps determine that the age of the men who mostly sign up in these websites is between 30 and 55 years. Likewise, the user population that predominates in these virtual spaces is female. In relation to the nationality of the users, they are from all over the world, however there is a greater presence of people from the Americas and Europe. The search criteria that segment the results according to the preferences of the users are: demographics, geolocation, hobbies, income level, level of education, number of children, physical characteristics, social and professional category. This aims to present the best profiles that meet the criteria defined by each user.

4 Conclusions

The virtual spaces represented through social networks provide ideal scenarios to establish channels of communication and interaction between groups of people. In recent years, the rise of websites for social interaction has progressively increased. This is because they are used in various aspects of life with the objective of sharing information and establishing professional, educational, affective or other links. In North and South America, for example, the most used online search networks are match, eharmony, bumble, edarling, latinamericancupid, okcupid, tinder, and grindr. These websites provide specific services according to the target audience. The results of the matches are guaranteed according to the requirements that are differentiated mainly by religious, races, nationalities, or sexual preference defined as heterosexual, bisexual, homosexual, lesbian, among other determining variables in the search for partners.

The psychological risk can be presented as an absolute necessity that coerces the person and leads to a mortifying disease, progressive, full of risks such as addiction and distortion of reality with the virtual; Depending on the user's objective, images of sexual nature, romantic idealistic relationships or sensual and sexual fantasies can be created. The need to interact with "real" people through virtual spaces can modify social behaviors, particularly due to the number of hours spent in the search for couples and their isolation from the real world. In this context, it is critical to consider the difference between hobby and addiction. A person can develop an addiction when the love of an object or activity becomes a need marked by intensity, frequency and loss in self-control.

Physical risk is determined by the presence of cross-border crimes that use technology as means of contact for their commitment. Among the most frequent crimes are human trafficking, kidnapping, rape, blackmail, identity theft, sexual harassment, pornography, fraudulent appropriation of personal data. In this context, it is necessary to apply computer security mechanisms that allow reducing the exposure of these threats and primarily, that users who interact with strangers in virtual spaces are aware of the information and level of trust they provide to these strangers.

5 Recommendations

For future research, it is significant to know the participation of the social and family environment regarding the use of social networks, as an indicator of social and cultural group. Moreover, it would be pivotal to establish comparative studies between professional and non-professional people segmented by their academic, economic and social level. In relation to the personality traits, it is necessary to continue investigating their participation in the love relationships through virtual spaces specifically social networks to seek online dates. This parameter is relevant to know if the personality determines the level of participation and social interaction through these virtual spaces. For this, it is vital to design complete personality instruments, as well as the information regarding to their self-introduction that people use in these social networks.

References

1. Fuchs, C.: Internet and Society: Social Theory in the Information Age. Routledge, Abingdon (2007)
2. Hinduja, S., Patchin, J.W.: Bullying, ciberbullying, and suicide. Arch. Suicide Res. **14**(3), 206–221 (2010)
3. Dominguez, D.C.: The social webs, typology, use and consumption of the webs 2.0 in today's digital society. Documentación de las Ciencias de la Información **33**, 45 (2010)
4. Fire, M., Goldschmidt, R., Elovici, Y.: Online social networks: threats and solutions. IEEE Commun. Surv. Tutor. **16**(4), 2019–2036 (2014)
5. Ledo, M.V., Vidal, M.N.V., García, L.H.: Redes sociales. Educación Médica Superior **27**(1), 146–157 (2013)
6. Del Moral, J.A.: Redes Sociales¿ Moda o nuevo paradigma. Asociación de usuarios de Internet, Madrid (2005)
7. Bergström, M.: La loi du supermarché? Sites de recontre et répresentation de l'amour. Ethnologie françaises **43**, 433–442 (2013)
8. Romero, C.L., de Amo, M.D.C.A., Borja, M.Á.G.: Adopción de redes sociales virtuales: ampliación del modelo de aceptación tecnológica integrando confianza y riesgo percibido. Cuadernos de Economía y Dirección de la Empresa **14**(3), 194–205 (2011)
9. Tobón, S., Guzmán, C.E., Silvano Hernández, J., Cardona, S.: Sociedad del Conocimiento: Estudio documental desde una perspectiva humanista y compleja. Paradigma **36**(2), 7–36 (2015)
10. Monsoriu, M.: Redes sociales profesionales: imprescindibles. Bit, pp. 74–77 (2008)
11. Glaser, M.: Your Guide to Social Networking Online. MediaShift (2007). Retrieved (2008)
12. García, B., López, M.C., García Jiménez, A.: Los riesgos de los adolescentes en Internet: los menores como actores y víctimas de los peligros de Internet. Revista latina de comunicación social **69** (2014)
13. Haz, L., Guarda, T., Zambrano, I., Sánchez, C.: Internet based parenting control application on teenagers. In: 2017 12th Iberian Conference on Information Systems and Technologies (CISTI), pp. 1–6. IEEE (2017)
14. García, B., Croker, L., Torrejón, C., del Rosario, T.M.: Exposición a las redes sociales y ciberviolencia de pareja en universitarios (2017)
15. Tello Navarro, F.: Emociones de computador. La experiencia sentimental de los usuarios chilenos de las páginas de citas (2017)
16. Tom, S., Walther, J.: Computer-mediated communication in personal relationship. In: EnWright, K., Webb, L. (eds.) Relational Maintenance and CMC, pp. 21–43. Peter Lang Publishing, Inc., New York (2011)

17. Peris, R., Gimeno, M.A., Carrero, V., Sanchiz, M.: Las ciberrelaciones: acercamiento psicosocial a las (os) internautas. Ponencia presentada al VII Congreso de Psicología Social y publicada en el libro: La mirada psicosociológica. Grupos, procesos, lenguajes y culturas, Madrid (2000)
18. Amaro López, J.A., Rodríguez Rodríguez, C.R.: Seguridad en internet. PAAKAT: revista de tecnología y sociedad 6(11), 00006 (2017)
19. Sadeghian, A., Zamani, M., Shanmugam, B.: Security threats in online social networks (2013)
20. McDowell, M., Householder, A.: Good Security Habits. US-CERT (2009)
21. Lardellier, P.: Les réseaux du coeur. Sexe, amour et séduction sur Internet. Francois Bourin Editeur, Paris (2012)
22. Lardellier, P.: El liberalismo a la conquista del amor. Algunas constataciones y reflexiones sobre el consumo sentimental y sexual de masa en la era de Internet. Revista de Sociología, Universidad de Chile, no. 29 (2014)

IoT Applications Using Blockchain and Smart Contracts

Rui Roriz[1] and José Luís Pereira[2(✉)]

[1] Mestrado Integrado em Engenharia e Gestão de Sistemas de Informação,
UMinho, Guimarães, Portugal
a64865@alunos.uminho.pt
[2] Departamento de Sistemas de Informação, UMinho and Algoritmi,
Guimarães, Portugal
jlmp@dsi.uminho.pt

Abstract. Blockchain is a relatively new technology, initially created for the Bitcoin's network to store transaction records happening in it. The system is redundant and distributed, making it difficult for fraudulent transactions. Beyond digital currencies, the blockchain concept has already demonstrated its potential in the insurance, health, digital identity, and many other areas. In order to deal with specific needs in those areas, a new technology has appeared – smart contracts – computational code programmed to meet and enforce certain conditions, like the ones seen in traditional contracts. Taking into account the benefits brought by blockchain and smart contracts, it is important to study their contributions to emerging technological solutions, like IoT (Internet of Things), in which security, connectivity and interoperability are major concerns. This paper aims to describe relevant blockchain and smart contract's concepts, discussing their contribution to support and improve modern IoT solutions.

Keywords: Blockchain · Smart contracts · Internet of Things (IoT)

1 Introduction

The financial system moves a lot of money every day, with many people making use of it. However, the system is full of problems, additional costs like fees and delays, bureaucracy and opportunities for fraudulent activities involving fraud and crime. For instance, 45% of financial intermediaries, like services for money transfer and stock exchange experienced some sort of crime every year [1]. For the entire economy the number is about 37% [2]. Therefore, this is an important concern for banks, as regulatory costs continue to increase. All these problems have costs to the institutions and, consequently, consumers have to pay for them [3].

Considering all these problems, in 2008, Satoshi Nakamoto proposed a solution based on a peer-to-peer payment system, using electronic coins – the Bitcoin [4]. Months later, the source code was published to allow any person to participate in a network of payments. Over the years, Bitcoin has risen in terms of popularity, although the technology which makes Bitcoin work remained relatively unknown. The technology is called *blockchain* and, in simple terms, consists in a network of computers

© Springer Nature Switzerland AG 2019
T. Antipova and A. Rocha (Eds.): DSIC 2018, AISC 850, pp. 426–434, 2019.
https://doi.org/10.1007/978-3-030-02351-5_48

(nodes), with the same set of records replicated by all (a *distributed ledger*). These records are divided into blocks, consisting in transactions occurred between the nodes. When a transaction is concluded the data corresponding to it is broadcasted to all the nodes, so they can add it to their distributed ledger.

A second generation of blockchain technologies introduced the *smart contract* concept. Smart contracts are self-executing contracts with the terms of the agreement between the parties being directly written as lines of code which are executed automatically when an event or a certain condition is met. Introduced in 1995 by Nick Szabo, his definition isn't too far from what smart contracts are being built for.

Nowadays, it is widely recognized that blockchain and smart contracts, as emergent technologies, came to stay and to revolutionize many business areas. Indeed, besides financial services, today there are noticeable efforts in business areas such as Healthcare, Media and Telecommunications, Energy, Retail and Consumer Goods, just to name a few, involving the use of these technologies to develop sophisticated business systems. With that reality in mind, in this paper we focus our attention on the use of blockchain and smart contracts to support the implementation of innovative technological solutions such as IoT (*Internet of Things*) systems and how these technologies might be used in combination with IoT to solve relevant aspects of daily life.

Regarding its structure, this paper begins by introducing some of the most relevant concepts about blockchain and smart contracts technologies. Following that, IoT solutions are briefly characterized, emphasizing the current major obstacles to their wider implementation. Then, we present a set of representative projects, describing how they use IoT with the blockchain and smart contracts technologies in order to support innovative requirements and functionalities. Finally, we conclude with some considerations about the use of blockchain and smart contracts technologies in the context of IoT implementations.

2 Blockchain Technology

At the end of October of 2008, someone under the name of Satoshi Nakamoto published a document entitled "Bitcoin: A Peer-To-Peer Electronic Cash System" [4]. In this document, some of the features of Bitcoin, which has the potential to revolutionize the financial sector, were described.

With these features, Satoshi Nakamoto, intended to create *"an electronic payment system based on cryptographic proof, instead of trust, allowing any two willing parties to transact directly with each other without the need for a trusted third party. Transactions that are computationally impractical to reverse would protect sellers from fraud, and routine escrow mechanisms could easily be implemented to protect buyers."* [4].

Through this initial idea, the bases for the creation of Bitcoin were established and later, on January of 2009, the first Bitcoin application was created. However, the success of the bitcoin comes from a cryptographic technology underlying it, namely the blockchain technology [5]. Since then, many authors have separated blockchain technology from the Bitcoin application, so they can use it in other industries [6]. *"This is much more than the financial services industry. Innovators are programming this*

new digital ledger to record anything of value to humankind – birth and death cer-
tificates, marriage licenses, deeds and titles of ownership, rights to intellectual prop-
erty, educational degrees, financial accounts, medical history, insurance claims,
citizenship and voting privileges, location of portable assets, provenance of food and
diamonds, job recommendations and performance ratings, charitable donations tied to
specific outcomes, employment contracts, managerial decision rights and anything else
that we can express in code." [7]. Because of all this potential, 10 years after the
creation of Bitcoin, it is estimated that more than 25 countries are investing in
blockchain technology, making more than 2500 patents and making a total of 1.3
billion dollars invested.

But, how do we define this technology? In simple terms, a blockchain is basically a
distributed database or ledger containing all transactions or digital events that have
been executed among participating parties [8]. All the transactions or digital events are
inserted into blocks and these are added to the blockchain in a linear, chronological
order [9].

To be added in a chronological order, the blocks need to be validated before being
inserted. A validation is made by a consensus protocol mechanism which, in bitcoin, is
called proof-of work. To make this mechanism work, there is a group of people in the
network called "miners". Their job consists on changing one variable until the network
accepts the solution [10]. The solution is met, the validation was made and the miner
who found the solution transmits the new block to all the nodes participating in the
network, so they can add it to their distributed ledger. For the effort of mining, the
miner is rewarded in bitcoins.

With this process, blockchain brings a new paradigm to the online business. Its
strengths are [11]:

- **Transparency** - blockchain's data are open to the parties;
- **Redundancy** - all the nodes of a blockchain have a copy of the data, thus protecting
  the system from malicious attacks and malfunctioning problems;
- **Immutability** - due to the consensus protocol, changing data in the blockchain is
  almost impossible as it needs the accordance of the majority of the nodes;
- **Disintermediation** - Eliminating intermediates, such as banks, from each transaction decreases the cost and risk from the transactions.

A distributed ledger is an asset database that can be shared across a network on
multiple sites, geographies, or institutions. All within the system can have access to the
ledger via copy or connection to the larger database. Any changes made on any one of
the ledgers will be reflected on all the ledgers that currently exist [12]. The ledger
produced can thus be considered authoritative, although its management is shared
among users with conflicting incentives [13].

A distributed ledger is a distributed database because every node has a synchronized copy of the database. However, there is a characteristic that differentiates it -
decentralization. The central authority (administrator) is eliminated and the integrity is
achieved using a consensus mechanism or validation protocol. The control of the
database (read/write access) is decentralized, this means that every node of the network
participates in it. There is no need for a central administrator to ensure the integrity of

the data or its consistency across nodes. Instead, this is achieved through some consensus mechanism or validation protocol [14].

3   Smart Contracts

In 1997, Nick Szabo published a paper called "The idea of smart contracts". He defined smart contract as *"A set of promises, including protocols within which the parties perform on the other promises. The protocols are usually implemented with programs on a computer network, or in other forms of digital electronics, thus these contracts are 'smarter' than their paper-based ancestors"* [15].

Despite the definition being two decades old, it stills makes sense in today's reality. If we dissect the definition and analyze the keywords, it will be present in other's definitions in our days. Nick Szabo said we could make a set of promises and according with Konstantinos Christidi *"We trigger a smart contract by addressing a transaction to it. It then executes independently and automatically in a prescribed manner on every node in the network, according to the data that was included in the triggering transaction."* [16]. Thus, we can program a smart contract to behave like we want, and the output will be like we promised.

Smart contracts have been designed to automate transactions and allow parties to agree with the outcome of an event without the need for a central authority [17]. Nick described smart contracts as *"protocols within which the parties perform"*, so when a transaction or an event occurs there is a set of rules who tells how the data should be processed, helping the process produce the right outcome [15].

This is accomplished by transforming a smart contract into lines of code with clauses and agreements embedded as computer code within the software. Smart contracts, seek to leverage the trustless, immutable nature of the blockchain to empower peer-to-peer, disintermediated agreements enforced automatically by code [18].

The life cycle of a smart contract typically consists of four broad phases: *creation of the smart contract, freezing of the smart contract, execution of the smart contract* and *finalization of the smart contract* [19]:

- **Create** - First, the parties involved need to agree on the objectives of the contract and this is like a classic contract negotiation. After everything is settled, the contract needs to be turned into code. Most of smart contract environments have a proper infrastructure to create, maintain and test the contract and with that we can validate the behavior and content of the contract. After the parties agree, the contract is then submitted to the distributed ledger;
- **Freeze** - After the smart contract has been submitted to the blockchain, it is persisted by a majority confirmation of the participating nodes. In exchange for this service, and to prevent a flooding of the ecosystem with smart contracts, a fee must be payed to the miners. From this point onward, the contract and all parties are public and accessible through the public ledger. During the freeze phase, any transfers made to the wallet address of the smart contract are being frozen and the nodes take on the role of a governance board, ensuring the preconditions for executing the contract are met;

- **Execute** - All the participant nodes can read the contracts stored in the distributed ledger. The code is executed by the inference engine of the smart contract environment after the contract's integrity is validated. The execution of the smart contract results in a set of new transactions that passed through all the conditions of the contract. The new state of information is validated through the consensus protocol and broadcasted to all the nodes;
- **Finalize** - After the smart contract has been executed, the new state of information is validated through the consensus protocol and broadcasted to all the nodes. The prior committed digital assets are transferred (unfreezing of assets) and with the confirmation of all transactions, the contract has been fulfilled.

4   Internet of Things

Internet of Things (IoT), also known as the Internet of Objects, is "*a foundation for connecting things, sensors, actuators, and other smart technologies, thus enabling person-to-object and object-to-object communications.*" [20]. Proposed in the end of the 1990's by MIT Auto-ID Labs, the primary purpose of IoT is to share information about objects, which reflects the manufacture, transportation, consumption and other details of people's life. It has three important characteristics [21]:

- Any objects can be involved in an IoT solution. Objects can be embedded with a chip, bar code, etc., and then we can address them;
- Terminals are interconnected, acting like autonomic network terminals;
- In such an extensively-interconnected network, every object participates in the service flow to make the pervasive service intelligent.

Despite all the potential of the IoT concept, it still has some major technological challenges, such as *security*, *connectivity* and *interoperability*. In relation to security, IoT endpoints are opening up vulnerabilities for systems around the world. Indeed, we all have heard stories about these endpoints being hacked and that creates a real threat to IoT solutions.

Regarding connectivity, companies all around the world are preparing their infrastructure for the internet age. Due to the high number of IoT devices, organizations need decentralized architectures, so their systems can handle the large amounts of data that are produced.

Concerning interoperability, it is a fact that IoT devices have been used by different industries for a long time. The interoperability of devices between brands can be a problem without standardized protocols so developing a common protocol where everyone would be involved can bring benefits for producers and consumers of IoT.

5 Examples of Blockchain and Smart Contracts Using IoT

In this section, we present some current development efforts, which are representative of the kind of innovative solutions that may be developed by integrating IoT technologies with blockchain and smart contracts.

- **Automatic Payments** [22]

Marie goes to work in her car. As soon as she enters the car, it synchronizes with Marie's smart phone's AutoPay service. This service gives security and trust to Marie through smart contracts on its blockchain interface, acting as a device for autonomous payments. AutoPay defines her office as her destination. The service interrogates the vehicle data about his fuel and if is low automatically finds a route which passes by a petrol station that is advertising competitive fuel price. After refueling the car, which was automatically paid by AutoPay's smart contract feature, Marie receives a message telling her that her work's car park is full and AutoPay, through smart contracts, paid for another car park, very close to her office. After work Marie goes home, and her daughter Ana asks to borrow her car. Ana's smart contract allows her to access her mother's car but don't let her make autonomous payments for everything. Ana can refuel using her mother's AutoPay service, but she can't use it on McDonald's drive to buy meals for her friends because those conditions are not present in the smart contract.

- **Supply Chain Traceability** [23]

Currently in the traditional fish business, there is a lot of problems, i.e., fishing practices are illegal, unregulated or unreported. All this lack of proper business management has impacts in the consumer's final product quality and there is no way for the consumer to know, creating a lack of vendor and consumer trust. To prevent these problems, combining IoT, blockchain and smart contracts could be the solution. Recording the trip of all sea-food since when it was fished until it gets to the final consumer. IoT sensors can be attached to any object entrusted to someone else for transport, with trackable ownership, possession, and telemetry parameters such as location, temperature, humidity, motion, etc. The final buyer can access a complete record of information and trust that the information is accurate and complete.

- **Smart City – Smart Homes** [24]

There is a panoply of household devices present in today's houses. Sensors to control the environment, refrigerators, dishwashers, cameras, door locks, alarms many of them are IoT devices. Homeowners or authorized parties can have access to these devices to control the smart home. All the information coming from these devices is stored in a central server and then it is presented to the controller in a device like a cell phone or a computer. This information cannot end up in other hand besides authorized person so maintaining the security is very important and blockchain can help in that field. Making use of distributed ledgers, the communication and control information can be recorded as transactions. Using symmetric cryptography to maintain confidentiality, hashing functions to integrity and asymmetric cryptography for authenticity a robust security can be ensured. Attempts to corrupt the system could be immediately detected.

- **Physical Theft** [25]

Multiple smart devices are stolen daily and most of them won't return to their rightful owner. A combination of IoT, blockchain and smart contracts could be an answer to help fighting this. After a smart device is stolen, its user could report the theft to the IoT blockchain network or even the network itself could report, based on a set of parameters that indicate the theft of the device. Marked as stolen, the device would forward its important data to the police, like, its current location or secret pictures taken of the thief, without the need for third party intervention. On a later stage, the device's unique manufacturing number could be written on a "blacklist" of stolen smartphones managed by the OS provider of the device, preventing the thief from performing a factory reset and use the device.

- **School Classes Management** [26]

In every class, teachers have to make the call for all the students, so they can mark them as present or not. If there are too many students this can be very time consuming, taking away time for what is most important - teaching. A system based on blockchain and IoT could be implement in schools to ensure students arrive at classes at the appropriate time. On a further approach, the parents could also be part of the school blockchain, verifying their child's attendance to classes. Teachers could also leave some feedback how students are performing, so parents can have insight about the progression being made. This kind of system could even avoid parent evening reunions that students, parents, and teachers don't appreciate much.

- **Automation and Control** [27]

The benefits of IoT opens multiple opportunities in automation, optimization and control. Having this idea, companies with industrial IoT solutions that provides drop-in decentralized wireless and radio mesh network technology, integrated their technology with blockchain. For instance, by connecting multiple taps sensors at different locations, sensors gather information about noise levels or ambient temperature. The benefits of deploying a solution like this one are many. Companies and cities can expect improvements like, minimization of manufacturing line's downtime, automate climate control in data centers, remotely control street lights across a city, avoid collisions between vehicles on remote work sites, etc.

6 Conclusions

There is no doubt that IoT has a great future, with more and more IoT solutions appearing in the landscape. However, as the previous examples have revealed, there are still some problems with IoT, which may be solved by combining it with the emerging blockchain and smart contracts technologies. Moreover, as the examples also demonstrate, by combining those technologies, an impressive list of innovative solutions might be envisaged and developed, with great impact in our daily lives. With this paper we aimed to clarify the concepts around the recent blockchain and smart contracts technologies and emphasize their potential when used in a IoT ecosystem.

References

1. PricewaterhouseCoopers: Threats to the Financial Services sector. Financial Services sector analysis of PwC's 2014 Global Economic Crime Survey (2014)
2. Medland, D.: Cost of regulation "Top Concern" for financial services. Forbes (2015)
3. Tapscott, A., Tapscott, D.: How blockchain is changing finance. Harv. Bus. Rev. (2017)
4. Nakamoto, S.: Bitcoin: a peer-to-peer electronic cash system. Cryptogr. Mail. (2008)
5. Pilkington, M.: Blockchain technology: principles and applications. In: Research Handbook on Digital Transformations (2015)
6. Nofer, M., Gomber, P., Hinz, O., Schiereck, D.: Blockchain. Bus. Inf. Syst. Eng. **59**, 183–187 (2017)
7. Tapscott, D., Tapscott, A.: Blockchain Revolution: How the Technology Behind Bitcoin is Changing Money, Business, and the World, Portfolio (2016)
8. Cosby, M., Nachiappan, P.P., Verma, S., Kalyanaraman, V.: Blockchain Technology: Beyond Bitcoin. Applied Innovation Review for Sutardja Center for Enterpreneurship & Technology Berkeley Engineering (2016)
9. Swan, M.: Blockchain: Blueprint for a New Economy. O'Reilly Media, California (2015)
10. Seffinga, J., Lyons, L., Bachmann, A.: The Blockchain (R)evolution - The Swiss Perspective White Paper, Deloitte, February 2017
11. Savelyev, A.I.: Copyright in the blockchain era: promises and challenges. High. Sch. Econ.: Comput. Law Secur. Rev. **34**(3), 550–561 (2017)
12. UK Government Chief Scientific Adviser: Distributed Ledger Technology: Beyond Blockchain. Government Office for Science (2016)
13. Pinna, A., Ruttenberg, W.: Distributed ledger technologies in securities post-trading revolution or evolution? European Central Bank (ECB) Research Paper Series - Occasional Papers (2016)
14. Benos, E., Garratt, R., Gurrola-Perez, P.: The Economics of Distributed Ledger Technology for Securities Settlement. Bank of England Working Papers, 23 Agosto 2017
15. Szabo, N.: Smart contracts glossary. Extropy (1995)
16. Christidi, K., Devetsikiotis, M.: Blockchains and smart contracts for the Internet of Things. IEEE Access **4**, 2292–2303 (2016)
17. Cant, B., Khadikar, A., Ruiter, A., Bronebakk, J., Coumaros, J., Buvat, J., Gupta, A.: Smart contracts in financial services: getting from hype to reality. Capgemini Consulting, pp. 1–24 (2016)
18. Brennan, C., William, L.: Blockchain: The Trust Disrupter. Credit Suisse Securities (Europe) Ltd., London (2016)
19. Sillaber, C., Waltl, B.: Life cycle of smart contracts in blockchain ecosystems. Datenschutz und Datensicherheit – DuD (2017)
20. Kuyoro, S., Osisanwo, F., Akinsowon, O.: Internet of Things: an overview. In: 3rd International Conference on Advances in Engineering Sciences & Applied Mathematics (2015)
21. Ma, H.: Internet of things: objectives and scientific challenges. J. Comput. Sci. Technol. **26** (6), 919–924 (2011)
22. Huckle, S., Bhattacharya, R., White, M., Beloff, N.: Internet of Things, blockchain and shared economy applications. Procedia Comput. Sci. **58**, 461–466 (2016)
23. Seafood Case Study in Supply Chain: Hyperledger. https://sawtooth.hyperledger.org/examples/seafood.html. Accessed 15 Mar 2018
24. Zahid, J., Hussain, F., Ferworn, A.: Integrating Internet of Things and blockchain: use cases. Newsletter (2016)

25. How to Secure the Internet of Things: DevTeam.Space. https://www.devteam.space/blog/how-to-secure-the-internet-of-things-iot-with-blockchain/. Accessed 15 Mar 2018
26. Schools are Using Bitcoin Technology to Track Students: CNBC. https://www.cnbc.com/2016/05/09/schools-are-recording-students-results-on-the-blockchain.html. Accessed 15 Mar 2018
27. Bahga, A., Madisetti, V.: Blockchain platform for industrial Internet of Things. J. Softw. Eng. Appl. **09**(10), 533–546 (2016)

Blockchain Technologies: Opportunities in Healthcare

Jorge Lopes[1] and José Luís Pereira[2](✉)

[1] Mestrado Integrado em Engenharia e Gestão de Sistemas de Informação,
UMinho, Braga, Portugal
a73263@alunos.uminho.pt
[2] Departamento de Sistemas de Informação, UMinho and Algoritmi,
Braga, Portugal
jlmp@dsi.uminho.pt

Abstract. Blockchain is a decentralized transaction and data management technology first developed for the Bitcoin cryptocurrency. The reason for the interest in this technology results from its decentralized nature, which provides security, anonymity and data integrity without any third-party organization in control of the transactions and, therefore, it creates interesting research areas. One of the business sectors that can benefit from this technology is healthcare. In this paper, we present and describe relevant aspects of the blockchain technology and we discuss some use cases in the healthcare area which might be improved by the use of this technology. To demonstrate its relevance to the healthcare sector, we present some real life examples of blockchain projects being developed to solve specific problems of this sector.

Keywords: Blockchain · Healthcare · Use cases · Blockchain technologies

1 Introduction

Being an emergent technology, blockchain has already conquered a noticeable place in the Information and Communication Technologies domain. While its origins can be traced back to the financial sector, in the present it is widely acknowledged that blockchain technologies have the potential to deeply transform the way technical solutions are developed in many business areas.

Indeed, companies in business areas such as Healthcare, Media and Telecommunications, Energy, Retail and Consumer Goods, among others, are using blockchain technologies to develop increasingly sophisticated technical solutions. Taking this fact into account, in this paper we focus our attention on the use of blockchain to support the implementation of innovative solutions in the healthcare sector. In order to do that, we briefly describe important aspects of the blockchain technology and we discuss some use cases in the healthcare domain, which might benefit from the use of this technology. To emphasize its relevance, we present some real life examples of blockchain projects being developed to solve specific problems in the healthcare domain.

© Springer Nature Switzerland AG 2019
T. Antipova and A. Rocha (Eds.): DSIC 2018, AISC 850, pp. 435–442, 2019.
https://doi.org/10.1007/978-3-030-02351-5_49

Regarding its structure, this paper commences by introducing some relevant concepts about blockchain technologies. After that, we discuss the use of this technology in the healthcare domain, presenting some important use cases and illustrative projects that make use of blockchain to solve particular problems in that sector. Finally, we conclude with some considerations about the use of these technologies in this specific context.

2   Blockchain Technology

Blockchain technology has indeed many applications that can influence significantly the way we approach many use cases in a vast list of business sectors. The purpose of this technology is the use of a distributed ledger for verifying and recording transactions. By doing so, it allows parties to send, receive, and record data through a Peer-to-Peer (P2P) network of computers.

2.1   Blockchain

In simple terms, a blockchain is a kind of distributed database that stores time ordered data in a continuously growing list of blocks. The blockchain is maintained using a network of computers with no central "master". Each *block* in the chain contains *transactions* which represent a change of state in the database; for example, the transfer of funds from one account to another. Transactions are verified by multiple nodes in the network and are eventually stored in blocks in the blockchain. Each block contains a signed hash of the contents of the preceding block, making it impossible for a block's contents to be altered [1].

Explaining the concepts in which the blockchain technology is based:

- **Network** - a group of computer nodes assembled together. Each node contains an address and a complete record of all the transactions that were ever recorded in that blockchain;
- **Decentralized** - means that no single entity has control over all the processing;
- **Distributed** - a model in which components located on networked computers communicate and coordinate their actions by passing messages. The key purposes of distributed systems can be represented by resource sharing, openness, concurrency, scalability, fault-tolerance, and transparency;
- **Transaction** - an exchange of value between two parties with their consent. Value cannot be transferred from an account without a digital signature only known by the owner of that account. That signature is like a private key;
- **Block** - list of transactions recorded over a given period and a hash pointer that converts the content of the block to a cryptographic key, linking the block to the previous one on the chain;
- **Chain** - a group of blocks linked together by hash pointers;
- **Time-Stamp** - instant when a block is generated. Used to implement a time order chain structure.

It is important to highlight that a transaction does not need to be coin-based. It can be an asset, which the owner can prove that he/she has it in his/her possession. In Fig. 1 a representation, borrowed from [2], of how a blockchain works.

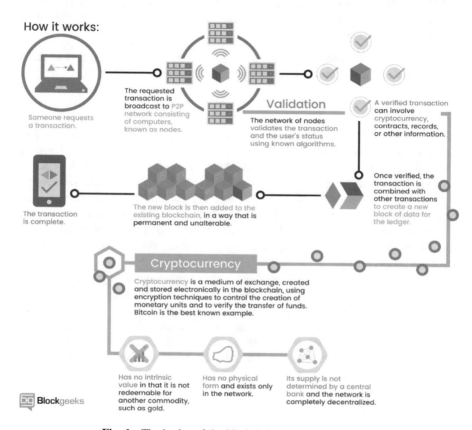

Fig. 1. The basics of the blockchain operation (from [2])

2.2 Decentralized Consensus

Blockchain aims to produce decentralized consensus, a specific state or set of information to be agreed upon by all agents via rules and protocols, without the need to trust or rely upon a centralized authority. This supposedly makes the consensus more secure and tamper-proof. Moreover, it rewards a community for properly maintaining the consensus, allowing greater recording and processing power in an incentive and typically competitive manner [3].

There are two main approaches to achieve consensus on a blockchain:

- **Proof-of-Work** (PoW) - rewards users who solve complicated cryptographical puzzles in order to validate transactions and create new blocks (i.e., mining). This system ensures that once a block is validated, it cannot be denied because doing so requires the malicious entity to have computing power that can compete with the entire existing network. Consequently, we achieve robust and tamper-proof consensus on the validity of these transactions that can prevent attacks such as Denial-of-Service (DoS attack);

- **Proof-of-Stake** (PoS) - the entity in charge of creating the next block is chosen in a deterministic manner, and the chance that an account is chosen depends on its wealth (i.e., the stake).

In both cases and many other consensus generation designs, the goal is to incentivize responsible and accurate recordkeeping, while reducing tampering.

2.3   Hash Pointers

If someone tries to change the contents of a block in the middle of the chain the hash generated for this block, which is stored in the previous node, also would need to change. So, even if the hash of the previous block and the data block were changed, the previous block next pointer is based on a hash of the previous block. Because the previous block is composed of both the data and the previous hash (which has a change) these would also indicate tampering. Thus, the only way to change a block is by changing the entire blockchain, but at that point the head of the blockchain would be incorrect and since this is the value that the users store, the users would be able to detect the tampering [4].

A hash pointer provides the blockchain with a tamper-evident system in a simple way. For example, if someone changes the contents of one block the hash of the next block will not mash up and we will notice the inconsistency.

Even if an attacker has enough computer power to change the block and all the hash pointers of the following blocks in the chain, he/she will arrive at a dead-end because the last hash pointer is the value we remember as being the head of the list and the inconsistency will be noticed inevitably.

2.4   Smart Contracts

A smart contract is like a program, which runs on the blockchain, and has its correct execution enforced by the consensus protocol. A smart contract can encode any set of rules represented in its programming language. For instance, it can execute transfers when certain events occur. Accordingly, smart contracts can implement a wide range of applications, including financial instruments and self-enforcing or autonomous governance [5].

A smart contract is linked to an account, is identified by an address and its code resides on the blockchain. After uploading a contract to the blockchain, it will react accordingly to the code implemented previously and is able to do transactions like a normal member of the blockchain. The main objectives are to satisfy common contractual conditions, minimize both malicious and accidental events, and minimize the need for trusted intermediaries [6]. Related economic goals include lowering fraud loss, arbitrations and enforcement costs.

3 Blockchain in the Healthcare Domain

In this section we present three important healthcare use cases which might be substantially improved by the use of blockchain technology: health information exchange, systems interoperability and data integrity and patient digital identities. Some recent blockchain development projects are also briefly described, as a demonstration of the highly dynamic nature of the healthcare sector.

3.1 Healthcare Use Cases

Using blockchain to support Health Information Exchange (HIE) might demonstrate the true value of systems interoperability (the ability of computer systems to exchange and make use of information). Blockchain-based systems have the potential to reduce or eliminate the friction and costs of current intermediaries, offering a promising new distributed framework to amplify and support the integration of healthcare information across a range of uses and stakeholders.

Table 1 addresses several existing difficulties regarding HIE and reveals how blockchain could make the healthcare system more efficient, disintermediated, and secure.

Table 1. HIE difficulties and blockchain opportunities

HIE difficulties	Blockchain opportunities
Trusting network is currently established by a third party whose function is to regulate and record all data shared	**Disintermediation** of trust its provided by giving access to the distributed leger to all participants, maintaining this way a secure and transparent exchange off data
Cost per transaction, outdated central systems for trading networks due to low transaction volumes	**Reduced transaction costs** and a more efficient system due to real-time transactions without intermediates
Master Patient Index needs to find a way to synchronize multiple patient's data between systems without jeopardize their privacy	**Distributed framework** using patients private and public identifiers secured through cryptography, providing this way a secure way to access the patient data
Varying data standards reduce compatibility between systems	**Shared data** between all parties in a near real-time updates
Few sources of integrated health records resulting on limited access to these records	**Distributed, secure access** to patient's continual data records on the distributed ledger
Inconsistent rules and bureaucracies hinder health organizations access to patient's data and increase the time for each process	**Smart contracts** able health organizations to create a consistent set of rules, regulating the access to HIE

Information stored on the blockchain could be universally available to specific individuals through the blockchain private key mechanisms, enabling patients to share their information with healthcare organizations much more seamlessly. This deployment of a transaction layer on the blockchain can help to accomplish interoperability goals while creating a trustless, and collaborative ecosystem of information sharing to enable new insights to improve the efficiency of the healthcare system.

As a transaction layer, there are two types of information that blockchain can store: "On-chain" data, which is directly stored in the blockchain or "Off-chain" data with links stored in the blockchain that act as pointers to information stored in separate, traditional databases (see Table 2).

Table 2. On-chain data vs off-chain data

	On-chain data	Off-chain data
Data types	Standardized text data fields from patients (e.g. age, gender)	Expandable medical data and binary data types (e.g. notes, MRI images, etc.)
Pros	Data contained on the blockchain is transparent and identical to all connected organizations	Possibility to store any data size and format
Cons	Restrictions regarding the size and types of data to be stored	In order to access data healthcare organizations have to require access to the system external

Storing medical information directly in the blockchain ensures that the information is fully secured by the blockchain's properties and is immediately viewable to those with permissions to access the chain. Unfortunately, at the same time, storing large data files in the blockchain slows block processing and presents potential challenges to scaling the system.

Creating interoperability requires frictionless submission and access to data. As such, the blockchain could serve as a transaction layer for organizations to submit and share data using one secure system. Once a standardized set of healthcare information is established, the specific data fields can be created in a smart contract to employ rules for processing and storing information on the blockchain, as well as stipulating required approvals prior to blockchain storage. Each time a patient interaction occurs healthcare organizations will pass information to the smart contract.

An interoperable blockchain can strengthen data integrity while better protecting patients' digital identities. According to Krawiec et al. [7], in 2015 there were 112 million healthcare record data breaches due to hacking/IT incidents and in 2016, it was estimated that one in three healthcare recipients was a victim of a data breach.

The blockchain's inherent properties of cryptographic public/private key access, proof of work, and distributed data creates a new level of integrity for healthcare information. Additionally, all health care organizations connected to the blockchain can maintain their own updated copy of the health care data. This feature improves security and can help limit the risk of the malicious activity, because changes are immediately broadcasted to the network, and distributed ledgers provide safeguard copies against harmful hacks.

3.2   Some Current Blockchain Projects

In face of the advantages mentioned above, there are already very interesting blockchain projects being developed in the healthcare area. These projects may be divided into three different groups, being them Electronic Medical Record Interoperability, Health Data Security and Supply Chain Management.

In the Electronic Medical Record Interoperability group, one of the projects deserving to be mentioned here was developed by the firm PokitDok [8] which involves the development of a Platform-as-a-Service (PaaS) that enables healthcare organizations to quickly build modern commerce experiences across the healthcare value chain. It provides a secure network for all sources of patient data, from Electronic Medical Records to medical devices and pharmacies. The platform's various APIs are detethered from the central hospital/pharmacy and instead connected by a distributed blockchain network. Another project, developed by the company Patientory [8, 9], uses blockchain technology to ensure end-to-end encryption while adhering to regulatory guidelines and compliance requirements, creating an easy way to securely store and manage health information in real time. This solution allows patients, clinicians, and healthcare organizations to access, store and transfer information safely, thus improving care coordination while ensuring data security.

In the Health Data Security projects group we highlight a project developed by GuardTime [8, 9], one of the largest providers of enterprise blockchain-based security systems. The company operates in multiple sectors including advertising, defence, healthcare, government, and financial services. In the healthcare domain, GuardTime partnered with the Estonian e-Health Authority in March 2016 to secure over 1 million patient records. Since Estonian citizens already carry unique identity credentials that link back to their health records, a blockchain layer is able to record and secure every interaction with their data. Another project worth mentioning is being developed by the company Nebula Genomics [9], which believes that the search for genomic data has similar characteristics and challenges as any market where data is involved, and they intent to use blockchain technology to solve data security issues and to ensure that data gets from the source to its end-user without any intermediaries. The company uses blockchain technology to enhance genomic data protection, enable buyers to efficiently acquire genomic data and address the challenges of genomic big data.

Finally, we have Chronicled [8, 9], a company developing projects focused on Supply Chain Management. Chronicled builds special-purpose blockchain-based applications for supply chain and IoT clients. The server synchronizes with Quorum, Hyperledger, and Ethereum blockchain protocols and has applications ranging from medical devices and pharmaceuticals to package tracking. A unique application of the Chronicled platform involves the tracking of products such as pharmaceuticals, blood, and human organs, which often need to be kept at low temperatures. Chronicled has developed a portable smart sensor that tracks and stores temperature readings on a blockchain back-end to ensure secure data transmission. Further, for pharmaceuticals and medical devices, Chronicled has developed "CryptoSeal," a tamper-proof adhesive sticker that provides a cryptographic identity for physical items. At each stop in the supply chain, the seal can be scanned and verified against the blockchain registry.

4 Conclusions

Nowadays, it is obvious that blockchain technologies have an enormous potential to impact the Information and Communication Technologies domain, being responsible for the appearance in the business landscape of increasingly innovative and technically sophisticated solutions. The healthcare sector is only one, although a very relevant one, of the many business sectors which might benefit from blockchain. Taking that into account, with this paper we aim to clarify the concepts around the recent blockchain technologies and emphasize their contribution to solve important problems still existing in the healthcare sector. To accomplish that we identified some representative use cases of the healthcare sector, describing how blockchain technologies might help to support their specific needs. In order to demonstrate that the healthcare sector is becoming very active in this context, continuously pursuing the development of technically advanced solutions, some recent blockchain development projects are also briefly described.

References

1. Bird, G.: Block chain technology, smart contracts and Ethereum. IBM (2016)
2. How does Bitcoin Blockchain work and what are the rules behind it? https://www.quora.com/How-does-Bitcoin-Blockchain-work-and-what-are-the-rules-behind-it. Accessed 21 Mar 2018
3. Cong, L.W., He, Z.: Blockchain disruption and smart contracts. SSRN Electron. J. (2018)
4. Hash Pointers – deltadeltaandmoredeltas. https://www.deltadeltaandmoredeltas.com/hash-pointers/. Accessed 21 Mar 2018
5. Luu, L., Chu, D.-H., Olickel, H., Saxena, P., Hobor, A.: Making smart contracts smarter. In: Proceedings of the 2016 ACM SIGSAC Conference on Computer and Communications Security, CCS 2016, pp. 254–269 (2016)
6. What is a Smart Contract? – Pactum – Medium. https://medium.com/pactum/what-is-a-smart-contract-10312f4aa7de. Accessed 21 Mar 2018
7. Krawiec, R.J., Housman, D., White, M., Filipova, M., Quarre, F., Barr, D., Nesbitt, A., Fedosova, K., Killmeyer, J., Israel, A.: Blockchain: opportunities for health care. In: Proceedings of the NIST Work Blockchain Healthcare, pp. 1–16 (2016)
8. 5 Blockchain Startups Working to Transform Healthcare. https://www.cbinsights.com/research/healthcare-Blockchain-startups-medicine/. Accessed 21 Mar 2018
9. Top 12 Companies Bringing Blockchain to Healthcare - The Medical Futurist. http://medicalfuturist.com/top-12-companies-bringing-Blockchain-to-healthcare/. Accessed 21 Mar 2018

Application of Matrix Fuzzy Logic in Machine Independent Temperature Controller

Ilya P. Seletkov$^{(\boxtimes)}$ and Leonid N. Yasnitsky

Perm State University, Bukireva St. 15, 614990 Perm, Russia
iseletkov@gmail.com, yasn@psu.ru

Abstract. The article deals with the problem of creating a temperature regulator, which does not require preliminary tuning for a specific production plant. Authors proposed to use a matrix approach of fuzzy logic for this purpose. It allows engineers to apply the linguistic rules formulated in the most general form for industrial processes control. It also allows building simple control algorithms for complex nonlinear systems. To verify the correctness of the algorithm, authors assembled installation with heaters of various types, powers and inertia. The full-scale experiment shown that the developed prototype of the device allows controlling the temperature in various settings without initial tuning.

Keywords: Matrix fuzzy logic · Temperature regulation · PID controller

1 Introduction

Currently, for control problems with the presence of feedback the proportional-integral-differential-regulators (PID) are most frequently used in industry [1]. They have a simple design and can easily be implemented by software or hardware. The main disadvantages of PID controllers are the need to select the coefficients for each particular installation and the lack of control over non-linear processes.

Several algorithms have been developed to adjust the regulators and to select the coefficients [2]. All these algorithms require the participation of a specialist installer, test runs of the installation, a large amount of time. Moreover, it is necessary to repeat the whole procedure of tuning when transferring a regulator already set up for one installation to another installation.

On the other hand, experts can formulate the rules on which temperature control should take place in a linguistic form. For example, "If the temperature of the object is low, it is necessary to heat up". These rules are not consider or take into account individual parameters (power, inertia, etc.) of the plants. With the use of this fact, we can try to develop a thermostat, which will not require a time-consuming preliminary tuning.

T. Antipova and A. Rocha (Eds.): DSIC 2018, AISC 850, pp. 443–449, 2019.
https://doi.org/10.1007/978-3-030-02351-5_50

2 Matrix Approach in Fuzzy Logic

The rules of management, formulated in a linguistic form, are most conveniently led to a numerical form by means of fuzzy logic [3]. To accurately match the generated control action with existing knowledge, and for greater flexibility in the settings, one can use the matrix approach of fuzzy logic [4].

2.1 Fuzzy Matrix Operations

Matrix approach [5] uses a two-dimensional vector x to describe the degree of truth of linguistic utterances. If we introduce the basis vectors $e^{(0)}$ and $e^{(1)}$, vector x will take the form (1):

$$x = x_0 \cdot e^{(0)} + x_1 \cdot e^{(1)}, \tag{1}$$

where the components of the vector x – numbers x_0, x_1 – satisfy the conditions (2):

$$0 \leq x_0, x_1 \leq 1; \quad x_0 + x_1 = 1. \tag{2}$$

The basis vectors $e^{(0)}$ and $e^{(1)}$ describe statements with a confidence level of 0 and 1 respectively.

With use of vectors, one can represent logical operations - a conjunction, a disjunction and an implication - in a matrix form. More specifically, a conjunction $C(x)$, a disjunction $D(x)$ and an implication $I(x)$ matrixes with size 2×2 look as follows [5]:

$$C(x) = \begin{pmatrix} 1 & x_0 \\ 0 & x_1 \end{pmatrix}; \; D(x) = \begin{pmatrix} x_0 & 0 \\ x_1 & 1 \end{pmatrix}; \; I(x) = \begin{pmatrix} x_1 & 0 \\ x_0 & 1 \end{pmatrix}. \tag{3}$$

Equations (4) show logical operations in term of matrix:

$$x \wedge y = C(x) \cdot y; \; x \vee y = D(x) \cdot y; \; x \to y = I(x) \cdot y, \tag{4}$$

where vector y is a matrix column 1×2, consists of its components $y = \begin{pmatrix} y_0 \\ y_1 \end{pmatrix}$, and dot stands for matrix multiplication. The result of logical operations, formulated in this form, will also be a fuzzy vector.

2.2 Fuzzy Matrix Inference

A fuzzy logical inference using matrix operation models is proposed in work [4]. Let fuzzy vector z determines the fuzzyness of rule $x \to y = z$. If fuzzy vectors $x = \begin{pmatrix} x_0 \\ x_1 \end{pmatrix}$ and $z = \begin{pmatrix} z_0 \\ z_1 \end{pmatrix}$ are known, then relations (3), (4) lead to the conclusion about the vagueness of the vector y. This problem reduces to solving a linear algebraic equation

with respect to an unknown vector y. The equation has the form $I(x) \cdot y = z$. In the case
of an implicative model of derivation or in more detail:

$$
\begin{pmatrix} x_1 & 0 \\ x_0 & 1 \end{pmatrix} \begin{pmatrix} y_0 \\ y_1 \end{pmatrix} = \begin{pmatrix} z_0 \\ z_1 \end{pmatrix} \text{ или } \begin{cases} x_1 y_0 = z_0 \\ x_0 y_0 + 1 = z_1 \end{cases}. \tag{5}
$$

It is known that a linear system has a solution if the determinant of the coefficient
matrix is not zero. In this case, we must require that condition

$$
\det(I(\mathbf{x})) = \det \begin{pmatrix} x_1 & 0 \\ x_0 & 1 \end{pmatrix} = x_1 \neq 0. \tag{6}
$$

In other words, degree of truth x_1 of vector x must not be equal to zero. Then
components of unknown vector y takes next values:

$$
y_0 = \frac{z_0}{x_1}; \quad y_1 = 1 - \frac{z_0}{x_1}. \tag{7}
$$

The solution obtained is meaningful only if both components of the vector y are
positive. So, in additional to requirement $x_1 \neq 0$ we obtain $x_1 \geq z_0$. Note then $x_1 = 1$
and $z_1 = 1$ we get well known result of modus ponens rule from (7) in classical logic:
$y_0 = 0$, $y_1 = 1$. Thus, we not only get a fuzzy logical conclusion, but also strictly
specify the scope of its implementation in this approach unlike other models of
implication.

3 Fuzzy Logic Finite-State Automata and Combinational Logic with Separate Memory Block

In case of using a PID controller a static error occurs in stationary mode [2]. It is
necessary to introduce the notion of a fuzzy state "Stable mode near the set point" to
monitor and prevent this situation in the case of a fuzzy controller (Fuzzy state
machine). A fuzzy set or a fuzzy predicate [4] numerically describes this concept. In
both cases expert have to specify truth/membership function [6].

A problem arises in the process of setting up real controllers in production. The
state of the fuzzy automaton describes the stage of the computational process and does
not explicitly relate to the subject area. Accordingly, it is difficult for experts to construct a graph of the truth function for a particular installation [7, 8].

To substantiate the possibility of using a combinational circuit, we well briefly
describe the operation of logical automata. As you know, one have to specify the tuple
$\{x, y, s, F, G\}$ to describe logical state machines with memory [9]. x is set of input
variables, y - is set of output signals, s - is set of sates, while F and G are transition and
output functions respectively: $s_{t+\Delta t} = F(s_t, x_t)$, $y_t = G(s_t, x_t)$. For a finite state automata sets x y, s are finite.

Expert does not need knowledge of the entire history of the signal changes to
determine the state of "Stable regime close to the installation" in this area. He can use

the machine with the final memory [9]. The operation of such automata is equivalent to the transformation of the form

$$y_t = h\left(x_{t-p\Delta t}, \ldots, x_{t-\Delta t}, x_t\right),\tag{8}$$

where $x_{t-p\Delta t}, \ldots, x_{t-\Delta t}, x_t$ – values of input variables vectors at discrete time moments $t, t - \Delta t, \ldots$; p – a finite positive integer which is the memory depth. It is easy to see that the Eq. (8) requires only combinational logic and external memory block.

Thus, the memory block content M is sequence of $p+1$ last values of input parameters $M(t) = \left\{x_{t-p\Delta t}, \ldots, x_{t-\Delta t}, x_t\right\}$. Algorithm updates memory content M by method FIFO (first input, first output). The combinational circuit analyzes a set of input variables $x_{t-p\Delta t}, \ldots, x_{t-\Delta t}, x_t$ and calculates the corresponding value of the output variable y_t.

In terms of the proposed model, one can reformulate the status of "Stable mode near the set point". Several sequential values of the input parameters "almost equal" to the setting value lead to this state.

4   Practical Application

For an experimental verification of the applicability of the matrix approach, we construct a test setup. Figure 1 shows the scheme of setup.

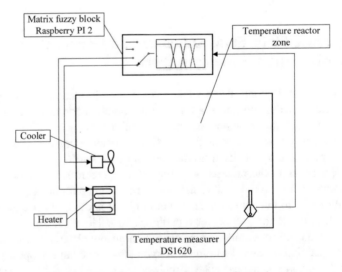

Fig. 1. Simplified scheme of the experimental setup.

Let us denote the parameters used in the formation of the control action as follows:

- T – current temperature in experimental setup working zone measured by DS1620, $T \in [0; 100]\ ^0C$;

- wc – power of cooler, $wc \in [0, 100]\%$;
- wh – power of heater, $wh \in [0, 100]\%$.

We will use T as input parameter for fuzzy algorithm and wc и wh as outputs.
To describe the subject domain, we introduce the following linguistic variables:

- α stands for "Object temperature" and $\alpha \in [a_0 = low, \; a_1 = near\,set, \; a_2 = high]$;
- βc stands for "Cooler power" and $\beta c \in [\beta c_0 = low, \; \beta c_1 = high]$;
- βh stands for "Heater power" and $\beta h \in [\beta h_0 = low, \; \beta h_1 = average, \beta h_2 = high]$.

Note we have to define $\beta h_1 = average$ to provide special level of heating to resist natural cooling in stable state.

Expert in testing of temperature regulators gave as set of linguistic rules looks as: IF current temperature is low WHEN heater power is high and cooler power is low. Let us rewrite them in terms of previously defined linguistic variables:

Table 1. Fuzzy rules in terms of linguistic variables.

$\alpha_{t-2\Delta t}$	$\alpha_{t-\Delta t}$	α_t	βc_t	βh_t
–	–	α_0	βc_0	βh_2
–	–	α_1	βc_0	βh_0
–	–	α_2	βc_1	βh_0
α_1	α_1	α_1	βc_0	βh_1

IF at the moment of time t $\alpha_t = \alpha_0$ WHEN $\beta c_t = \beta c_0$ and $\beta h_t = \beta h_2$.
Table 1 contains short formulations of all rules.

As one can see in most cases controller will act as nonlinear without history dependency. However, when sequence of three input temperatures "near set point" $\alpha_{t-2\Delta t} = \alpha_1$, $\alpha_{t-\Delta t} = \alpha_1$, $\alpha_t = \alpha_1$ arrives controller switches to "stable state near set point" and produces "average heater power".

Figures 2 and 3 show membership/truth functions numerically specifying all values of linguistic variables.

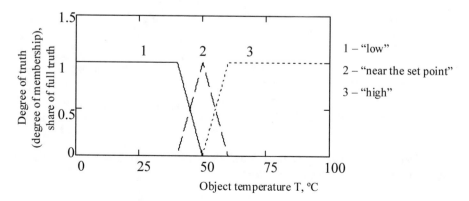

Fig. 2. Truth functions for temperature linguistic variable.

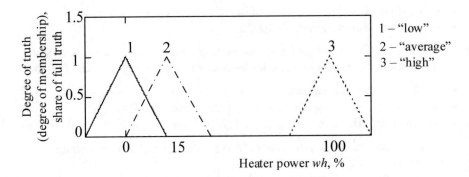

Fig. 3. Truth functions for heater power.

We carried out experiments with three different heating elements (700 W, 1500 W, 2200 W) and with two objects: 0.5 and 1.5 L of water. Figure 4 show respective step change processes.

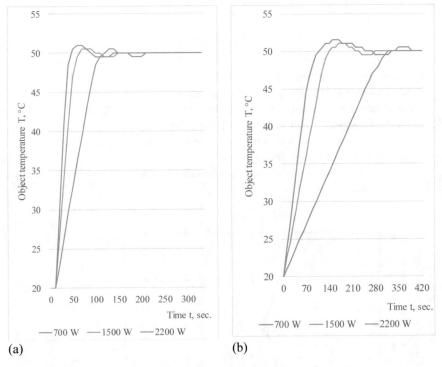

(a)   (b)

Fig. 4. Step change processes of (a) 0.5 and (b) 1.5 L of water between initial and set temperatures with heaters with different powers.

As one can see then controller detects temperature is near set point it switches on special heater power value and resists against natural cooling and stable state error. In addition, it does not need special tuning of controller parameters for each individual production-heating reactor.

5   Conclusion

The problems of industrial processes control in case of using the most popular industrial controllers - PID: required customization for each individual plant and stable state error problem can be resolved by using fuzzy inference.

The conducted experiments show that matrix approach of fuzzy logic make it possible to construct the device independent controller without tuning.

Analysis of history of object state changes with combination of fuzzy state definition allow engineers to solve stable state error problem.

Additionally paper eliminates problem of numerical specification of hard understandable concept of fuzzy state. Proposed combinational logic with separate memory block make fuzzy system design process much more easy.

Actually, the developed controllers are at the stage of introduction into the processes of temperature control and frequency control of electric motors in oil mining organizations.

References

1. Blevins, T.L.: PID advances in industrial control. In: IFAC Conference on Advances in PID Control, pp. 23–28. Elsevier, Brescia (2012)
2. Tehrani, K.A., Mpanda, A.: PID control theory. In: Panda, R.C. (ed.) Introduction to PID Controllers - Theory, Tuning and Application to Frontier Areas, pp. 213–228. InTech, Den Haag (2012)
3. Zadeh, L.A.: Fuzzy sets. Inf. Control **8**, 338–353 (1968)
4. Martsenyuk, M.A., Polyakov, V.B., Seletkov, I.P.: Matrix implementation of fuzzy inference algorithms. St. Petersbg. State Polytech. Univ. J. Comput. Sci. Telecommun. Control Syst. **6**, 133–141 (2012)
5. Martsenyuk, M.A.: Matrix representation of fuzzy logic. Fuzzy Syst. Soft Comput. **2**, 7–35 (2007)
6. Piegat, A.: Fuzzy Modeling and Control. Physica-Verlag, Heidelberg (2001)
7. Zhang, X., Li, L., Dai, Y.: Fuzzy state machine energy management strategy for hybrid electric UAVs with PV/fuel cell/battery power system. Hindawi Int. J. Aerosp. Eng. **2018**, 1–16 (2018)
8. Gadelhag, M., Lotfi, A., Langensiepen, C., Pourabdollah, A.: Unsupervised learning fuzzy finite state machine for human activities recognition. In: PETRA 2018 Proceedings of the 11th Pervasive Technologies Related to Assistive Environments Conference, pp. 537–544. ACM New York, Corfu, Greece (2018)
9. Brauer, W.: Automatentheorie. Vieweg + Teubner Verlag, Brunswick (1984)

The Problem of Solution Restoration by Measurements for the Laplace Equation

Dmitry A. Tarkhov⊙, Maksim A. Migovan⊙, Kirill A. Ivanenko⊙,
Sergey A. Smirnov$^{(\boxtimes)}$⊙, and Aleksandra M. Kobicheva⊙

Peter the Great St. Petersburg Polytechnic University, Saint Petersburg, Russia
dtarkhov@gmail.com, wap_migovan@mail.ru,
kaivanenko1971@gmail.ru, serg.a.smir@gmail.com,
kobicheva92@gmail.com

Abstract. In this paper, we test the methods and algorithms for constructing neural network models from equations and data using the example of the problem on the restoration of the Laplace equation solutions according to the measurements in a unit square. We estimate the quality of approximate solutions constructed with the help of neural networks for different sets of system parameters (the number of points in which the operator is calculated, the number of test points on one side of the square). During the experiment, all test points inside the square are regenerated. Selection of the solution is carried out by optimization of the error functional. Optimization is carried out using the algorithm of training neural networks Resilient Propagation (RProp). The algorithms considered can be applied to a wide range of practically interesting problems, since they practically do not depend on the form of the differential equation, its linearity, the geometry of the region, etc.

Keywords: Neural network modeling · Differential equation
Laplace equation · Boundary value problem

1 Introduction

When solving real technical problems, there often arises the need to construct a mathematical model of an object for a set of initial data that is insufficient for an unambiguous solution. Investigation of new methods for solving such problems is necessary, because the creation of a universal method, which covers a wide range of tasks on the one hand, and on the other hand retains the simplicity of application remains relevant to this day. We propose to use the apparatus of neural networks in similar tasks. There are other methods that allow such a class of problems to be solved, but they usually require a significant modification when changing the differential equation, type of conditions, etc.

There are many papers on the application of neural networks to solving differential equations. For example, in [1], authors consider methods for solving differential equations with initial and boundary conditions with the help of neural networks. The trial solution of the differential equation is written as the sum of two parts. The first part satisfies the initial/boundary conditions and does not contain adjustable parameters.

© Springer Nature Switzerland AG 2019
T. Antipova and A. Rocha (Eds.): DSIC 2018, AISC 850, pp. 450–455, 2019.
https://doi.org/10.1007/978-3-030-02351-5_51

The second part is constructed so as not to affect the initial/boundary conditions. This part includes a direct neural network containing configurable parameters (weights). Consequently, by construction, the initial/boundary conditions are satisfied, and the network is trained to satisfy the differential equation. These methods allow solving both ordinary differential equations and partial differential equations.

The relevance of the neural networks application for constructing models by differential equations and experimental data is described in [2]. In [3], fractional delay differential equations of variable order are implemented using an approach based on artificial neural networks. The book [4] presents various neural network methods for solving differential equations arising in science and technology. In [5], the solution of partial differential equations is investigated using the Chebyshev single-layer neural network.

Our papers [6–9] are distinguished by a number of unique features, in particular, the use of regeneration of test points in the process of training a neural network, which significantly improves the accuracy of the problem solution.

In this paper, we test the methods [6–9] on the problem of the solution restoration of the Laplace equation from the data. We are looking for a solution in the form of an RBF network. Selection of the solution is carried out by optimization of the error functional. In the western papers, the test points do not regenerate. In our work, all test points within the unit square in which the equation will be considered will be regenerated.

2 Methods

In this paper we present the results of computational experiments for the Laplace equation $\frac{\partial^2 u}{\partial x^2} + \frac{\partial^2 u}{\partial y^2} = 0$ in the unit square $[0; 1] \times [0; 1]$.

In the basis of the model for "measurements" we took the well-known solution of the Laplace equation $v(x, y) = \log((x + a)^2 + y^2)$. For specific calculations, we chose $a = 0.1$.

We search the solution in the form of an RBF network $y = \sum_{j=1}^{n} W_j \varphi_j(\|\mathbf{x} - \mathbf{c}_j\|)$, in which we use standard Gaussians as basis functions

$$\varphi_i(x, y, a_i, x_i, y_i) = exp\left\{-a_i\left[(x - x_i)^2 + (y - y_i)^2\right]\right\}.$$

The solution is selected by optimizing the error functional $I = I_1 + \delta I_2$. As the first summand I_1 responsible for the fulfillment of the equation, we used

$$\sum_{j=1}^{M}\left(\Delta u\left(x_j', y_j'\right)\right)^2 \qquad (1)$$

where $\left(x_j', y_j'\right)$ - test points from a square $[0; 1] \times [0; 1]$.

As data on the sides of the square, we chose the values of the function $v(x, y)$ at fixed points of the boundary with a random error.

As a component of the error functional I_2 responsible for the data, we used

$$\sum_{j=1}^{M_1}\left(u\left(x_j'',0\right)-v\left(x_j'',0\right)-\xi_{1j}\right)^2+\sum_{j=1}^{M_1}\left(u\left(0,y_j''\right)-v\left(0,y_j''\right)-\xi_{2j}\right)^2$$
$$+\sum_{j=1}^{M_1}\left(u\left(x_j'',1\right)-v\left(x_j'',1\right)-\xi_{3j}\right)^2+\sum_{j=1}^{M_1}\left(u\left(1,y_j''\right)-v\left(1,y_j''\right)-\xi_{4j}\right)^2,$$

(2)

where the test points in which the values of the function are calculated, we choose on the boundaries of the square at regular intervals, $\xi_{i,j}$ - random variables that simulate measurement errors and are uniformly distributed across the gap $[-\varepsilon, \varepsilon]$. The optimization was carried out using the RProp method. The results were compared with $v(x, y)$.

3   Results

In each series of computational experiments, the number of neurons n is constant. J_1 denotes the root-mean-square error at the boundary points, J_2 denotes the error at all points. The quality of the neural network model was evaluated by $J_1 = 0.01\sqrt{I_1}J$ and $J_2 = 0.005\sqrt{I_2}$ calculated at 10,000 random test points inside the square and 10,000 randomly distributed at equal intervals on each of its boundaries. With each set of parameters, we started the learning process 10 times and based on the results we calculated the minimum of errors and their average value.

Below are the results of computational experiments 1. In these experiments, the number of neurons $n = 10$, $M = 100$ and $M_1 = 10, 30, 100$ (Table 1).

Table 1. Results of computational experiments for n = 10 neurons.

M_1		$\varepsilon = 0.1$		$\varepsilon = 0.3$		$\varepsilon = 1$	
		J_1	J_2	J_1	J_2	J_1	J_2
10	Minimum	0.0964	0.0375	0.117	0.0501	0.208	0.0964
	Average	0.1374	0.0559	0.1535	0.0653	0.285	0.183
30	Minimum	0.0600	0.0242	0.0699	0.0294	0.120	0.0599
	Average	0.0823	0.0337	0.0938	0.0421	0.156	0.0905
100	Minimum	0.0509	0.0191	0.0487	0.0170	0.0564	0.0234
	Average	0.0688	0.0273	0.0654	0.0264	0.102	0.0588

A series of computational experiments 2. In these experiments, the number of neurons $n = 30$ and $M_1 = 10, 30, 100$ (Table 2).

A series of computational experiments 3. In these experiments, the number of neurons $n = 100$ and $M_1 = 10, 30, 100$ (Table 3).

The results of the calculations show that the root-mean-square error in most cases is substantially less than the error in the initial data. In all cases, the mean value of the

Table 2. Results of computational experiments for n = 30 neurons.

M_1		$\varepsilon = 0.1$		$\varepsilon = 0.3$		$\varepsilon = 1$	
		J_1	J_2	J_1	J_2	J_1	J_2
10	Minimum	0.0910	0.0349	0.103	0.0526	0.202	0.109
	Average	0.117	0.0474	0.143	0.0759	0.269	0.149
30	Minimum	0.0486	0.0192	0.0712	0.0290	0.110	0.0632
	Average	0.0638	0.0252	0.0860	0.0384	0.160	0.0877
100	Minimum	0.0373	0.0118	0.0388	0.0166	0.0723	0.0383
	Average	0.0448	0.0163	0.0533	0.0216	0.0972	0.0518

Table 3. Results of computational experiments for n = 100 neurons.

M_1		$\varepsilon = 0.1$		$\varepsilon = 0.3$		$\varepsilon = 1$	
		J_1	J_2	J_1	J_2	J_1	J_2
10	Minimum	0.0997	0.0344	0.105	0.0535	0.212	0.135
	Average	0.130	0.0515	0.137	0.0671	0.294	0.201
30	Minimum	0.0511	0.0203	0.0632	0.0220	0.108	0.0636
	Average	0.0642	0.0255	0.0808	0.0361	0.165	0.0970
100	Minimum	0.0309	0.0116	0.0389	0.0132	0.0677	0.0348
	Average	0.0447	0.0165	0.0549	0.0229	0.0911	0.0585

errors does not greatly exceed their minimum, from which we conclude that the training procedure is stable. In all cases, the increase in the number of points in which information about the function at the boundary is known leads to a reduction in the error not only at the boundary, but also within the region from which we conclude that the neural network processes the information as a whole, and not locally. In this case, for large errors in the initial data, $\varepsilon = 1$, the errors with the number of neurons $n = 10$ and with the number of neurons $n = 100$ practically coincide. With a small error $(\varepsilon = 0.1)$, the effect of increasing the number of neurons and the number of test points on the boundary of the square is more noticeable. Let's illustrate the results with graphs Fig. 1.

Comparing the graphs for different values of the parameter ε, we notice an increase in the approximation error for a larger value of the parameter ε. At the same time, this increase in the error does not lead to catastrophic consequences - the quality behavior of the solution is monitored correctly, while the relative error inside the square where there were no "measurement" points is substantially less than the error at the boundary.

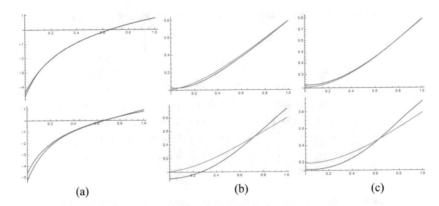

Fig. 1. (a) Graph $u(x,x)$ and $v(x,x)$, (b) graph $u(x,1)$ and $v(x,1)$ (c) graph $u(1,x)$ and $v(1,x)$. Parameter values $n = 100; M_1 = 30$. For the first line - $\varepsilon = 0.1$. For the second line - $\varepsilon = 1$.

4 Conclusion

Considered methods and algorithms practically do not depend on the form of the differential equation, its linearity, the geometry of the domain, etc. We expect that similar results will be obtained for a wide range of practically interesting problems of constructing mathematical models of real objects on heterogeneous information.

Acknowledgment. This paper is based on research carried out with the financial support of the grant of the Russian Scientific Foundation (Project No. 18-19-00474).

References

1. Lagaris, I.E., Likas, A., Fotiadis, D.I.: Artificial neural networks for solving ordinary and partial differential equations. IEEE Trans. Neural Netw. **9**(5), 987–1000 (1998)
2. Rostami, F., Jafarian, A.: A new artificial neural network structure for solving high-order linear fractional differential equations. Int. J. Comput. Math. **95**(3), 528–539 (2018)
3. Zuniga-Aguilar, C.J., Coronel-Escamilla, A., Gomez-Aguilar, J.F., Alvarado-Martinez, V.M., Romero-Ugalde, H.M.: New numerical approximation for solving fractional delay differential equations of variable order using artificial neural networks. Eur. Phys. J. Plus **133**(2), 1–16 (2018)
4. Yadav, N., Yadav, A., Kumar, M.: Introduction to Neural Network Methods for Differential Equations. SpringerBriefs in Applied Sciences and Technology, Computational Intelligence. Springer, Dordrecht (2015)
5. Mall, S., Chakraverty, S.: Single layer Chebyshev neural network model for solving elliptic partial differential equations. Neural Process. Lett. **45**(3), 825–840 (2017)
6. Tarkhov, D.A., Vasilyev, A.N.: New neural network technique to the numerical solution of mathematical physics problems. I: Simple problems. Opt. Memory Neural Netw. (Inf. Opt.) **14**(1), 59–72 (2005)
7. Lazovskaya, T.V., Tarkhov, D.A., Vasilyev, A.N.: Parametric neural network modeling in engineering. Recent Patents Eng. **11**(1), 10–15 (2017)

8. Gorbachenko, V.I., Lazovskaya, T.V., Tarkhov, D.A., Vasilyev, A.N., Zhukov, M.V.: Neural network technique in some inverse problems of mathematical physics. In: Cheng, L., Liu, Q., Ronzhin, A. (eds.) Advances in Neural Networks. ISNN 2016. LNCS, vol. 9719, pp. 310–316. Springer, Cham (2016)
9. Budkina, E.M., Kuznetsov, E.B., Lazovskaya, T.V., Tarkhov, D.A., Shemyakina, T.A., Vasilyev, A.N.: Neural network approach to intricate problems solving for ordinary differential equations. Opt. Memory Neural Netw. **26**(2), 96–109 (2017)

Comparison of Reconstruction Strategies of Compressive Sensing Applied to Ultrasound Images

Erick Toledo Gómez[(⊠)], Humberto de Jesús Ochoa Domínguez,
Soledad Vianey Torres Argüelles,
and Leandro José Rodríguez Hernández

Departamento de Ingeniería Industrial y Manufactura,
Universidad Autónoma de Ciudad Juárez, Ciudad Juárez, Mexico
`etg.pds@gmail.com`

Abstract. Ultrasound medical images are important for medical diagnose. The method allows the real-time visualization of organs of the body and it is not invasive. In this study, a comparison of reconstruction greedy search methods, used in compressive sensing, is performed. The methods and the algorithms are explained and experiments are carried out in synthetic and measured data. Result show that the orthogonal matching pursuit outperforms the other methods in the greedy search classification.

Keywords: Compressive sensing · Sampling matrix · Sparsity
Ultrasound images

1 Introduction

Medical images are important for the diagnose of human beings. An important area of study is the ultrasound (US) images, which have the special feature of being captured in real time. The ultrasound signals are acquired through a transducer that sends ultrasonic waves at a frequency higher than 20 kHz [1], spreading across of the body, until colliding with the soft tissues which causes the wave to be reflected.

The compressive sensing (CS) area integrates different stages such as sampling, reduction of the dimensionality, compression and optimization, and it has been used to introduce improvements in the reconstruction of these images. The CS [2–4] aims at reconstructing signals by a number of measures significantly lower than the necessary when using the Shannon/Nyquist sampling theory [5, 6]. To apply the CS to signals, a fundamental property called sparseness [7, 8] must be fulfilled.

The reconstruction of the US image is usually computationally expensive, hence, the reconstruction algorithms are an important step in CS. These algorithms are divided into five groups: Bayesian methods, convex relaxation, greedy search, non-convex relaxation and brute force [9]. In this paper, a comparison of greedy algorithms for reconstructing US images is performed. The metrics used are the structural similarity index (SSIM), which is a quality metric for measuring the similarity between the original and the recovered images and peak signal to noise ratio (PSNR).

© Springer Nature Switzerland AG 2019
T. Antipova and A. Rocha (Eds.): DSIC 2018, AISC 850, pp. 456–464, 2019.
https://doi.org/10.1007/978-3-030-02351-5_52

The paper is organized as follows: Sect. 2 presents the conditions that must be met both the measurement vector and the sampling matrix to be able to apply the CS. In Sect. 3, the greedy algorithms (OMP, CoSaMP, HTP and IHT [10–13]) are explained. Section 4 provides the results of the algorithms when retrieving the US images in simulated data. Section 5 provides the results of measured data. The paper concludes in Sect. 6.

2   Compressive Sensing

The goal of CS is to reconstruct a vector $x \in \mathbb{R}^N$ that satisfies the linear equation $y = \Phi x$, where $\Phi \in \mathbb{R}^{MxN}$ is the sensing matrix and the vector $y \in \mathbb{R}^M$ has a reduction of dimensionality with respect to the input sparse vector $x \in \mathbb{R}^N$, that is $M \ll N$.

Sparsity property allows to obtain compressed samples, which can be reconstructed with precision [7, 8]. A signal is sparse if it has only a few non-zero coefficients, compared to the signal length, for a vector $x \in \mathbb{R}^N$ the sparsity can be expressed as follows:

$$\|x\|_0 \le k. \tag{1}$$

A sparse vector $x \in \mathbb{R}^N$ can be represented through with a linear combination of few coefficients of a known base or dictionary Ψ. If this representation is exact then the signal is sparse.

$$x_i = \sum_{i=1}^{n} z_i \psi_i \quad \Rightarrow \quad x = \Psi z \quad \text{with} \quad \|z\|_0 \le k. \tag{2}$$

Ψ is an array of $N \times N$ with $[\psi_1, \psi_2, \ldots \psi_N]$ column vectors and z_i the sequence of coefficients of x [14]. The sparse signals can be recovered using CS if they have been contaminated with noise $y = \Phi x + \eta$, where η is the noise component. In order to reconstruct the signal the Restricted Isometry Property (RIP) [8, 14] must be fulfilled.

Theorem 1 [8]. *If for any positive number L there exists a Constant Restricted Isometry (RIC) $\delta_L \in (0, 1)$, it is said that the matrix Φ satisfies the L-order RIP, in other words,*

$$(1 - \delta_L)\|x\|_2^2 \le \|\Phi x\|_2^2 \le (1 + \delta_L)\|x\|_2^2 \quad \text{for all } x \text{ such that} \quad \|x\|_0 \le L. \tag{3}$$

If $\Phi \in \mathbb{R}^{MxN}$ satisfies RIP of order $2k$ implies that the distances between all vectors k-sparse are preserved. Then, the sampling matrix assign a single vector k-sparse at the same point. To recover any sparse signal $x \in \mathbb{R}^N$, that satisfies $y = \Phi x$ with $\Phi \in \mathbb{R}^{MxN}$, $y \in \mathbb{R}^M$ and $M \ll N$ it is required to solve the following optimization problem,

$$min \ \|x\|_0 \quad s.t \quad \mathbf{\Phi}x = y. \tag{4}$$

When the signal is contaminated with noise, the model of the signal becomes $y = \mathbf{\Phi}x + \eta$ and the optimization problem changes to solve,

$$min \ \|x\|_0 \quad s.t \quad \|y - \mathbf{\Phi}x\|_2 \leq \sigma_n^2. \tag{5}$$

where σ_n^2 is a measure of the power of the noise. For the two cases of optimization problems to solve previously mentioned with the l_0, they are algorithms of the type NP-HARD [15]. That is, solving this type of algorithms for any measurement matrix $\mathbf{\Phi}$ is computationally intractable.

3   Greedy Search Algorithms

Greedy algorithms present a simple analysis and low complexity [16]. In these algorithms for each iteration the residue vector is calculated from the projection of the sparse vector on the sampling matrix ($\mathbf{\Phi}$), until the stopping condition is satisfied, throwing an approximation to the original vector as output. Within this group are the Orthogonal Matching Pursuit (OMP) [10] and the Compressive Sampling Matching Pursuit (CoSaMP) [11]. Algorithms based on threshold methods are another variety that give us an approximation to the original vector, among them give the Iterative Hard Thresholding (IHT) [13] and Hard Thresholding Pursuit (HTP) [12]. The following algorithms are used to find the solution of (6).

$$x^* = argmin \ \|x\|_0 \quad s.t \quad \|y - \mathbf{\Phi}x\|_2 \leq \sigma_n^2. \tag{6}$$

3.1   Orthogonal Matching Pursuit (OMP)

The OMP is characterized by its simplicity and high speed [10]. The algorithm initializes a residual vector r equal to the vector y and in each iteration the function is orthogonally projected on all the vectors that have been selected from the sampling matrix, until the stop condition is met, resulting in an approximation to the original vector. The Algorithm 1 implements the OMP.

3.2   Compressive Sampling Matching Pursuit (CoSaMP)

In each iteration, the indices of the $2k$-sparse vectors for which the correlation between the sampling matrix and the residual is maximum. Subsequently, the minimum square is searched using the selected vectors of the sampling matrix. Leaving only the k largest components of this solution and updating the residual vector. When the stopping conditions are met, an approximation to the original vector is obtained. The Algorithm 2 implements the CoSaMP.

The non-linear operator $H_k(*)$ is a hard thresholding of order k which retains the k highest absolute values of x and makes zero the remaining.

Algorithm 1 : Orthogonal Matching Pursuit (OMP)

In: Φ , y , σ_η^2

Initialize: $r \leftarrow y$; $\Omega \leftarrow \varnothing$; $j \leftarrow 0$; $i \leftarrow 0$; $x^* \leftarrow \varnothing$

while $\|r\|_2^2 > \sigma_\eta^2$ **do**

    $x_j \leftarrow \phi_j^T r$ for all $j \notin \Omega$

    $i \leftarrow \underset{j \notin \Omega}{argmax} |x_j|$

    $\Omega \leftarrow \Omega \cup i$

    $x_\Omega^* \leftarrow \underset{x}{argmin} \|\Phi_\Omega x - y\|_2^2$

    $r \leftarrow y - \Phi_\Omega x_\Omega^*$

    $j \leftarrow j + 1$

end while

Out: x^*

Algorithm 2 : Compressive Sampling Matching Pursuit (CoSaMP)

In: Φ , y , k , σ_η^2

Initialize: $r \leftarrow y$; $\Omega \leftarrow \varnothing$; $j \leftarrow 0$; $S \leftarrow \varnothing$; $x \leftarrow \varnothing$

while $\|r\|_2^2 > \sigma_\eta^2$ **do**

    $\Omega_{j+1} \leftarrow supp(H_{2k}(\Phi^T r_j))$

    $S_{j+1} \leftarrow \Omega_{j+1} \cup supp(x_j)$

    $\widetilde{x}_{j+1} \leftarrow \underset{x}{argmin} \|\Phi_{S_{j+1}} x - y\|_2^2$

    $(\widetilde{x}_{j+1})_i \leftarrow 0$ for all $i \notin S_{j+1}$

    $x_{j+1} \leftarrow H_k(\widetilde{x}_{j+1})$

    $r_{j+1} \leftarrow y - \Phi x_{j+1}$;

    $j \leftarrow j + 1$

end while

Out: $x^* \leftarrow x_{j+1}$

3.3 Iterative Hard Thresholding (IHT)

Thresholding algorithms use the operator $H_k(*)$ to maintain the k highest absolute values of $x \in \mathbb{R}^N$. In each iteration, a better approximation to the vector k-sparse is sought through the Eq. (7). Where the residue vector is projected onto the sampling matrix and added to the approximation of the previous iteration and then only the k higher absolute values of the solution thrown are maintained [17]. The Algorithm 3 implements the IHT.

$$x_{n+1} = H_k\big(x_n + \Phi^T(y - \Phi x_n)\big). \tag{7}$$

3.4 Hard Thresholding Pursuit (HTP)

In each iteration the residue vector is projected on the sampling matrix, the result added to the approximation of the previous iteration and the supreme of the k highest absolute values obtained is sought. Solving a minimum square taking only the selected vectors columns of the sampling matrix [18]. The Algorithm 4 implements the HTP.

Algorithm 3 : Iterative Hard Thresholding (IHT)

In: Φ , y , k , σ_η^2
Initialize: $r_0 \leftarrow y$; $d_0 \leftarrow \varnothing$; $j \leftarrow 0$; $x_0 \leftarrow \varnothing$
while $\|r\|_2^2 > \sigma_\eta^2$ **do**
$\quad d_{j+1} \leftarrow x_j + \Phi^T r_j$
$\quad x_{j+1} \leftarrow H_k(d_{j+1})$
$\quad r_{j+1} \leftarrow y - \Phi x_{j+1}$
$\quad j \leftarrow j+1$
end while
Out: $x^* \leftarrow x_{j+1}$

Algorithm 4 : Hard Thresholding Pursuit (HTP)

In: Φ , y , k , σ_η^2
Initialize: $r_0 \leftarrow y$; $d_0 \leftarrow \varnothing$; $j \leftarrow 0$; $x_0 \leftarrow \varnothing$; $S \leftarrow \varnothing$
while $\|r\|_2^2 > \sigma_\eta^2$ **do**
$\quad d_{j+1} \leftarrow x_j + \Phi^T r_j$
$\quad S_{j+1} \leftarrow supp(H_k(d_{j+1}))$
$\quad x_{j+1} \leftarrow \underset{x}{argmin} \left\| \Phi_{S_{j+1}} x - y \right\|_2^2$
$\quad (x_{j+1})_i \leftarrow 0 \ for \ all \ i \notin S_{j+1}$
$\quad r_{j+1} \leftarrow y - \Phi x_{j+1}$
$\quad j \leftarrow j+1$
end while
Out: $x^* \leftarrow x_{j+1}$

Table 1 shows the RIC conditions a be fulfilled [19] for the strategy tested.

Table 1. RIC conditions for the different algorithms [19].

OMP	CoSaMP	HTP	IHT
$\delta_{13K} < 0.1666$	$\delta_{4K} < 0.4782$	$\delta_{3K} < 0.5773$	$\delta_{3K} < 0.5773$

4   Results of Synthetic Data

In this investigation, the image of US of a phantom of cysts [20] was used. The image was divided into blocks of 8 × 8 samples and the orthonormal basis of the discrete cosine transform (DCT) was used to obtain the sparse vector. The results show that as we take fewer samples it becomes more difficult to obtain a reconstructed image similar to the original one.

Table 2 shows the quantitative results of the cyst phantom. The performance of the algorithms was measured using the Structural Similarity Index Metric (SSIM) and the Peak Signal to Noise Ratio (PSNR).

Table 2. Results of the simulation using the cyst phantom [20].

Transformed	Algorithm	(%) Coefficients	SSIM	PSNR
DCT-II	OMP	15.63	**0.92063**	**31.1298**
DCT-II	CoSaMP	15.63	0.53382	20.2263
DCT-II	HTP	15.63	0.62673	23.8842
DCT-II	IHT	15.63	0.62660	23.8387
DCT-II	OMP	31.5	**0.98245**	**36.3830**
DCT-II	CoSaMP	31.5	0.74486	24.9654
DCT-II	HTP	31.5	0.87370	28.9667
DCT-II	IHT	31.5	0.87362	28.9565
DCT-II	OMP	50	**0.99847**	**46.1240**
DCT-II	CoSaMP	50	0.84211	27.5830
DCT-II	HTP	50	0.84633	27.9177
DCT-II	IHT	50	0.84606	27.9021

Figure 1 shows the original and recovered cyst phantom image. Notice that visually, the recovered images with CoSaMP (Fig. 1(b) and (g)) have dark point that are not part of the original image.

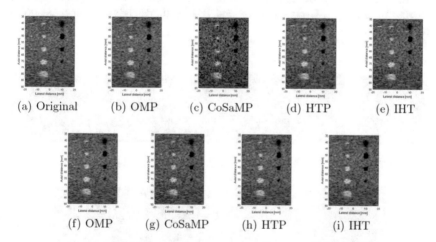

(a) Original   (b) OMP   (c) CoSaMP   (d) HTP   (e) IHT

(f) OMP   (g) CoSaMP   (h) HTP   (i) IHT

Fig. 1. Synthetic US images (a) original cyst phantom [20] and reconstructed phantoms with k = 4 and 15.6% of coefficients using (b) OMP, (c) CoSaMP, (d) HTP, (e) IHT, k = 15 and 50% of coefficients using (f) OMP, (g) CoSaMP, (h) HTP and (i) IHT.

5   Results of Measured Data

Tests on measured data were performed under the same test conditions as used in the cyst phantom [20]. Table 3 shows the quantitative results using measured data.

Table 3. Results with real US image

Transformed	Algorithm	(%) Coefficients	SSIM	PSNR
DCT-II	OMP	15.63	**0.77571**	**25.5944**
DCT-II	CoSaMP	15.63	0.57632	19.6994
DCT-II	HTP	15.63	0.66668	23.2074
DCT-II	IHT	15.63	0.66651	23.1992
DCT-II	OMP	31.5	**0.88170**	**28.6785**
DCT-II	CoSaMP	31.5	0.70914	23.2360
DCT-II	HTP	31.5	0.71074	24.4908
DCT-II	IHT	31.5	0.71047	24.4806
DCT-II	OMP	50	**0.95891**	**32.5981**
DCT-II	CoSaMP	50	0.80196	25.1387
DCT-II	HTP	50	0.81063	25.9815
DCT-II	IHT	50	0.81009	25.9670

In Fig. 2, recovered images keeping the 15.6% of the transform coefficients exhibit blocks artifacts at the edges. The recovered image using the CoSaMP presents the dark points as in the case of the Fig. 1(b). In the recovered images keeping 50% of the transform coefficients OMP does not exhibit block artifacts in the edges as the rest of the methods. However, CoSaMP, HTP and IHT methods show less speckle noise in smooth regions.

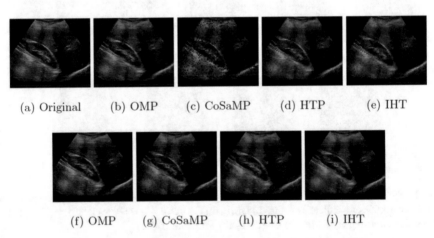

(a) Original   (b) OMP   (c) CoSaMP   (d) HTP   (e) IHT

(f) OMP   (g) CoSaMP   (h) HTP   (i) IHT

Fig. 2. Measured US data (a) original image and reconstructed images with k = 4 and 15.6% of coefficients using (b) OMP, (c) CoSaMP, (d) HTP, (e) IHT, and k = 15 and 50% of coefficients using (f) OMP, (g) CoSaMP, (h) HTP and (i) IHT.

6 Conclusions

In this study, we compared four algorithms for the reconstruction of signals with CS applied to US images. The comparisons were made on a cyst phantom and using the SSIM and PSNR metrics. The discrete cosine transform was used to find the sparse representation of the image. The results showed that the OMP algorithm has a better performance in terms of PSNR and SSIM compared to the other algorithms in the greedy search classification. Therefore, to maintain a good image quality with fewer samples than the required using the Shannon/Nyquist theorem, the OMP results the best option algorithm.

References

1. Herrick, J.F., Krusen, F.H.: Ultrasound and medicine (a survey of experimental studies). In: Transactions of the IRE Professional Group on Ultrasonic Engineering, vol. PGUE-1, pp. 4–13 (1954)
2. Donoho, D.L.: Compressed sensing. Stanford University, Technical report (2004)
3. Candes, E.J., Romberg, J., Tao, T.: Robust uncertainty principles: exact signal reconstruction from highly incomplete frequency information. Technical report (2004)
4. Lustig, M., Donoho, D.L., Santos, J.M., Pauly, J.M.: Compressed sensing MRI. IEEE Signal Process. Mag. 25(2), 72–82 (2008)
5. Shannon, C.E.: Communication in the presence of noise. Proc. IRE 37(1), 10–21 (1949)
6. Nyquist, H.: Certain topics in telegraph transmission theory. Trans. Am. Inst. Electr. Eng. 47(2), 617–644 (1928)
7. Donoho, D.L.: Compressed sensing. IEEE Trans. Inf. Theory 52(4), 1289–1306 (2006)
8. Candes, E.J., Romberg, J., Tao, T.: Robust uncertainty principles: exact signal reconstruction from highly incomplete frequency information. IEEE Trans. Inf. Theory 52(2), 489–509 (2006)
9. Tropp, J.A., Wright, S.J.: Computational methods for sparse solution of linear inverse problems. Proc. IEEE 98(6), 948–958 (2010)
10. Tropp, J.A., Gilbert, A.C.: Signal recovery from random measurements via orthogonal matching pursuit. IEEE Trans. Inf. Theor. 53(12), 4655–4666 (2007)
11. Needell, D., Tropp, J.A.: CoSaMP: iterative signal recovery from incomplete and inaccurate samples. Appl. Comput. Harmonic Anal. 26(3), 301–321 (2009)
12. Foucart, S.: Hard thresholding pursuit: an algorithm for compressive sensing. SIAM J. Numer. Anal. 49(6), 2543–2563 (2011)
13. Blumensath, T., Davies, M.E.: Iterative hard thresholding for compressed sensing. Appl. Comput. Harmonic Anal. 27(3), 265–274 (2009)
14. Candes, E.J., Wakin, M.B.: An introduction to compressive sampling. IEEE Signal Process. Mag. 25(2), 21–30 (2008)
15. Natarajan, B.K.: Sparse approximate solutions to linear systems. SIAM J. Comput. 24(2), 227–234 (1995)
16. Mallat, S.G., Zhang, Z.: Matching pursuits with time-frequency dictionaries. IEEE Trans. Signal Process. 41(12), 3397–3415 (1993)
17. Liu, L., Xie, Z., Feng, J.: Backtracking-based iterative regularization method for image compressive sensing recovery. Algorithms 10(1), 7 (2017)

18. Bouchot, J.L., Foucart, S., Hitczenko, P.: Hard thresholding pursuit algorithms: number of iterations. Appl. Comput. Harmonic Anal. **41**(2), 412–435 (2016)
19. Foucart, S., Rauhut, H.: A Mathematical Introduction to Compressive Sensing. Birkhauser, Cambridge (2013)
20. Jensen, J.A.: Field II simulation program. https://field-ii.dk/

Virtual Tutor: A Case of Study in University Aberta

Elizabeth Carvalho$^{(\boxtimes)}$ and Adérito Marcos

CIAC/UAb, Universidade Aberta, Rua da Escola Politécnica,
141-147, 1269-001 Lisbon, Portugal
{elizabeth.carvalho,aderito.marcos}@uab.pt

Abstract. The project VIRTUAL TUTORING – the virtual tutor as learning mediating artifact in online university education, is an ongoing project, with the main goal of analyzing the pedagogic impact of an anthropomorphic user interface on a typical distance learning environment targeted to support online higher education. It implies the development of 3D rigged avatars that should perform typical online tutor activities. The virtual tutor should mimic a human tutor, being a kind of emphatic interface between the student and the course module in Moodle. But more than this, the virtual tutor should give support in the learning process of the student, working as much as a guide inside the contents offered by the e-learning course. This paper gives an overview of the project present development status.

Keywords: Virtual tutor · E-learning · Avatar · Embodied pedagogical agents

1 Introduction

In this project, we want to analyze the pedagogic impact of anthropomorphic user interfaces, also named Embodied Conversational Agents (ECA) or "avatars", on online learning (OL) environments that are based on learning management systems (LMS) and targeted for use in university level courses. The main idea behind the project is to understand how ECAs may be modeled/adapted as Virtual Tutors in LMS and make available online help and guidance to each individual student. A Virtual Tutor is analogous to a human tutor in that it can autonomously interpret each individual learning situation and effectively intervene, in face of the online knowledge domain and the student profile, in the learning process according to a given tutoring plan that encompasses a set of instructions given by the human teacher/tutor.

Generally, it is not well understood which properties an avatar must possess, in terms of its visual appearance, behavior, emotion, nonverbal expression, and function, to appropriately respond to the students' requests and needs, when engaged in an online learning environment. In this project, an adaptable and interactive anthropomorphic interface will be implemented that is capable of supporting natural human-computer interaction, and which will be modeled as a Virtual Tutor, and evaluated for its efficiency and pedagogic impact as an online teaching-learning mediator artifact in the context of trial scenarios.

© Springer Nature Switzerland AG 2019
T. Antipova and A. Rocha (Eds.): DSIC 2018, AISC 850, pp. 465–471, 2019.
https://doi.org/10.1007/978-3-030-02351-5_53

This project will encompass a set of three trial scenarios for virtual tutoring, one for each of the following areas: natural sciences; computer sciences; and social sciences. In these teaching-learning scenarios, the students will be engaged in virtual classrooms in either formal university graduation or post-graduation courses, where each student will have access to a particular instance of the Virtual Tutor. In these trials we will analyze the pedagogic impact of the application of virtual tutoring, taking special attention to the following central features:

- User-centered learning, measured in terms of the level of student autonomy within the learning process (e.g. the quality of the student's answers/contributions that were generated by suggestion of the virtual tutor; the self-exploitation of content that followed from the virtual tutor suggestions);
- Collaborative learning, measured in terms of the level of interaction within the virtual classroom and the quality of the student intervention and participation in group activities and online discussions that clearly resulted from the direct/indirect intervention or influence of the virtual tutor's actions;
- Teaching effort, measured in terms of the effective time spent by the human teacher/tutor to configure the virtual tutor plan, launch the virtual tutor and monitor it against the overall teaching-learning process;

It should be noticed that we conceive virtual tutoring as a complement to other actual online teaching-learning instruments/tools such are online forums, chatting, quizzes or virtual simulations. One of the main expected outcomes from this project is the design of a virtual pedagogic model for online teaching-learning that explicitly integrates application scenarios of the virtual tutors as a valuable mediating artifact/instrument.

2 Background

Kokane et al. [1] implemented a Learner Centered Design Approach of E-learning System using 3D Virtual Tutors. Further they enhanced their system using WebRTC Based Multimedia Chat system focusing especially young learners to interact with human tutors. They added live video lecture session by which students can interact with the tutor, besides the presentation of 3D virtual tutors' narrations of articles in the text form of live transcriptions of avatars' speech and timed quiz by which a real-life objective examination can be mimicked and thereby evaluating the performance of students. The general architecture of their system is illustrated in Fig. 1.

Soliman and Guelt [2] developed a prototype of an intelligent pedagogical (IPA) agent interacting with a learner and a learning object in natural science experiment in a virtual world while providing supporting multimodal communication abilities. The IPA has features of text chat based on the Artificial Intelligence Markup Language (AIML) and non-verbal communication abilities through gesture animation. According to them, a multimodal communication module is central to the IPA since IPA is the focal point of interaction with the learner and helps in improving engagement and believability. Figure 2 shows one of the available environments to the learners.

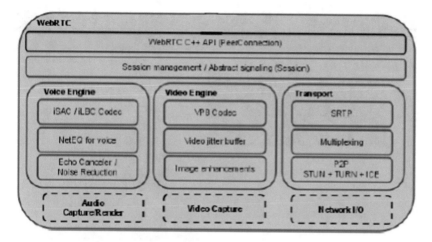

Fig. 1. System architecture [1]

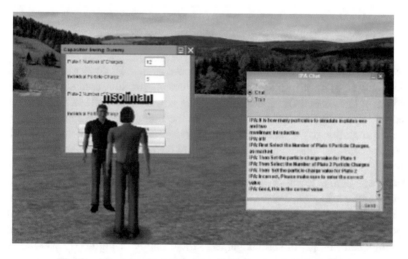

Fig. 2. A pedagogical agent **monitors** learner interaction [2]

Currie et al. [3] developed an Avatar-based system to guide students through the materials provided by a university student employability service. Based on the assessment, they realized that some of the parts of the information provided by the Avatar were considered by the employability representative to be too long for the hearing delivery. This was a highly relevant observation, which was the crux of nature and the role of Avatar-based interfaces. Clearly, a great advantage of the avatar is the ability to guide the end user to solutions through the interaction process, and to conduct this dialogue orally, helps to give a similar tone to human-to-human conversation. The environment is shown in Fig. 3.

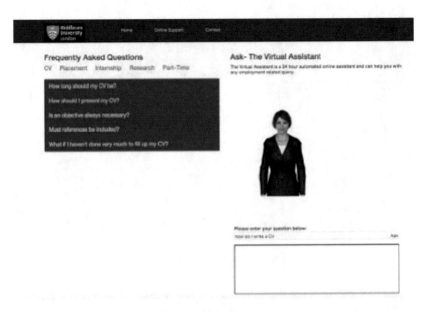

Fig. 3. Virtual assistant interface [3]

Online learning (OL) has grown in importance as a direct consequence of the rapid development taking place in information and communication technology (ICT). This development has pushed OL agents into finding new methods of teaching and learning that could explore the technological media to the limits that ICT could actually offer. Due to the evolution of OL, it is difficult to find a precise and current definition. Nichols [4] describes OL as "education that occurs only through the Web", that is, it does not involve any physical learning materials issued to the students or an actual face to face contact. Pure online learning is essentially the use of eLearning tools in a distance education mode using the Web as the sole medium for all student learning and contact." Even though this statement is still valid, OL has evolved to include aspects such as collaborative learning [5], connectivist learning [6], online participation [7], massive open online learning [8] or serious games, virtual reality or digital storytelling [9, 10].

The use of Embodied Conversational Agents (ECAs) as Virtual Tutors in real online teaching-learning scenarios is still a ground-breaking challenging since OL itself is still in its infancy. In spite of the achievements registered in the area of ECAs there is still a lack of virtual entities that can effectively give support to online teaching-learning. Very few experiments are referred in the literature. They have remained at a stage of limited demonstration and basic proof-of-concept, or expressive talking-heads where no actual pedagogic impact analysis has been carried on [11, 12].

3   Virtual Tutor Development

To start the development of the Virtual Tutor project, we made a detailed identification of the relevant requirements that it should satisfy, besides making an appraisal of existing solutions and/or approaches. In terms of pedagogical requirements, the most significant indicated that the tutor should not be intrusive, although it should behave much as an advisor or reminder of significant activities. Because its deployment aimed to attend the requirements of the pedagogical model of University Aberta, several aspects must be regarded in order to guarantee the compliance of the Tutor's behavior with it.

We also evaluated what technological platform might be used to develop the Virtual Tutor interaction, considering that a 3D model was going to be created to represent the physical Tutor. Unity 3D^{TM} [13] was the first choice because of its technological characteristics, making straightforward available the development for both a desktop and mobile platforms.

Further, it was decided to develop two distinct versions of the virtual tutor: one specific to be used in Moodle platform and other to be uploaded as a mobile app. The difference between these two approaches is sensitive. The first one is supposed to work as much as an adviser of what is available inside the e-learning course, alerting the student of his lack of activity in the forums or important evaluation milestones, besides highlighting any important resource or existing topic. The second one is much more oriented to work just as a reminder of relevant things that are happening inside the course. It does not include the presentation of the Moodle platform, running as a standard alone application on the mobile platform. Presently, only the Android^{TM} [14] system operation is able to run the app.

In both cases, there are two tutors that appear: Mary and John, two young adult people. Both were obtained by digitizing the face and the bust of two real people. This procedure was done, in order to ensure that the aspect of the avatar is the closest to the real one of a person. We assumed, that in the future, the teachers themselves can be used as the basis for the virtual 3D model creation. The Avatars has 8 basic facial

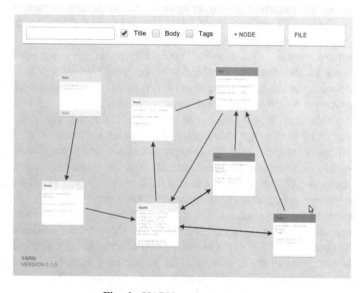

Fig. 4. YARN environment [15]

expressions (rigs) to express feelings such as approval, reprisal or neutrality. We also tried to make the transitions between facial expressions look as natural as possible, interpolating some casual gests as nodding or gazing around.

The version of the virtual tutor that is available in Moodle is made up of a pop-up window with dialogs written in balloons and buttons, which allow the navigation through the area of the online course in question. You can also ask some questions to search for space content. The color of the buttons varies depending on the activation or relevance, or a kind of post-it may appear to draw attention to something in particular in the course area.

The mobile phone version works as an app that provides warnings regularly for the student. The simulation of an empathic behavior is one of the goals in both cases, in addition to the behavior of the avatar being modulated according to, for example, the notes that the student has, their participation in forums or based on a dialogue tree defined by the teacher.

4   Conclusions and Future Work

The Virtual Tutor project is still an ongoing one. We have to work on the artificial intelligence module in order to give much more support to automatic answers according the e-learning course content and the student's performance. Some analytics should also be included to provide smart evaluations on students' interactions in the e-learning course area, based on Moodle's logs, for instance. Integration of databases used by both versions of the virtual tutor and the UAb's Moodle environment is also one of our priorities at this moment.

Presently, we are running tests with two different groups of students from University Aberta. The students are from the first cycle of the bachelor degree in psychology and environment sciences. In order to test the present version of the tutor, usability questionnaires were developed. They focused on evaluating the simplicity or not of the end-users interaction with the tutor.

Besides these questionnaires targeting on usability issues, there were also produced others, focusing on the empathic capabilities of the tutor. We wanted to know if the student, just looking at the facial expression of it, would be able to deduce if the virtual tutor was sending a positive, negative or neutral visual signal.

Finally, at this stage, it was also a goal to find out if the elaboration of the virtual tutor's dialogues by the teachers (real ones) was easy or not to do. Yarn Spinner, an open source implementation of the Yarn language was adopted to make dialogue in our virtual tutors (Fig. 4, illustrates the environment). It adopts a tree-node-tag approach to design the dialogue. We proposed that it was used to create dynamic dialogues to be used by the virtual tutor, according to the teacher and course needs.

The results of all these tests will help us to proceed in some corrections and enhancements of what we have accomplished so far. One of the significant goals and outputs of this project is to enrich the Moodle e-learning platform of University Aberta, and thus, to contribute to the evolution of its pedagogical learning model [16]. The continuation of the development of projects that prioritize the quality of the interaction and attendance to the individualities of the eLearning student is a growing line in the investigations of the area inside University Aberta.

Acknowledgement. This project is financed by FCT - Fundação para a Ciência e Tecnologia, under the Grant FCT- PTDC/IVC-PEC/3963/2014.

References

1. Kokane, A., Singhal, H., Mukherjee, S., Reddy, G.R.M.: Effective e-learning using 3D virtual tutors and WebRTC based multimedia chat. In: 2014 International Conference on Recent Trends in Information Technology (ICRTIT), pp. 1–6. IEEE (2014)
2. Soliman, M., Guetl, C.: Implementing intelligent pedagogical agents in virtual worlds: tutoring natural science experiments in OpenWonderland. In: 2013 IEEE Global Engineering Education Conference (EDUCON), pp. 782–789. IEEE (2013)
3. Currie, E., Harvey, P., Daryanani, P., Augusto, J.C., Arif, R., Ali, A.: An investigation into the efficacy of avatar-based systems for student advice. In: EAI Endorsed Transactions on e-Learning, vol. 3, no. 11 (2016)
4. Nichols, M.: A theory for eLearning. Educ. Technol. Soc. **6**(2), 1–10 (2003)
5. Garrison, R.: Implications of online and blended learning for the conceptual development and practice of distance education. J. Dist. Educ./Revue De L'ÉDucation à Dist. **23**(2) (2009). ISSN 1916-6818
6. Anderson, T., Dron, J.: Three generations of distance education pedagogy. Int. Rev. Res. Open Dist. Learn. **12**(3), 80–97 (2011). ISSN 1492-3831
7. Hrastinski, S.: A theory of online learning as online participation. Comput. Educ. **52**(1), 78–82 (2009)
8. Teixeira, A.M., Mota, J.: Innovation and openness through MOOCs: Universidade Aberta's Pedagogic model for non-formal online courses. In: Proceedings of the EDEN 2013, pp. 332–339. EDEN, Oslo (2013)
9. Leitão, R., Rodrigues, J.M.F., Marcos, A.F.: Game-based learning: augmented reality in the teaching of geometric solids. Int. J. Art Cult. Des. Technol. (IJACDT) **4**(1), 63–75 (2014)
10. Rodrigues, P., Bidarra, J.: Transmedia storytelling and the creation of a converging space of educational practices. Int. J. Emerg. Technol. Learn. (iJET) **9**(6), 42–48 (2014)
11. Urbano, P., Balsa, J., Ferreira, P., Baptista, J.: The importance of ties in the efficiency of convention emergence. In: Proceedings of ICAART 2011, pp. 321–329 (2011)
12. Nascimento, L., Leal, M., Morgado, L.: A tutoria e o Tutor em EaD: retratos do presente versus visões do futuro. Atas Conferência Internacional AIESAD, Rio de janeiro (2014)
13. Unity 3D. https://unity3d.com/pt. Accessed 27 June 2018
14. Android. https://www.android.com/. Accessed 27 June 2018
15. YARN Spinner. https://www.secretlab.com.au/blog/2016/2/10/yarn-spinner. Accessed 27 June 2018
16. Pereira, A., Mendes, A., Morgado, L., Amante, L., Bidarra, J.: Modelo Pedagógico Virtual da Universidade Aberta: para uma universidade do futuro, pp. 1–112. Universidade Aberta, Lisboa (2007)

Author Index

© Springer Nature Switzerland AG 2019
T. Antipova and A. Rocha (Eds.): DSIC 2018, AISC 850, pp. 473–475, 2019.
https://doi.org/10.1007/978-3-030-02351-5